Gary T. Bender, Ph.D.

Professor of Chemistry
University of Wisconsin, La Crosse

Principles of
Chemical
Instrumentation

1987

W.B. SAUNDERS COMPANY

Philadelphia   London   Toronto   Mexico City
Rio de Janeiro   Sydney   Tokyo   Hong Kong

W. B. Saunders Company: West Washington Square
Philadelphia, PA 19105

Library of Congress Cataloging-in-Publication Data

Bender, Gary.
  Principles of chemical instrumentation.

  1. Instrumental analysis. I. Title.
QD79.I5B46 1987   543'.07   85-27721
ISBN 0-7216-1834-0

Editor: Harry Benson
Designer: Terri Siegel
Production Manager: Bob Butler
Manuscript Editor: Martha Tanner
Illustrator: Karen McGarry
Illustration Coordinator: Peg Shaw
Page Layout Artist: Patti Maddaloni

Principles of Chemical Instrumentation   ISBN 0–7216–1834–0

Last digit is the print number: 9   8   7   6   5   4   3   2

This book is dedicated to the memory of the late Mrs. Raden "Dee" Wiriaatmadja Situmeang, B.S., M.S. Mrs. Situmeang served as my proofreader on all but six chapters. She was a constant source of encouragement and a good friend.

PREFACE

The objective of this book is to survey the theory and practice of instrumental analysis as it is applied in clinical chemistry and molecular biology. The text is written for students who have a background in quantitative chemical analysis and algebra. The author has not assumed that the student has a background in either physics or calculus.

The text is organized on an "as it is needed" basis. Chapter 1, **Basic Concepts of Electronics**, covers the most basic principles of electronics. The material is needed as background for the chapters that follow. More advanced principles of electronics are covered in Appendix C, **Electronics**.

Chapter 3, **Analytical Applications of the Absorption of Radiant Energy; Absorption Spectroscopy**, Chapter 4, **Visible Absorption Spectroscopy**, and Chapter 5, **Deviations from Beer's Law and Errors in Absorption Spectroscopy**, are used as background for chapters 6 through 12, which cover ultraviolet absorption spectroscopy, molecular fluorescent spectroscopy, flame emission spectroscopy, atomic absorption spectroscopy, turbidimetry, nephelometry, refractometry of liquids, reflectance photometry, and reflectance densitometry. Chapter 13, **Potentiometric Methods of Analysis**, provides considerable background for chapters 14 through 16, which discuss voltammetry, constant-current coulometry, coulometric titrations, and electrical conductance measurements. Chapter 17, **Liquid Chromatography**, provides extensive background for both Chapter 18, **Ion-Exchange Chromatography**, and Chapter 19, **Gas Chromatography**. The entire book serves as background for Chapter 25, **Automation in Clinical Chemistry**.

Appendix A, **Immunochemistry and its Applications to Quantitative Analysis**, is included for students who have no background in immunochemistry.

A solutions manual for the problems in this text is available from the author at nominal expense. Send inquiries directly to the chemistry department at the University of Wisconsin–La Crosse in La Crosse, Wisconsin 54601.

A number of registered trademarks and trade names are included in the text. An alphabetized list of these along with the names of their owners is provided at the end of the book.

GARY T. BENDER

ACKNOWLEDGMENTS

I would like to gratefully acknowledge the help of my editor, Baxter Venable. Mr. Venable's experience in publishing and his constant encouragement were very helpful to me during the writing and production of this book. I wish Mr. Venable the best during his years of retirement.

I would like to acknowledge the contributions of the many W. B. Saunders employees who worked on the book. In particular, I acknowledge Debra Vickery, my editor's administrative assistant; Martha Tanner, my copy editor; Bob Butler, production manager; and Harry Benson, my new editor.

I would especially like to acknowledge the personnel of the St. Francis Medical Center of La Crosse, Wisconsin, for their never-ending cooperation in the production of this work. Many of the photographs that have not been attributed to other sources were photographed in the St. Francis Laboratory. Their library was always available to me as I sought information for inclusion in this work. I want to thank Abbas Rahimi, M.D.; Kathy Thrower, M.S.M.T. (ASCP);* Silvia Kraus, B.S.M.T. (ASCP);* Barbara Horstman, B.S.M.T. (ASCP);* Elroy Sondreal, B.S.M.T. (ASCP);* Donald Faas, B.S.M.T. (ASCP);* Debbie Lee, B.S.M.T. (ASCP); Karen Adams, B.S.M.T. (ASCP); Madeleine McDonald, M.S.; Rita Scheubel, B.S.M.T. (ASCP);* and the laboratory administrator, Gordon Triebs, M.S.M.T. (ASCP).* All of these individuals played a part in making this book possible.

I would also like to acknowledge the assistance of the Marshfield Clinic of St. Joseph's Hospital Joint Venture Laboratory in Marshfield, Wisconsin. The laboratory administrator, James Kosmicki, B.S.M.T. (ASCP), M.B.A.; James Mikula, M.S.M.T. (ASCP); Michael Arndt, B.S.M.T. (ASCP); and Virginia Narlock, M.S.M.T. (ASCP), all contributed background information for the book.

The Laboratory of St. Joseph's Hospital in St. Paul, Minnesota was very helpful to me. The chemistry supervisor, Dennis Modline, B.S.M.T. (ASCP), and Dawn Erlandson, B.S.M.T. (ASCP),* provided a lot of background information and helpful suggestions. Mr. Modline provided a very helpful review of much of the book.

I would like to acknowledge the help of the staff of the Mayo Clinic Laboratory. Harold Markowitz, M.D.; Garry Mussmann, B.A.; Larry Ebnet, B.S.; and David Squillace, B.S.,* all provided background information. David Squillace was very helpful in the preparation of the mass spectroscopy section (in Chapter 19) of the book.

I would like to acknowledge the assistance and cooperation of Walter Mason, Ph.D., who is the director of Quality Assurance of Ortho Pharmaceutical Corporation, Raritan, New Jersey.

I would like to acknowledge the assistance of the Laboratory of the Lutheran Hospital. Marian Johnson, B.S.M.T. (ASCP);* Arvin Ellefson, B.S.M.T. (ASCP); Joann Toftne, B.S.M.T. (ASCP);* Kay Case, B.S.M.T. (ASCP); Ken David, CLT; Robert Mordin, B.S.; and Russel Ducherschein, B.S.M.T. (ASCP), all provided background information for my work. Mr. Ducherschein provided many helpful suggestions for Chapter 25, **Automation in Clinical Chemistry**.

The Laboratory of St. Mary's Hospital, in Madison, Wisconsin, was very helpful to me. The laboratory supervisor, Shirley Armstrong, B.S.M.T. (ASCP); and Eunice Hardy, B.S.M.T. (ASCP),* provided a lot of background information and helpful suggestions.

I would like to acknowledge the assistance of Robert Lavine, M.S., and Professor Dennis Evans, Ph.D., of the University of Wisconsin–Madison. Mr. Lavine provided many suggestions and helped with the production of many of the illustrations. Professor Evans provided suggestions relating to Chapter 13, **Potentiometric Methods of Analysis**.

Merl Evenson, Ph.D., of the University Hospital Laboratory, Madison, Wisconsin, also provided background information for this text.

Grace Smith, B.S.M.T. (ASCP), M.S., the director of our Medical Technology Program, provided many suggestions for the hematology chapters (Chapters 23 and 24).

William Nieckarz, Ph.D., of our campus, and Donald Showalter, Ph.D., of the University of Wisconsin–Stevens Point, both provided suggestions for the material on nuclear chemistry (Chapter 22).

Linda Heisler, B.A., Anita Morstead, Janice Haynie, and Sue Butterfield all worked on the typing of the manuscript. Ms. Heisler was also my final proofreader.

I would also like to acknowledge my students from both the La Crosse and the Madison campuses for their assistance in catching errors. In particular, I want to thank Teresa Schmidt, B.S.M.T. (ASCP);* Carl Gorsk, B.S.;* Ruth Davis, B.S.;* and Cathy Snell, B.S.,* for their careful examination of the manuscript as viewed by university juniors.

I would like to acknowledge the cooperation of the many instrument manufacturers that have provided background information and illustrations for this book.

And lastly, I would like to acknowledge the State of Wisconsin. The state's commitment to education has provided me with the surroundings that have made this work possible.

*Former students of the University of Wisconsin–La Crosse.

CONTENTS

Basic Concepts of Electronics

POTENTIAL DIFFERENCE

Potential difference, sometimes called EMF (electromotive force) or voltage, is the most basic concept of electronics. It is a measure of the electrical potential energy that exists between two points. It is best understood in terms of a hypothetical experiment. Visualize two isolated small metal spheres that are separated from each other in a vacuum. Initially let us assume that both spheres are electrically neutral—they possess no excess positive or negative charge (Fig. 1–1).

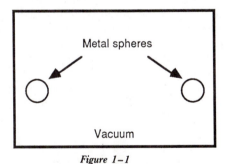

Figure 1–1

Next visualize electrons from one of the spheres being transported and deposited on the other sphere. This action would create a difference in the electrical charge between the two spheres. A charge separation would be said to exist. This separation of charge would create an electrical force field. This would be true because the sphere made positive by the removal of its electrons would be attracting the negative charge now held by the other sphere. Such a force field is depicted in Figure 1–2.

When an electrical force field exists, there is said to be a *potential difference* across the field. The unit that is used to express the magnitude of a potential difference is the volt. A volt is defined as the potential difference necessary to move 1 coulomb of electrical charge in 1 second. A coulomb of charge is equivalent to the charge on 6.24×10^{18} electrons.

In summary, the potential difference or voltage between two points is caused by the interaction of separated charge. The value of the potential difference is directly proportional to the amount of charge that is separated. The distance that separates two charged objects does not affect the potential difference.

The force field that results from the interaction of charged bodies is measured in units of volts per centimeter. Thus the distance between charged objects does have an effect on the electrical field strength. At constant potential, the farther apart two objects reside, the lower the field strength.

In electronics, one important source of potential difference is the battery. The following discussion will be helpful in understanding the origin of the potential difference in batteries. This background will also be used extensively in Chapter 13.

BATTERIES

Some of man's earliest experiences with electronics resulted from the discovery of various batteries. You will recall from your previous scientific training that batteries are used to convert chemical energy to electrical energy. One of the early bat-

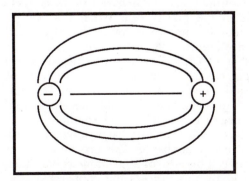

Figure 1–2. "Lines of force" in an electrical field.

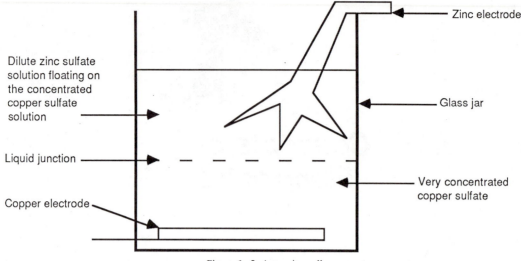

Figure 1–3. A gravity cell.

teries in widespread use was the gravity cell, depicted in Figure 1–3. Let us consider the gravity cell in some detail to see how the chemical energy produces a potential difference.

The potential difference that occurs in a gravity cell is caused by the development of charge separations across the solution-electrode interfaces. Such charge separations develop whenever a metal electrode is immersed into a solution of its ions. Consider, for example, the behavior of the zinc electrode in the gravity cell. When the cell is constructed, an uncharged bar of zinc (Zn) is placed in a solution of zinc ions. When the bar and solution meet, a small amount of zinc metal dissolves from the bar to become zinc ions. The electrons from this reaction

$$Zn \rightarrow Zn^{+2} + 2e$$

remain behind on the zinc electrode. This reaction occurs because of a favorable chemical energy change. This charge separation is typified by the bar having an excess of electrons and the solution having more zinc cations than anions to counter their charge. This charge separation is responsible for the existence of a potential difference at the zinc electrode-solution interface.

Consider next the behavior of the copper electrode in the gravity cell. When the cell is constructed, an uncharged copper bar (Cu) is placed into a solution of copper ions. In this case, a small number of copper ions attach themselves to the copper electrode, making the electrode positively charged. The loss of copper ions leaves the solution side of the solid-liquid interface negatively charged. The negative charge on the solution side of the interface is due to anions that no longer have cations to balance their charge. This process occurs in the gravity cell because of a favorable chemical energy change. By this process the copper bar–solution

interface develops a charge separation and hence a potential difference.

Thus we see that in the gravity cell and, for that matter, in any battery, there are at least two solid-liquid interfaces. One interface has a charge separation with electrons in excess on the electrode, while the other interface has an excess of positive charge on the electrode. A potential difference exists then at both interfaces; the potential difference that the battery supplies is the sum of the two potential differences at the two electrodes.

The value of the potential difference supplied by a battery does not change with time as long as the battery temperature and the concentration of electroactive ions in the battery remain constant. A battery is said to be a DC (direct current) voltage source. The electrical symbol for a battery is depicted in Figure 1–4. Often the potential difference,

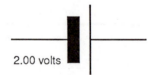

Figure 1–4. Electrical symbol for a battery.

or voltage, of a battery is listed in close proximity to the symbol.

Batteries in modern instruments sometimes serve as a source of voltage. More commonly, electrical devices, called power supplies, are built to replace batteries. In this text, power supplies will be represented as rectangular boxes (Fig. 1–5).

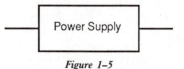

Figure 1–5

A power supply contains many electrical components. Its purpose is to convert the alternating voltage supplied by utilities companies into a direct voltage of proper magnitude.

ELECTRICAL CURRENT

Consider once again the gravity cell in Figure 1–3. Until now we have not connected the electrodes of the gravity cell to an external circuit. We have only discussed the spontaneous generation of the potential difference that is available at the cell's two electrodes.

If we connect the terminals of the gravity cell with a wire, the electrons that have accumulated on the zinc electrode *will travel under the influence of the potential difference* (voltage) down the wire to the copper electrode. This is illustrated in Figure 1–6.

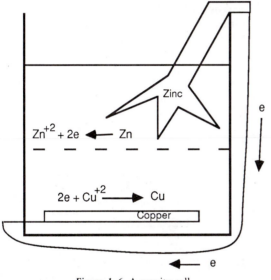

Figure 1–6. A gravity cell.

When this occurs, the zinc undergoes the following reaction:

$$Zn \rightarrow Zn^{+2} + 2e$$

This reaction keeps the voltage of the zinc electrode at its proper value. When the electrons arrive at the copper electrode, copper ions are converted to copper metal. More copper ions then attach to the copper electrode. In this way, the battery maintains its potential difference (voltage), but it does so at the expense of the chemicals in the cell. As long as there is a conductive path for the electrons, the process outlined will continue. The movement of charge is defined as an electrical current. Two types of current flow in the circuit are illustrated in Figure 1–6. The first type is an electron current; it is a movement of electrons through the wire. The second type of current flowing is an ion current. As Cu^{+2} ions are removed from the solution, the sul-

fate ions (SO_4^{-2}) migrate through the solution, across the solution boundary and on into the zinc sulfate solution where they counterbalance the charge of newly formed Zn^{+2} ions. In this way, charge is carried through the battery back to the zinc half cell. Notice that the current flow is cyclic. In a series circuit, the value of the current is the same at all places in the circuit.

The unit of measure that is applied to current is the ampere or amp. It is defined as the movement of 1 coulomb of charge per second.

ELECTRICAL CONDUCTION AND RESISTANCE TO CURRENT FLOW

In the preceding section we saw that electrons flow down a wire from one terminal of the battery back to the other terminal. One might ask why a wire is necessary: Why doesn't the current flow through the air, or why isn't a glass rod used for the conduction of the electrons? The answer to these questions can be found by a systematic discussion of conductance and resistance.

From the experimental point of view, materials are divided into three classes of electrical current conductors—conductors, semiconductors, and insulators (nonconductors). Most metals are classified as conductors, many metaloids as semiconductors, and many nonmetals as insulators. The characteristic of metals that makes them conductors is their loosely held valence electrons. These loosely held valence electrons can move from one metal atom to the next. Thus, if an electron enters a wire on one end, a second electron is pushed out the other end of the wire. The electrons of insulators, on the other hand, are tightly held in covalent bonds. These tightly held electrons are not free to move; thus the conductivity of insulators is poor.

Metals, which are highly conductive to the flow of electrons, are said to be of low electrical resistance. Insulators, which have poor conduction of current, are said to have high electrical resistance. Thus conductance and resistance are the inverse of each other:

$$R = \frac{1}{S}$$

R is resistance measured in ohms

S is conductance measured in mhos

It has been found to be convenient to discuss electrical resistance of materials in terms of the resistance of a cube of material measuring 1 cm on a side. If the resistance is measured between two parallel faces, the value obtained is the resistivity (ρ). A partial listing of resistivity data is given in Table 1–1.

TABLE 1-1. Resistivity Data

Material	ρ in microhm cm
Carbon	350.0
Copper	1.7
Mercury	94.1
Nichrome	100.0
Platinum	11.0
Silver	1.5
Bakelite 588	2×10^{16}
Beeswax	5×10^{14}
Glass	9×10^{13}
Quartz	5×10^{18}
Germanium	8.9×10^{4}
Selenium	8×10^{6}

TABLE 1-2. Temperature vs. Resistivity for Copper

Temperature	ρ microhm cm
−150°C	0.567
−100°C	0.904
20°C	1.7
100°C	2.28
500°C	5.08

The resistance of conductors other than 1-cm cubes is related to the resistivity by the following equation:

$$R = \rho \frac{l}{A}$$

where R is resistance in ohms (Ω)

ρ is the resistivity in ohm cm

l is length of a conductor in cm

A is the cross-sectional area of the conductor in cm²

Thus a strand of copper wire 10 cm long and 0.2 cm in diameter will have a resistance of

$$R = \rho \frac{l}{A}$$
$$R = \frac{1.7 \times 10^{-6} \Omega \text{ cm } (10 \text{ cm})}{(3.14)(10^{-1}\text{cm})^2}$$
$$R = \frac{1.7 \times 10^{-5}}{3.14 \times 10^{-2}} \Omega$$
$$R = 5.4 \times 10^{-4} \Omega$$

Because the resistance of such a wire is quite low, electrical current passes through the wire easily. Next consider a glass rod 10 cm long with a diameter of 0.2 cm:

$$R = \frac{9 \times 10^{13} \times 10^{-6} \Omega \text{ cm } (10 \text{ cm})}{3.14 \times 10^{-2} \text{ cm}^2}$$
$$R = 2.9 \times 10^{10} \Omega$$

Thus a rod of glass has a high resistance to the flow of electrons through its structure. Air is also a poor conductor and hence is quite resistant to the flow of current.

As a further note of interest, the value of the resistivity varies with temperature. Some data for copper are presented in Table 1-2. One can calculate the resistance of a given conductor at some temperature (T) in terms of its resistance (R_0) at 0°C and its temperature coefficient of resistivity (α)

$$R = R_0 (1 + \alpha T)$$

Hence for metals, which have positive temperature coefficients, resistance increases with temperature. This variation of resistance with temperature is one of the reasons that instruments must be warmed up in order to get stable operation.

In summary, we have seen that some materials pass electrons, that is, current, through their structure with ease; others do not. The resistivity for most conductors increases with temperature. Hence the resistance of these conductors increases with temperature.

RESISTORS

Electrical resistors are made from several materials of high resistivity so that a small device may provide the desired resistance. A common resistor is illustrated in Figure 1-7. The plastic case is filled

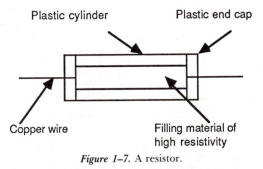

Figure 1-7. A resistor.

with a material of high resistivity. For example, carbon-filled resistors are common; carbon's resistivity is 350 microhm cm. Resistors can be manufactured to have a resistance value anywhere between a few ohms and several hundred thousand ohms.

Colorbands painted on a resistor indicate its resistance value (Fig. 1-8). The color code for these

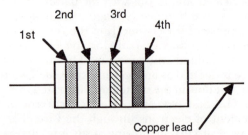

Figure 1-8. Color coding indicates the resistance value of a resistor.

TABLE 1–3. Color Code (Key)

Black	0	Yellow	4	Gray	8
Brown	1	Green	5	White	9
Red	2	Blue	6	Silver	tolerance of $\pm$ 10%
Orange	3	Violet	7	Gold	tolerance of $\pm$ 5%

bands is given in Table 1–3. The first two bands give the resistance to two significant figures. The third band is the power of 10 by which the first two bands are multiplied. Thus a resistor that has orange, green, red, and silver bands would be a 3500-ohm resistor with a tolerance around that nominal value of $\pm$ 10 per cent.

In addition to carbon, Nichrome wire, silicon, and selected ceramic materials are used to make resistors. Resistors come in different physical sizes. The larger they are, the more heat they can dissipate. All resistors heat up to some extent when current is passing through them.

OHM'S LAW

Ohm's law defines the interrelationship between potential difference, current, and resistance. Consider the electrical circuit in Figure 1–9. It is a

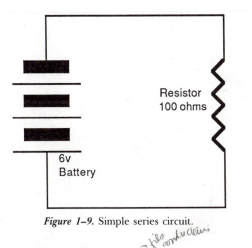

Figure 1–9. Simple series circuit.

simple series circuit consisting of three simple batteries in series applying a 6-volt potential difference across the copper wire leads of a carbon-filled electrical resistor. As explained in the previous discussion, the electrical resistance of the leads is low compared with that of the 100-ohm resistor. It is a common practice in electronics to disregard the resistance of the lead wires and to consider only the resistance of the components. A long time ago the mathematical relationship between applied potential (V), resistance (R), and current (I) was discovered. The relationship is Ohm's law and is stated

$$V = IR$$

If we ignore the resistance of the copper wire leads that connect the battery and resistor, we may easily calculate the current through the circuit.

$$V = IR$$

$$I = \frac{V}{R}$$

$$I = \frac{6v}{10^2 \ \Omega}$$

$$I = 6 \times 10^{-2} \text{ amps}$$

$$I = 60 \text{ milliamps}$$

The 60-ma current that flows through the resistor is also flowing through the leads and the battery. This is true because this is a series circuit.

When current passes through a resistor, heat is produced. It has been found experimentally that the heat dissipated in a resistor is equal to the product of the applied voltage and the current flow.

$$\text{Heat} = IV$$

If Ohm's law is used to substitute for the voltage, the equation becomes

$$\text{Heat} = I^2R$$

SOME USEFUL RESISTIVE CIRCUITS

Many electrical circuits used in the electronics of instruments have several resistors. An example is the following simple voltage divider (Fig. 1–10).

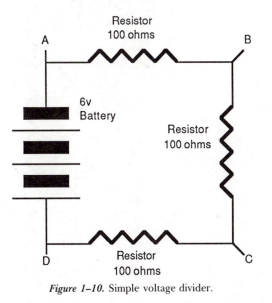

Figure 1–10. Simple voltage divider.

This circuit is a series circuit; thus we know that the current through each resistor is the same. The value for the current may be readily calculated because the total resistance in a series circuit is the sum of

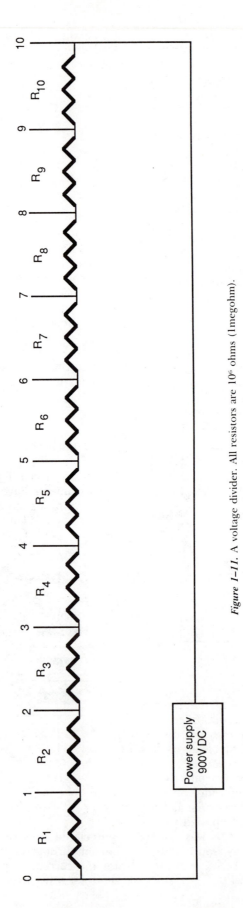

Figure 1-11. A voltage divider. All resistors are 10^6 ohms (1megohm).

the individual resistances. Thus, for our circuit the resistance is

$$R_T = R_1 + R_2 + R_3$$
$$= 100 \; \Omega + 100 \; \Omega + 100 \; \Omega$$
$$R_T = 300 \; \Omega$$

The current through the circuit is

$$I = \frac{V}{R} = \frac{6v}{300\Omega} = 2 \times 10^{-2} \text{ amps}$$

Note that this value is only one third of the current that flowed in the circuit in Figure 1–9.

Now that the current through the resistors is known, we can calculate the potential across each resistor. The voltage or potential difference that is available across the leads A and B is

$$V = IR$$
$$V = (2 \times 10^{-2} \text{ amps}) \; 10^2 \; \Omega$$
$$V = 2 \text{ volts}$$

The voltage across terminals B and C as well as C and D is also 2 volts.

The fact that the applied voltage (6v) can be found distributed across the series resistors in a circuit is called Kirchhoff's voltage law. This law is useful in the analysis of more complex circuits. Consider the voltage between points A and C.

$$V = IR = (2 \times 10^{-2} \text{ amps})(R_1 + R_2)$$
$$V = (2 \times 10^{-2} \text{ amps}) \; (10^2 \; \Omega + 10^2 \; \Omega)$$
$$V = (2 \times 10^{-2} \text{ amps}) \; (2 \times 10^2 \; \Omega)$$
$$V = 4v$$

The voltage available across terminals B and D is also 4 volts. The voltage available across terminals A and D is 6 volts, of course, as they are the battery terminals. This type of circuit is called a voltage divider because the applied voltage can be "divided up" by the series resistors in order to produce various values.

A second example of a voltage dividing circuit is the circuit used in powering photomultiplier light detectors. This light detector must have 10 different potential difference values supplied to its electrodes. In order to do this most economically, the series circuit illustrated in Figure 1–11 is often employed. In this circuit a potential difference source of 900 volts is placed across the 10 series 1 megohm* resistors. The resistance in the complete series circuit is

$$R_T = R_1 + R_2 + R_3 + R_4 + R_5$$
$$+ R_6 + R_7 + R_8 + R_9 + R_{10}$$

*A megohm is 10^6 ohms.

In this case, the value is 10 MΩ or 10×10^6 Ω. The current through each resistor is the same and can be calculated by Ohm's law.

$$I = \frac{V}{R} = \frac{9 \times 10^2 v}{10^7 \; \Omega}$$
$$= 9 \times 10^{-5} \text{ amps or .09 milliamps}$$

Now let us calculate the voltage available between terminals 0 and 1

$$V = (9 \times 10^{-5} \text{ amps}) \; 10^6 \; \Omega = 90 \text{ volts}$$

and the voltage available between terminals 0 and 2.

$$V = (9 \times 10^{-5} \text{ amps})(R_1 + R_2)$$
$$V = (9 \times 10^{-5} \text{ amps})(2 \times 10^6 \; \Omega) = 180 \text{ volts}$$

The voltage available between terminals 0 and 3 is

$$V = (9 \times 10^{-5} \text{ amps})(R_1 + R_2 + R_3)$$
$$V = 270 v$$

Thus we see that the circuit divides the 900-volt input into 10 outputs, each of which is 90 volts greater than the output before it. The potential difference available between leads 0 and 10 is the full 900 volts.

In voltage dividing networks, it is not necessary for all of the resistors to have the same value. The values used are those that produce the potentials desired. The actual values used in the division depend on the demand for current. You can see that low resistance values in a series circuit allow for increased current flow. The resistance values used in Figure 1–11 are high because the demand for current is small in this application.

THE D'ARSONVAL METER

A common device used to measure electrical current is the d'Arsonval meter (Fig. 1–12). It functions by making use of the magnetic field that is always associated with an electrical current flow.

The meter has a permanent magnet surrounding a coil of resistance wire that is wound on an iron mandrel. Attached to the mandrel is a pointer that is associated with the meter face scale. The iron mandrel and its coil are supported on a low friction pivot system. The suspension spring returns the mandrel and its pointer to zero in the absence of current. When a current is to be measured, the wire through which it flows is interrupted and the ends are hooked to the meter coil. The current passing

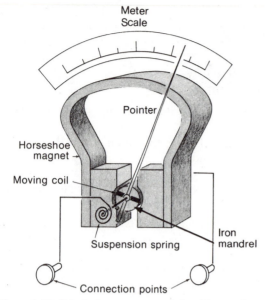

Meter Scale

Pointer

Horseshoe magnet

Moving coil

Suspension spring

Iron mandrel

Connection points

Figure 1–12. Principal components of the d'Arsonval meter. (Adapted from: A.J. Diefenderfer, Principles of Electronic Instrumentation, 2nd Ed., Philadelphia, Saunders College Publishing, 1979.)

through the meter coil sets up a concentrated magnetic field that causes a twisting force on the mandrel and thus moves the mandrel and pointer until the spring force counteracts the magnetic force. The current can then be read from the meter face.

The resistance of the meter is a design consideration depending on the range of current to be measured. The resistance value is easily controlled by selecting the proper wire from which to make the moving coil. It is easy to build meters that measure high currents. Practical limitations do limit the measurement of low currents. The most sensitive of the d'Arsonval meters can measure currents in the microamp range.

The symbols used for a d'Arsonval meter are depicted in Figure 1–13.

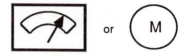

Figure 1–13. Symbol used for a d'Arsonval meter.

The Volt-Ohm Meter

The electronics of modern instrumentation is built up from such components as power supplies, amplifiers, and microcomputers. These devices are generally built on printed circuit boards that plug into the instrument's electronic chassis. An example of such a board is shown in Figure 1–14.

Most service work these days involves the detection and replacement of the malfunctioning circuit board. Many instruments are able to diagnose themselves with the help of their dedicated computer. Others require that the voltage between two test points or the resistance between two test points be measured by the instrument operator. These measured values are then used to determine which boards must be replaced.

The instrument used to make these measurements is a volt-ohm meter (VOM). The common VOM is a d'Arsonval meter that has been configured to measure voltage or resistance.

The d'Arsonval meter can be used to measure voltage even though it is inherently a current meter. This can be done because the meter resistance is a constant and Ohm's law applies. Thus the indicated current multiplied by the meter's internal resistance gives the voltage. As this multiplication is an inconvenience, d'Arsonval meters that are set up to measure voltage have a voltage scale placed under

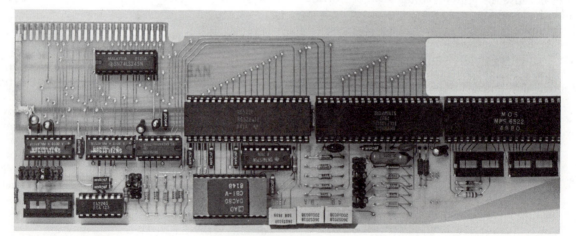

Figure 1–14

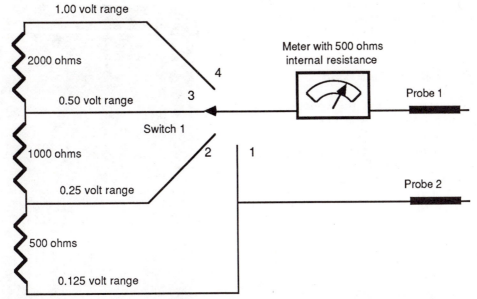

Figure 1–15. A multirange voltmeter.

the pointer. This scale takes the meter's internal resistance into account.

Multirange voltage meters can be made as well (Fig. 1–15). A switched voltage dividing network is used in order to keep the voltage on scale. Consider a multimeter that is built with a 250-microamp, 500-ohm meter. The highest voltage that the meter can measure directly is 0.125 volts. If a voltage between 0 and 0.125 volts is expected, the best measurement is made by placing switch 1 in position 1. If a voltage between 0 and 0.25 volts is expected, the best measurement is made by placing the switch in position 2. If a voltage between 0 and 0.50 volts is expected, position 3 is used. And if a voltage between 0 and 1.00 volts is expected, position 4 is used. These voltmeters generally have multiple meter faces in order to expedite the measurements.

The d'Arsonval Meter as Ohmmeter

The d'Arsonval meter can be used in conjunction with a battery to measure resistance. In essence,

the battery is connected through the meter to the resistor of unknown value (Fig. 1–16). Ohm's law is then used to calculate the value of the resistance. The circuit in Figure 1–16 contains a variable resistor for adjusting the meter to zero ohms. In order to set this circuit up for the measurement of the resistor, probes 1 and 2 are connected together. The variable resistor is then adjusted to make the meter read 250 microamps of current, which corresponds to zero ohms. The meter reads zero microamps when probe 1 is not connected to anything; this corresponds to a resistance of infinity. In this way the 250-microamp d'Arsonval meter covers the resistance range from zero to infinity. So calibrated, probes 1 and 2 can be connected to the unknown resistance. The current that flows can be measured and correlated to the resistance. Unfortunately, this correlation of resistance and current is nonlinear. The high resistance values become crowded at the low-current end of the meter scale. Still, the meter is useful in situations in which high accuracy is not

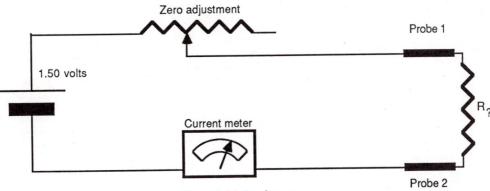

Figure 1–16. An ohmmeter.

a requirement. Commonly the meter face of a VOM is provided with a nonlinear resistance meter face. This allows for the direct readout of the resistance.

DIGITAL VOLTMETERS

Digital voltmeters convert the voltage to be measured to a binary number, which in turn is used to actuate a set of light-emitting diodes that present the voltage to the user. No mechanical meter movement is involved. The design of the digital converter depends on the special needs of the user, but, in general, high-input resistance is obtainable. This allows the voltage to be read without a flow of significant current from the source of voltage. Field effect transistors (FET) are generally used for the input stage for digital voltmeters.

THE OSCILLOSCOPE

The cathode ray tube oscilloscope is useful in the study of voltage versus time. The device uses a cathode ray tube (CRT) to display voltage on the y-axis and time on the x-axis. Consider the CRT illustrated in Figure 1–17.

The voltage to be studied is connected to the vertical deflection electrodes. The electrical field associated with the voltage on these electrodes deflects the electron beam in a manner proportional to the magnitude of the applied voltage. The time during which the applied voltage is to be studied is selected by the operator of the oscilloscope. This is done by applying a linearly increasing voltage from a saw-toothed generator to the horizontal deflection electrodes. The magnitude of the voltage is just enough to move the electron beam from the left of the tube to the right.

Oscilloscopes are frequently used on blood cell counters to monitor the pulse heights that result when blood cells pass through the conductivity cell.

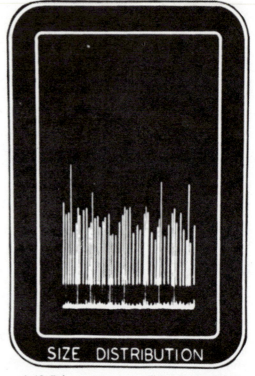

Figure 1–18. Pulse pattern on oscilloscope screen. (Courtesy of Coulter Electronics, Inc., Hialeah, Fla. Reprinted with permission.)

The readout shows voltage pulses that occur over several seconds (Fig. 1–18).

AMPLIFIERS

Electrical Amplifiers

In the operation of many instruments, the currents or voltages produced by the analyte are so small that it is necessary to increase their size to a much higher proportional value. This is done with an electrical circuit called an amplifier. Amplifiers

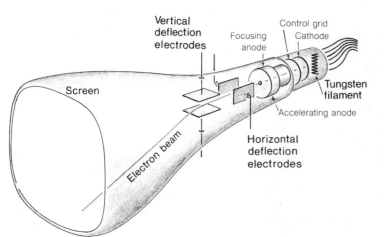

Figure 1–17. Schematic diagram of cathode ray tube. (Adapted from: A.J. Diefenderfer, Principles of Electronic Instrumentation, 2nd Ed., Philadelphia, Saunders College Publishing, 1979.)

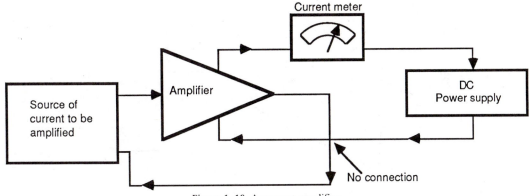

Figure 1–19. A current amplifier.

can be configured to amplify either current or voltage, but more often than not it is current that is amplified. In common electrical amplifier circuits the small current that is to be amplified is placed in control of a higher current circuit. This is shown symbolically in Figure 1–19.

The current to be amplified is flowing through the amplifier and back to its source of origin. The current flowing in this loop is frequently on the order of a few microamps. The amplifier is built so that this independent input current has complete control of the current in the second circuit loop.

The second circuit loop consists of a direct current power supply, a meter, and the amplifier. The current flowing in this loop is frequently on the order of a few milliamps and is in this range because of the power that is supplied by the power supply. The current flowing in this loop could be called the dependent current. The meter is provided for the measurement of the dependent current. The relationship between the independent current and the dependent current is shown in Figure 1–20. Thus this amplifier has an amplification factor of 1000. This is true because an input current of 5 microamps is associated with an output current of 5 milliamps. Amplification factors both lower and higher are possible.

It is important to realize that an amplifier is not a something-from-nothing device. The 1000-fold amplification illustrated in Figures 1–19 and 1–20 is obtained by supplying power (from a power supply). The amplification takes place only because the independent low-current loop controls the dependent high-current loop that also contains the power supply.

The triangular symbol in Figure 1–19 is a symbol for a complex circuit. Such a circuit contains transistors, resistors, capacitors, and associated wiring. An instrument's amplifier could easily have 30 components.

Difference Amplifiers

A second type of common amplifier is the difference amplifier. It is called a difference amplifier because it amplifies the difference between two currents. Its symbol is shown in Figure 1–21.

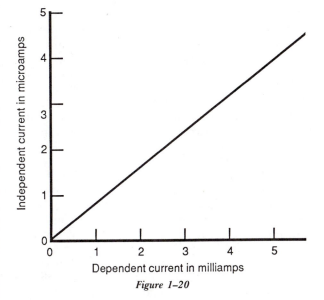

Figure 1–20

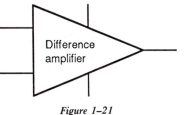

Figure 1–21

The difference between the current passing into input 1 and the current passing into input 2 controls the current in the high-current loop of the circuit. The overall circuit is shown in Figure 1–22.

Once again, the difference amplifier is a relatively complicated circuit but its function is fairly straightforward.

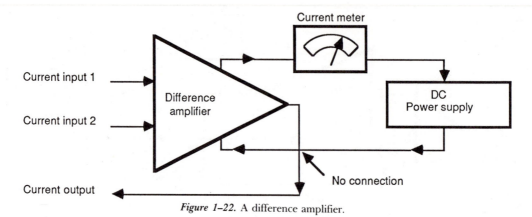

Figure 1–22. A difference amplifier.

SOME PRACTICAL CONSIDERATIONS

The power cord supplied with most instruments is the type that terminates with a three-conductor plug (Fig. 1–23). Notice that the plug has

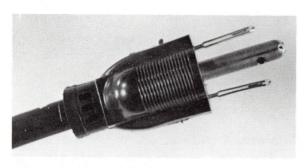

Figure 1–23

two parallel blade-shaped conductors. The supplied potential is provided to the instrument across this pair of conductors. Also notice that there is a rounded conductor below the other two. The rounded conductor is connected through a wire to the metal chassis of the instrument. It is through this rounded conductor pin that the instrument is grounded. A wire runs from the wall receptacle for this pin to an earth ground rod. When the instrument is plugged in, the grounding connection is made. This grounding provision is there to protect the instrument operator from electrical shock or electrocution. If any electrical failure were to cause the instrument cabinet or chassis to become electrified, the ground wire would conduct the current to ground and blow a fuse or trip a circuit breaker. Never cut off the grounding pin and plug an instrument into a two-prong receptacle. This could endanger the instrument's circuits and you. Do not use a three-pin to two-pin adapter. This can lead to the operation of an ungrounded instrument.

All instruments are provided with electrical fuses. A fuse is a special segment of conductor placed in the power conductor running to a circuit. Two such devices are illustrated in Figure 1–24.

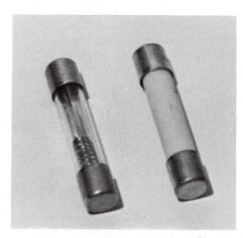

Figure 1–24

The conductor is visible on some fuses through the glass that is connected between the end caps. The conductor is made from a low melting, fairly resistive metal alloy. If the current being drawn by a circuit becomes excessive, the heat generated by the current as it passes through the fuse melts the conductor and disconnects the circuit. Some fuses have a ceramic insulator between the end caps.

If an instrument suddenly loses electrical power, it is likely that the instrument has blown a fuse. Fuses are held in fuse holders (Fig. 1–25). Fuse holders are frequently located on the back of an instrument. Before changing a fuse, one should shut down and unplug the instrument. The cap of the fuse holder should be removed and the fuse pulled out and inspected. One can sometimes tell by visual inspection that a fuse is blown. A fuse can also be checked with an ohmmeter. A fuse should have nearly zero resistance. If a fuse has been blown,

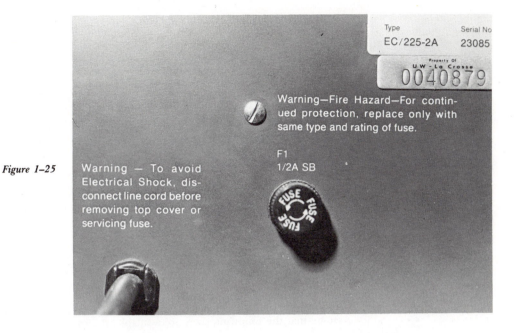

Figure 1–25

a new fuse of identical rating should be installed. The instrument can then be plugged in and powered up in the normal manner. When this is done, the instrument frequently will blow the new fuse. This occurs because of the ongoing electrical problem that was responsible for the loss of the first fuse. This is the time to call a service person. Do not put in a larger fuse. Do not bypass the fuse holder. Such procedures will leave an already troubled instrument with no protection. Performing such procedures generally leads to a seriously damaged instrument or an electrical fire.

When doing periodic maintenance on lab instruments, inspect for damaged or frayed insulation on power cables and exposed wires. A service person should replace any damaged wires.

Limit your service work on electrical systems to those tasks that the manufacturer has delegated to the instrument owner. The trouble-shooting section of the instrument manual should be the limit. If you are measuring voltages at test points on an instrument, avoid touching the instrument and the voltmeter probes with both hands. Touching electrical circuits with both hands can lead to passage of current through the chest. If a lot of current is available, such as experience can be terminal. I always keep one hand in my pocket and use the other hand to connect the meter to the check points.

Problems

1. What is the resistance of a Nichrome wire 0.020 cm in diameter and 37 cm long?

2. What is the resistance of a Bakelite 588 rod 1.5 cm² in area and 10 cm long?

3. Consider the circuit

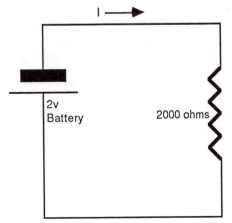

What is the current, I?

4. Consider the circuit

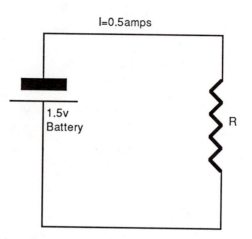

(a) What is R?
(b) What is the heat dissipated in R?
(c) If the resistor must be replaced with one of equal resistance, what size should be used? Assume that the following sizes are available: 0.25 watts, 0.50 watts, and 1.0 watts.

5. Consider the circuit

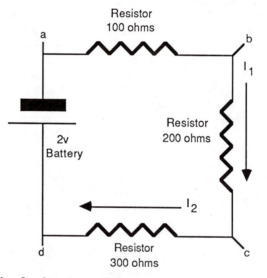

(a) What is the value for the current, I_1?
(b) What is the voltage between points a and b?
(c) What is the voltage between points b and c?
(d) What is the voltage between points c and d?
(e) What is the voltage between points a and c?
(f) What is the voltage between points b and d?
(g) What is the voltage between points a and d?
(h) What is I_2?
(i) How much heat is dissipated in the 300-ohm resistor?

6. Consider the circuit

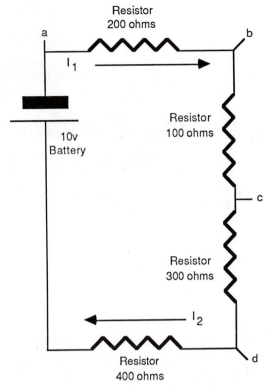

Using the circuit in the preceding figure, answer questions a through i from problem 5.

7. If a 10-milliamp meter of 1000 ohms internal resistance were to be used to measure voltage, what would be the highest readable voltage?

8. What is the voltage of five 2-volt batteries connected in series?

9. What is the voltage of five 2-volt batteries connected in parallel?

Electromagnetic Radiation and Some of Its Interactions with Matter

Many instruments in the clinical laboratory use electromagnetic radiation to measure concentration and/or composition of samples. These measurements are of extreme significance in the practice of medicine; it would be hard to overstate their importance. Because of the wide use of these methods of analysis, a thorough introduction to electromagnetic radiation is provided. The background here will be useful to you in your study of the next ten chapters.

ELECTROMAGNETIC RADIATION

Electromagnetic radiation is a form of energy that has some very unusual properties. First of all, its speed of propagation, or travel, is extremely great. Within the context of laboratory work one can assume that its speed is so great that it can traverse an instrument instantaneously. Secondly, much of the behavior of electromagnetic radiation can be explained in terms of wave propagation. Thirdly, electromagnetic radiation interacts with matter in ways that require that it be particulate in character. These properties define the boundaries of a dichotomy that has not as yet been fully reconciled. Because of the dual nature of radiant energy, we need to use both the wave theory and the particulate theory to explain its behavior.

The Wave Theory of Electromagnetic Radiation

Let us consider the wave nature of electromagnetic radiation. We can explain a lot of the behavior of electromagnetic radiation by assuming that radiant energy is propagated in sine waves as illustrated in Figure 2–1. The energy carried in a

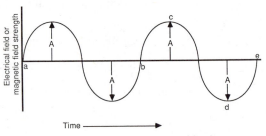

Figure 2–1. Radiant energy propagated in sine waves.

beam of radiant energy causes an electrical force field and, at 90° to the electrical, a magnetic force field. These fields vary with time in a cyclic manner. The distance a to b is called the wavelength (λ) of the radiant energy; the value of A is the wave amplitude and is related to the radiation intensity. The illustration shows two complete cycles of the electromagnetic radiation. The number of cycles that occur per second is the frequency of the radiant energy. We can picture a beam of radiant energy as being composed of many of these propagating sine waves.

Electromagnetic radiation of any wavelength can be produced. The wavelengths are somewhat arbitrarily divided into regions. Thus, we speak of the visible region of the electromagnetic spectrum as 380 to 800 nm in wavelength, the infrared region as about 1000 to 25,000 nm, and the ultraviolet region as about 200 to 380 nm. X-rays are shorter in wavelength than 10 nm, whereas γ rays are

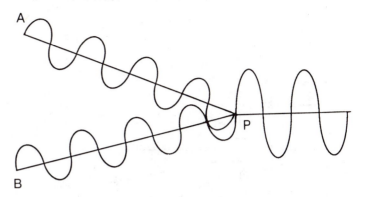

Figure 2–2. Constructive interference.

shorter than 0.1 nm in wavelength. The frequency of radiant energy is controlled by the source that produces it and is in no way affected by the medium through which it is propagated. The speed of propagation of radiant energy is affected by the density of the material through which it is being propagated. The speed of propagation is 3×10^{10} cm per second in a vacuum but is slower in more dense media.

The intensity of a beam of radiant energy is defined as the energy per unit area per second and is related to the square of the amplitude (A). The term intensity is often interchanged with the term power.

The wave model of radiant energy is particularly useful in explaining the properties of electromagnetic radiation called diffraction and interference. These topics are of great importance in the study of laboratory instruments, so they will be covered in some detail.

Constructive and Destructive Interference

When two beams of electromagnetic radiation arrive at point P, they may interact with each other either constructively or destructively. For example, if the distances AP and BP are the same, and if the electromagnetic radiation beams leaving A and B have the same frequency, and, further, if the electromagnetic radiation beams are in phase at points A and B (phase coherent), then, when they arrive at point P, the electrical and magnetic fields will combine, and the combined beam will have the same frequency but higher intensity. This process is called constructive interference and is illustrated in Figure 2–2.

By contrast, if the two beams of electromagnetic radiation are 180° out of phase at points A and B, the electrical and magnetic fields will destroy each other at point P. This is called destructive interference, and it is illustrated in Figure 2–3.

Diffraction of Electromagnetic Radiation

Diffraction of electromagnetic radiation is indeed a strange phenomenon. Diffraction is the name applied to the changes in the direction of propagation that occur when radiant energy interacts with the edge of a solid object. The phenomenon of diffraction is shown in Figure 2–4. Figure 2–4 shows a narrow slit illuminated by phase coherent monochromatic radiant energy. The radiant energy, rather than expanding in the shadow of the slit as one would expect, is found at much reduced intensity at many points on the screen. The smaller

Figure 2–3. Destructive interference.

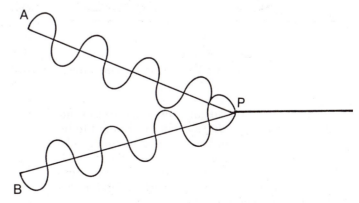

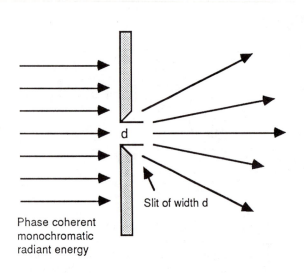

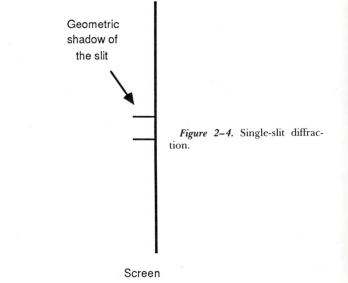

Figure 2–4. Single-slit diffraction.

Screen

the slit width, the more pronounced this phenomenon is; the larger the slit, the less pronounced. When the slit is larger than ten times the wavelength of the radiant energy, the diffraction is hardly noticed, but when d is equal to the wavelength, the diffraction is very great (Fig. 2–5).

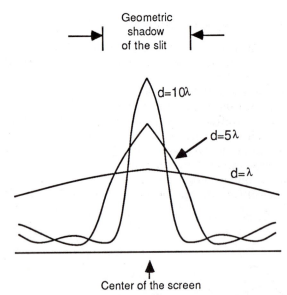

Figure 2–5. A plot of intensity versus position on the screen.

The property of radiant energy that causes diffraction is not well understood but is thought to be responsible for a number of experimentally observed phenomena. Examples include the Tyndall effect, the x-ray diffraction patterns for ionic materials, and the diffraction of electromagnetic radiation by diffraction gratings.

Polarized Electromagnetic Radiation

In Figure 2–1 electromagnetic radiation is depicted as a sine wave propagating through space. For many applications this simplistic model of electromagnetic radiation is sufficient; it allows for a reasonable explanation of the behavior of electromagnetic radiation. On the other hand, some applications require a more complex model.

Experiments have shown that a more accurate model of electromagnetic radiation is the one depicted in Figure 2–6. The figure shows two complete cycles of the electrical field associated with the electromagnetic radiation. The magnetic fields are not illustrated, but the student should remember that the magnetic fields are present. The illustration shows that the radiant energy is composed of two components at 90° to one another. The components are in phase and of equal amplitude. One component undulates in the XZ plane, while the other undulates in the YZ plane. This type of radiant energy is said to be unpolarized radiation and is typical of the radiation produced by most sources.

If the radiant energy depicted in Figure 2–6 were passed through a device that absorbed all of the radiation in the XZ plane, the transmitted radiation would be polarized (Fig. 2–7). Radiant energy of this type is frequently said to be polarized vertically. Polarized electromagnetic radiation has some important applications. Horizontally polarized radiation can of course also be isolated. The device that absorbs the unwanted component is called a dichroic filter.

The Particulate Theory of Light

For many years, the wave theory of electromagnetic radiation provided a good model for ex-

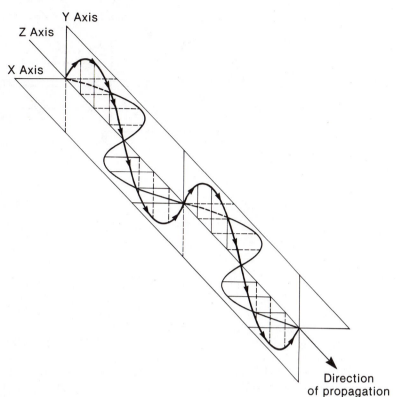

Figure 2–6. Monochromatic electromagnetic radiation. Only electrical field shown. (Adapted from: D.A. Skoog, Principles of Instrumental Analysis, 3rd Ed., Philadelphia, Saunders College Publishing, 1985.)

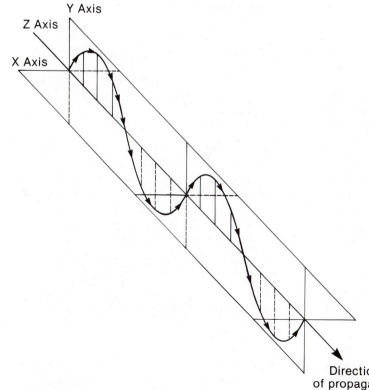

Figure 2–7. Polarized monochromatic electromagnetic radiation. (Adapted from: D.A. Skoog, Principles of Instrumental Analysis, 3rd Ed., Philadelphia, Saunders College Publishing, 1985.)

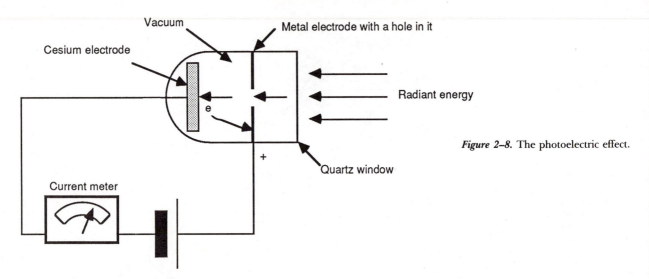

Figure 2–8. The photoelectric effect.

plaining all of the known properties of light. However, in the 1920s R. A. Millikan studied the photoelectric effect, which Albert Einstein then interpreted by postulating the particulate theory of electromagnetic radiation. A brief explanation of the photoelectric effect follows.

Millikan found that electromagnetic radiation can eject electrons from the surface of alkali metals. He studied this phenomenon with a tube somewhat like the one depicted in Figure 2–8. The cesium surface was given a negative charge, while the metal electrode with a hole in it was made positive. The hole was provided to let in the radiant energy. Because of the charge on the electrodes, any electrons ejected by radiant energy could be collected by the metal electrode and would cause electrical current to be detected on the d'Arsonval galvanometer. Millikan found that electrons could be ejected by the radiant energy only if its wavelength were short enough.

Einstein postulated that the reason for Millikan's results is that the electromagnetic radiation is composed of *photons,* or particles of energy, and that those photons of longer wavelength do not possess enough energy to knock the electrons free. No two smaller photons can combine to eject the electrons because they cannot interact simultaneously with the cesium atoms. Other workers have studied Einstein's theory and have found widespread applicability. The theory is useful for understanding atomic and molecular spectroscopy.

The Interaction of Photons and Atoms

Further support for the particulate nature of electromagnetic radiation can be found in the study of atoms and their interaction with electromagnetic radiation. For example, if sodium vapor is introduced into a vacuum tube and the tube is provided either heat energy or electrical energy, the sodium vapor will glow with a yellow appearance. If one

carefully studies the emitted radiant energy, one finds that it is composed of only a few wavelengths. For example, radiation of wavelength 589 nm is the most predominant, but radiant energy of wavelengths 285 nm, 330 nm, 568 nm, 819 nm, and 1139 nm are also present. Careful study has further shown that the emitted radiant energy can be correlated with changes in the energy of the valence electrons of sodium atoms. Thus, when electrical energy excites a sodium atom, the 3s electron is increased in energy to some other orbital, and when the excited atom loses the excess energy, a photon of radiant energy is emitted. When the radiant energy is emitted, the electron returns to the 3s level. Consider Figure 2–9.

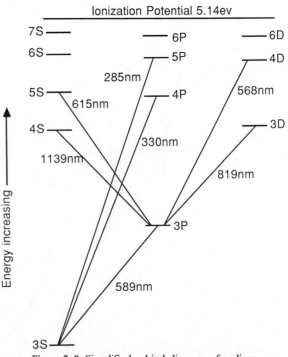

Figure 2–9. Simplified orbital diagram of sodium.

When an electron that has been driven to the 3p energy level (orbital) reverts back to the 3s level, a particle of radiation of wavelength 589 nm is emitted. When an electron has been excited to the 4p level by the electrical energy, it will eventually (10^{-12} to 10^{-15} sec) return to the 3s level and emit a 330 nm photon in the process. It is by this type of process that all atoms and molecules emit electromagnetic radiation.

It is also interesting to note that although sodium vapor can absorb photons of radiant energy, it can only absorb photons of particular wavelengths. Thus, sodium vapor will not absorb a 580 nm photon, nor will it absorb a 590 nm photon; they have the wrong energy. The sodium atoms will, however, absorb 589 nm photons very well.

Although the emission and absorption of electromagnetic radiation by molecules are quantitatively more complex, the principles are much the same. Molecules have electrons in various energy levels, and there are empty molecular orbitals in the molecular energy levels. Hence transitions of the electrons between full and empty molecular orbitals account for the absorption and emission of radiation by molecules.

In summary, the absorption and emission processes involve discrete, known transitions between higher and lower energy levels in specific atoms or molecules. They are known as quantized transitions that require photons of just the right energy. The wavelengths at which they occur are characteristic of the atoms or molecules that are undergoing the transitions; the wavelengths emitted or absorbed can frequently be used for qualitative analysis.

Analytical Applications of the Absorption of Radiant Energy; Absorption Spectroscopy

If one were to consider all the samples analyzed in the world, one would very probably find that well over 50 per cent of the samples had been analyzed by some form of absorption spectroscopy. This is especially true if one considers samples of biological origin. Although absorption spectroscopy can be performed using any of the wavelength ranges of the electromagnetic spectrum, those most commonly used in the clinical laboratory are the visible and ultraviolet regions. This chapter has been written as introductory material for the chapters on visible, ultraviolet, and atomic absorption spectroscopy; the information in this chapter has general application to an understanding of those forms of absorption spectroscopy. It is also applicable to infrared spectroscopy.

LAMBERT'S EXPERIMENT

In about 1760 Lambert performed experiments concerning the interaction of monochromatic electromagnetic radiation with varying thicknesses of absorbing material. In relating his work to modern absorption spectroscopy, it is best to think of the absorbing material as a solution prepared from a radiation-absorbing compound and a nonabsorbing solvent. The resulting solution is held in a cell made of a nonabsorbing material. Consider Figure 3–1.

In Figure 3–1 we see a cell b centimeters on a side. The cell is filled with a solution prepared from an absorbing solute and a nonabsorbing solvent. A beam of monochromatic radiation of an absorbable wavelength is incident on the cell. The angle of incidence is 90° relative to its face. The small reflective losses are not illustrated. The intensity of the incident beam is Po; the intensity of the transmitted beam is P. As the compound in the solution absorbs some of the monochromatic radiation, P must be less than Po. The ratio P/Po is called the transmittance and is given the symbol **t**. Assume for the sake of discussion that the value of the transmittance in Figure 3–1 is 0.5. Lambert found that the same solution when placed in a cell 2b centimeters thick produces a transmittance of 0.25, and if the same solution were placed in a cell 3b centimeters on a side, the transmittance would be 0.125. The mathematical equation that predicts this behavior is

$$(2.303) \log P/Po = -kb \qquad \text{Equation 3–1}$$

where k is a constant related to the concentration of the solute and the solute's intrinsic ability to absorb radiation at the wavelength supplied.

Thus, as b increases linearly, the transmittance decreases exponentially. Moreover, as b increases to large values, P approaches zero. As the transmittance is equal to P/Po, the transmittance also approaches zero as b increases to very large values.

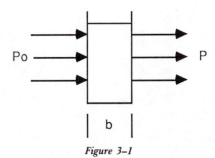

Figure 3–1

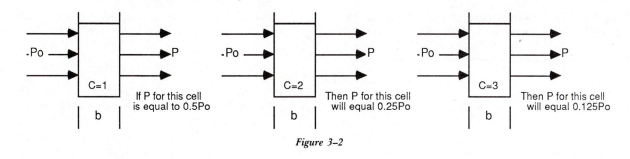

Figure 3–2

BEER'S EXPERIMENT

In about 1850 Beer studied the transmittance of solutions of a given compound under differing concentration conditions. He kept the cell path constant and used monochromatic radiation of an absorbable wavelength. His work showed that for a given cell path and compound, linear increases in concentration are accompanied by exponential decreases in transmittance (Fig. 3–2). If P/Po were equal to 0.5 for the solution on the left, then the transmittance of the solution in the middle would be 0.25, and that of the one on the right would be 0.125. Just as expected from the similarity to Lambert's work, the equation that relates this behavior is

$$(2.303) \log P/Po = -hbC \qquad \text{Equation 3-2}$$

where $h = k/C$

In order to simplify the foregoing equation, the constants are combined to form a new constant. Thus

$$-\log P/Po = \epsilon bC$$

$$\text{or} \qquad\qquad \text{Equation 3-3}$$

$$-\log t = \epsilon bC$$

$$\text{where } \epsilon = \frac{h}{2.303}$$

As $-\log t$ is a lot to write or say, it has been given a simpler name and symbol; it is called the absorbance (A). Thus

$$-\log P/Po = A = \epsilon bC \qquad \text{Equation 3-4}$$

where $A = -\log t = -\log P/Po$

A second name applied to A is the optical density. It is significant to note that as b or C increases linearly, so does A. This is in direct contrast to the exponential decrease one finds for t as b or C increases linearly. This fact is illustrated in Figures 3–3 and 3–4. As the transmittance changes from 0 to 1, absorbance varies from infinity to zero. Transmittances, then, are decimal numbers between 0 and 1. As people don't like numbers less than 1 very well, transmittance values are often multiplied by 100 in order to make them larger numbers. The multiplied equivalent of transmittance is the per cent transmittance, symbolized as T. T varies from 0 to 100 per cent, whereas t varies from 0 to 1. The interrelationship of t, T, and A is given in Table 3–1.

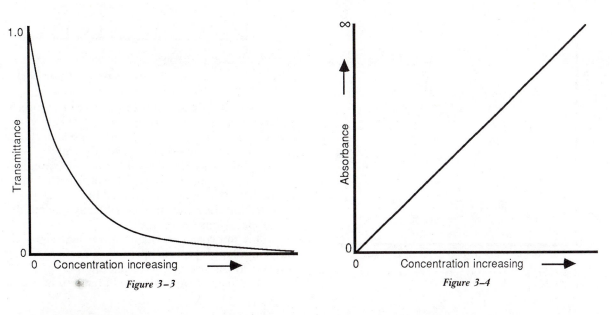

Figure 3–3   *Figure 3–4*

TABLE 3–1. Comparison of Transmittance, Per Cent Transmittance, and Absorbance for Selected Values of Transmittance

t	T	A
0	0	∞
0.001	0.1%	3
0.01	1%	2
0.1	10%	1
1	100%	0

Referring to Equation 3–4, let us briefly consider the properties of the constant ϵ. Epsilon is called the molar absorptivity and can be thought of as the absorbance of a 1-molar solution of an absorbing solute when held in a 1-cm cell. As concentration is expressed in moles per liter (mol/l) and the cell path b in centimeters, in order for the absorbance to be unitless, epsilon must be expressed in liters per mole centimeters (l/mol cm). For a given compound, epsilon varies greatly with wavelength. The constant ϵ may have values as low as zero and as high as several million.

Molar absorptivity is somewhat predictable, as similar compounds have somewhat similar molar absorptivities. This relationship is very approximate, however. The value of epsilon may be used in attempts to perform qualitative analysis. This is possible because the experimentally determined epsilon values produced by a compound should be the same as the published values. This statement assumes that the values of ϵ are determined on high-quality equipment. Thus, if an unknown and a sample of the pure material have the same epsilon values at all wavelengths, they *may* be the same compound. If the molar absorptivities are not the same at all wavelengths, the two compounds cannot be the same. It is important to remember that ϵ does vary with wavelength (Fig. 3–5).

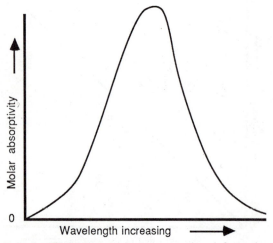

Figure 3–5. Molar absorptivity versus wavelength for a selected compound.

Now that the molar absorptivity is more fully appreciated, let us consider the use of Equation 3–4.

$$A = \epsilon bC = -\log P/P_o$$

This equation is called Beer's law and illustrates the fundamental relationship between the experimentally measured values of P_o, P, and the molar concentration of the absorbing solute. The equation is the mathematical basis of quantitative analysis by absorption spectroscopy.

QUANTITATIVE ANALYSIS

Quantitative analysis by absorption spectroscopy is done in three ways. The first way is based on a graphical display of Beer's law. In this approach, the absorbance values for several standard solutions of known concentration are determined and graphed. The resulting graph is called a Beer's law plot or a standard calibration curve (Fig. 3–6).

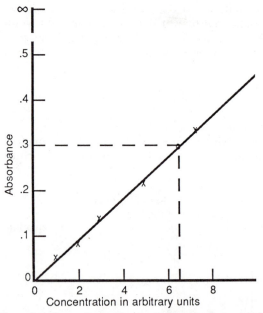

Figure 3–6. A standard calibration curve (Beer's law plot).

In the Beer's law plot illustrated in Figure 3–6, five standards have been plotted. Their values are marked with "x." An unknown with absorbance of about 0.3 is marked on the plot with an "o." One can see that the concentration that corresponds to an absorbance of 0.3 is about 6.3 arbitrary units. In this way the standard curve or Beer's law plot is used to find the concentration of analyte. Note that the slope of this curve is ϵb.

This approach to quantitative analysis is the only popular graphical approach. No one has much interest in the use of curves like the one in Figure 3–3. Too many standards would have to be run in order to gain an intimate knowledge of the curve. At least 15 standards would be required before an unknown could be analyzed, whereas curves like the one in Figure 3–6 are frequently established on the basis of three points.

The second way quantitative anazlysis is approached involves a simple calculation. If ϵ and b are known values, the concentration of the analyte may be calculated from the absorbance of the unknown solution. The instrument being used may provide the absorbance directly, or the analyst may have to calculate it from the per cent transmittance. When this approach is used, the calculation takes on the form

$$C = \frac{A_u}{\epsilon b} \qquad \text{Equation 3–5}$$

where C is measured in moles per liter and A_u is the absorbance of the unknown. Consider an example problem.

Example 1. A solution of vanillin has an absorbance of 0.13 at its wavelength of maximum absorbance (λ_{max}). The molar absorptivity of the compound at the λ_{max} is 1.19×10^4 l/mol cm. The solution is held in a 1.10-cm cell. What is the solution concentration in moles per liter?

$$A = \epsilon b C$$

$$C = \frac{A}{\epsilon b}$$

$$C = \frac{A}{(1.19 \times 10^4 \text{ l/mol cm})(1.10 \text{ cm})}$$

$$= \frac{0.13}{(1.19 \times 10^4 \text{ l/mol cm})(1.10 \text{ cm})}$$

$$C = 9.93 \times 10^{-6} \text{ mol/l}$$

At times, the user of the data prefers that the concentration of the analyte be supplied in units other than moles per liter. In that case a unit correction factor must be used. Beer's law becomes

$$C' = \frac{A_u f}{\epsilon b}$$

where C' can have any concentration units

A_u is the absorbance of the unknown solution

f is a unit correction factor

ϵ is the molar absorptivity

b is the cell path

Consider an example problem.

Example 2. Reconsider example 1, but calculate the results in mg per dl. The molecular weight of vanillin is 152 g/mol.

$$C' = \frac{A_u f}{\epsilon b}$$

$$C' = \frac{A_u}{(1.19 \times 10^4 \text{ l/mol cm}) 1.10 \text{ cm}}$$

$$(1.52 \times 10^2 \text{ g/mol})(10^3 \text{ mg/g})\left(\frac{1}{10 \text{ dl}}\right)$$

where 10^3 mg/g is the conversion from grams to milligrams and 1/10 dl is the conversion from liters to deciliters.

Thus $f = 1.52 \times 10^4 \dfrac{\text{mg l}}{\text{mol dl}}$

$$C' = \frac{A_u\ 1.52 \times 10^4 \text{ mg l}}{1.19 \times 10^4 \text{ l/mol cm } (1.10 \text{ cm}) \text{ mol dl}}$$

$$C' = (1.16)A \text{ mg/dl}$$

$$C' = .15 \text{ mg/dl}$$

Sometimes an experimental procedure affects the concentration of the analyte. This can happen when a sample is carried through a cleanup procedure; this type of cleanup is sometimes necessary to remove interfering compounds. It can also happen when dilution of the sample is unavoidable. When this is the case, the factor f contains corrections for the experimental procedure. Consider the next example.

Example 3. Consider the following procedure for the analysis of porphobilinogen in urine (see reaction below). In this procedure, 25.0 ml of urine from a 24-hour sample is treated by a separation procedure that isolates the porphobilinogen in 25.0 ml of water. Two ml of the water solution is treated with 2.0 ml

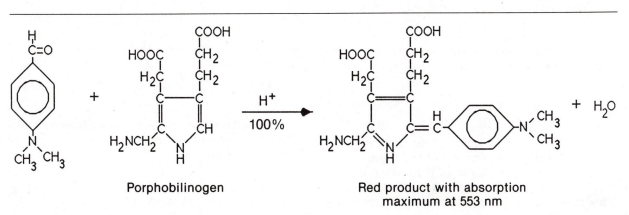

Porphobilinogen   Red product with absorption
maximum at 553 nm

of p-dimethyl-aminobenzaldehyde solution. The resulting solution turns red; the absorbance of the solution is directly proportional to the original porphobilinogen concentration. It is measured in a 1-cm cell.

The molar absorptivity of the red product has been estimated as 3.80×10^4 l/mol cm. The molecular weight of the porphobilinogen is 226.2 g/mol. What is the equation that gives the porphobilinogen concentration in μg per dl of urine? Let's start with Equation 3–5.

$$C = \frac{A_u}{\epsilon b}$$

Equation 3–5 gives the concentration, in moles per liter, of the red product in the cell. One must provide a modification to the equation in order to get the concentration in μg per dl.

$$C' = \frac{A_u f}{\epsilon b}$$

where f = (unit correction)(dilution factor)

C' = the concentration in the μg/dl

$$C' = \frac{A_u f}{(3.8 \times 10^4 \text{ l/mol cm})(1.00 \text{ cm})}$$

Substituting for f, ϵ, and molecular weight,

$$C' = \frac{A_u \text{mol}}{3.80 \times 10^4 \text{l}} \left(\frac{226 \text{ g}}{\text{mol}}\right)\left(\frac{10^6 \mu\text{g}}{\text{g}}\right)\left(\frac{1}{10 \text{ dl}}\right)\left(\begin{matrix}\text{dil.}\\\text{factor}\end{matrix}\right)$$

This equation gives the concentration in μg/dl of porphobilinogen that would be in the cell had it not reacted with p-dimethylaminobenzaldehyde. The concentration in the urine is 2C′ because the color development reagent cut the concentration in half. Thus the dilution factor is 2.
Plugging in the dilution factor

$$C_{urine} = 2C' = \frac{A_u(226)(10^5)2}{3.80 \times 10^4}$$
$$= 1.19 \, A_u \times 10^3 \, \mu\text{g/dl}$$

The third way that quantitative analysis is done by absorption spectroscopy is by the analysis of a standard of known concentration followed by the analysis of the unknown. The concentration of the unknown can be calculated

$$C = \left(\frac{A_u}{A_s}\right) C_s \qquad \text{Equation 3–6}$$

where C is the concentration of the unknown
$\quad$ C$_s$ is the concentration of the standard
$\quad$ A$_u$ is the absorbance of the unknown
$\quad$ A$_s$ is the absorbance of the standard
This approach to the determination of concentra-

tion depends on the fact that the standard and unknown are carried through identical procedures. The assumption is made that the method is linear over the needed range of concentrations; that is, an unknown twice as concentrated as the standard will have twice the absorbance of the standard. The use of Equation 3–6 is common in the clinical laboratory. I have illustrated the approach by the uric acid method from *Fundamentals of Clinical Chemistry,* 2nd Ed. This volume, edited by N. W. Tietz, is one of the best clinical analysis texts in the world.

Please consider the method and its calculations; it has been simplified for teaching purposes.

Example 4.
1. For plasma or serum, prepare a protein-free filtrate by mixing 1.0 ml of plasma or serum with 8.0 ml of water, 0.5 ml of sulfuric acid, and 0.5 ml of sodium tungstate. Mix and filter. This is a 1/10 dilution that is protein free.
2. Measure the following into test tubes.

	Standard	Unknown	Blank
Protein-free filtrate (1/10)	0	3.0 ml	0
Distilled water	0	0	3.0 ml
Uric acid standard (1 mg/dl)	3.0 ml	0	0
Sodium carbonate (14%) (pH adjusting reagent)	1.0	1.0	1.0
Phosphotungstic acid (color- producing reagent)	1.0	1.0	1.0

3. Let the colored product form for 15 minutes.
4. Measure the absorbance of the standard and unknown at 710 nm. Use the blank to set the instrument to 0 absorbance.

$$C = \left(\frac{A_u}{A_s}\right) 1 \text{ mg/dl} \times 10$$

The 10 in the previous equation compensates for the 10-fold dilution in step 1. Otherwise the unknown, blank, and standard solutions were treated exactly the same.

Each of these three approaches to quantitative analysis has its place in the laboratory; each method has its advantages. The graphical approach serves to check method linearity and give warning of a bad standard or absorbance. The use of

$$C = \frac{A_u}{\epsilon b} \qquad (3–5)$$

is quick and dirty but provides for no internal check of the instrument or of the linear range of the method. The use of

$$C = \frac{A_u}{A_s} C_s \qquad (3–6)$$

while popular, is not much better than the use of Equation 3–5. The determination of A_s on the same instrument as is used to determine A_u can lead to some error reduction. The use of either Equation 3–5 or Equation 3–6 is a compromise of the internal checks of the graphical method in the interest of saving time.

THE ABSORPTION SPECTRUM

In Figure 3–5 you saw that the molar absorptivity of a compound changes with wavelength. Thus, for a solution of a given concentration, when held in a cell of a given thickness, the absorbance must also vary with wavelength. This fact is evident from the equation

$$A = \epsilon bC \qquad (3-4)$$

A plot of a compound's absorbance versus the wavelength is called an absorption spectrum (Fig. 3–7).

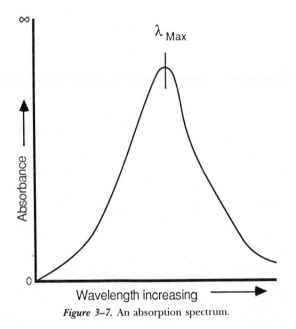

Figure 3–7. An absorption spectrum.

Keep in mind that the actual absorbance is affected by concentration and cell path but that the spectral shape is not changed by these variables. For example, with cell path held constant, three solutions of concentrations 1×10^{-4} mol/l, 2×10^{-4} mol/l, and 3×10^{-4} mol/l would produce the spectra illustrated in Figure 3–8. Notice that both λ_{max} and the curve shape for the three solutions have stayed the same. The molar absorptivity of the compound, at λ_{max}, is the same for each solution. The molar absorptivity of the compound in each of the three solutions is the same at any given wavelength.

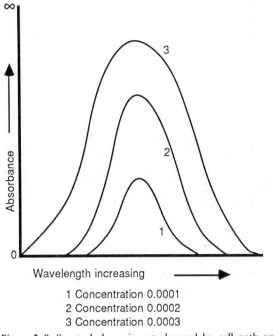

1 Concentration 0.0001
2 Concentration 0.0002
3 Concentration 0.0003

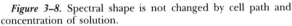

Figure 3–8. Spectral shape is not changed by cell path and concentration of solution.

The absorption spectrum of a compound is useful for two reasons. First, most compounds have at least small differences in their spectra. These variations can range from large differences in peak shape to small differences in λ_{max} or molar absorptivity. In either case, the spectrum of a bona fide sample can be compared with that of an unknown compound in an attempt to do qualitative analysis. If the spectra are obtained experimentally on solutions of equal concentration, identical spectra suggest that the two compounds are the same. If the spectra are different, the compounds cannot be the same. This approach to qualitative analysis is very popular in toxicology.

Second, the spectrum of a compound must be used to select the wavelength that will be used for purposes of quantitative analysis. Consider Figure 3–9.

When an analyst wants to do quantitative analysis of a compound, he or she must select a wavelength at which the absorbances of unknowns, standards, and blanks will be measured. Here are some of the considerations that go into the selection of the wavelength. First, λ_1 is no good at all because the molar absorptivity at this wavelength is zero. Under such conditions all solutions of the analyte would have an absorbance of zero. As a possible wavelength λ_2 has three disadvantages. The first problem with λ_2 is that the absorbance is changing rapidly with wavelength. If a standard curve were determined at λ_2 and then a slight error in wavelength setting were incurred, the values obtained for unknowns would be badly affected. A second

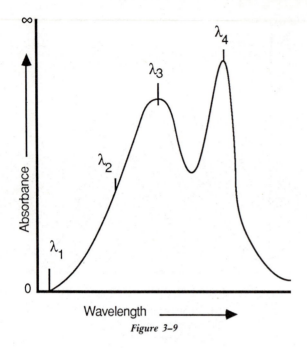

Figure 3–9

problem with λ_2 is that a small amount of compound would produce only one-half the absorbance that the same solution would produce at λ_3. A third problem with λ_2 is that on instruments of poor quality, a nonlinear Beer's law plot could occur (this problem will be discussed in Chapter 5). On the other hand, λ_3 would produce much greater absorbance for a given amount of compound and would be least sensitive to absorbance error from small errors in the setting of wavelength. The molar absorptivity at $\lambda_3 \pm 1$ nm has about the same value through the specified range. The nonlinear Beer's law plot problem associated with wavelengths in which absorbance changes rapidly with wavelength is not a problem at λ_3. On the other hand, λ_4 would produce the greatest absorbance for a given amount of compound, but it would be subject to this type of error and would also be susceptible to absorbance errors caused by slight errors in the setting of the wavelength. Thus, of all the wavelengths on the spectrum in Figure 3–9, λ_3 is the best and would be the appropriate choice for use in quantitative analysis. The only exception to this statement would be an analysis requiring the lowest limit of detection. In this case λ_4 would be best.

SIMULTANEOUS SPECTROPHOTOMETRIC ANALYSIS

From time to time an analyst finds that the analyte is not the only compound in the sample that absorbs at a particular wavelength. When this is the case, the absorbing second compound is an interference in the analysis. Interferences can be handled

in several ways. Sometimes a separation procedure can remove the interfering compound from the solution. This procedure leaves the analyte in solution but free from the interfering compound. Sometimes a new wavelength can be selected. If the analyte absorbs well at some new wavelength but the interfering compound does not, then the problem with the interfering compound is solved. When the interfering compound is chemically similar to the analyte, it is frequently difficult to remove the interfering compound by a separation step. The chemical similarity also means that the absorption spectra of the two compounds are likely to be similar. In such a case, the use of simultaneous spectrophotometric analysis is a good solution to the problem.

Simultaneous spectrophotometric analysis is based on the fact that absorbing solutes absorb independently of each other. The sum of the absorbances of the absorbing compounds is the absorbance of the mixture. For a two-component mixture, this can be written in equation form

$$A_{\lambda_1} = (\epsilon_A)_{\lambda_1}bC_A + (\epsilon_I)_{\lambda_1}bC_I \qquad \text{Equation 3–7}$$

where C_A is the analyte concentration
C_I is the concentration of the interfering compound
$(\epsilon_A)_{\lambda_1}$ is the molar absorptivity of the analyte at λ_1
$(\epsilon_I)_{\lambda_1}$ is the molar absorptivity of the interference at λ_1

It is apparent that Equation 3–7 contains two unknowns, C_A and C_I. When one has two unknowns but only one equation, it is necessary to find a second independent equation, which can be done by measuring the absorbance of the mixture at a second wavelength. One may write

$$A_{\lambda_2} = (\epsilon_A)_{\lambda_2}bC_A + (\epsilon_I)_{\lambda_2}bC_I \qquad \text{Equation 3–8}$$

The two equations can then be used to find C_A and C_I. The only requirement is that both of the selected wavelengths must be wavelengths where both compounds absorb. The use of the two equations does require a knowledge of $(\epsilon_A)_{\lambda_1}$, $(\epsilon_A)_{\lambda_2}$, $(\epsilon_I)_{\lambda_1}$, and $(\epsilon_I)_{\lambda_2}$. These values can be determined by measuring the absorbance at λ_1 and λ_2 of a solution of known concentration of the analyte. If the process is repeated by determining the absorbance at λ_1 and λ_2 of a solution of known concentration of the interference, one can calculate $(\epsilon_I)_{\lambda_1}$ and $(\epsilon_I)_{\lambda_2}$. These four ϵ values can then be substituted into equations 3–7 and 3–8. The resulting equations can then be used to find C_A and C_I on the basis of A_{λ_1} and A_{λ_2} for the mixture. Consider an example.

Vanillylmandelic acid and *p*-hydroxymandelic acid are two urinary phenolic compounds. The vanillylmandelic acid concentration is of diagnostic sig-

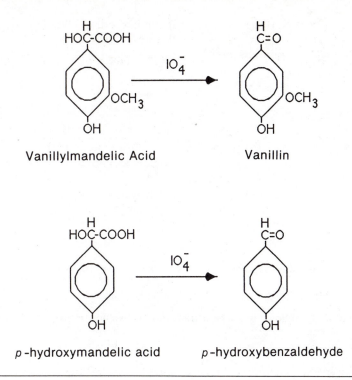

Vanillylmandelic Acid   Vanillin

p-hydroxymandelic acid p-hydroxybenzaldehyde

nificance; p-hydroxymandelic acid is an interfering compound. The two compounds are difficult to separate, but the mixture of the two can be extracted from the urine. In the course of the procedure, the two compounds are coextracted from the urine and finally find their way into a sodium carbonate solution. For reasons beyond the limit of this discussion both compounds are then oxidized with periodate to form vanillin and p-hydroxybenzaldehyde (see reaction above). These two compounds are then determined by ultra-violet absorption spectroscopy. The vanillylmandelic acid concentration is found by ascertaining the vanillin concentration and working mathematically back through the procedure. With this background, let's consider some representative data.

Vanillin is the analyte; its λ_{max} is 355 nm. A 1.00×10^{-5} molar vanillin solution produces an absorbance of 0.269 at 355 nm. It also produces an absorbance of 0.187 at 340 nm. The cell is 1.00 cm in all cases. p-Hydroxybenzaldehyde is the interference; its λ_{max} is 340 nm. A 1.10×10^{-5} molar solution produces an absorbance of 0.175 at 355 nm. The same solution also produces an absorbance of 0.311 at 340 nm.

Calculating the ϵ_A value for vanillin at 355 nm,

$$A = \epsilon bC$$

$$0.269 = (\epsilon_A)_{355} \ (1 \ cm) \ 1.00 \times 10^{-5} \ mol/l$$

$$(\epsilon_A)_{355} = 2.69 \times 10^4 \ l/mol \ cm$$

Calculating the ϵ_A value for vanillin at 340 nm,

$$A = \epsilon bC$$

$$0.187 = (\epsilon_A)_{340} \ (1 \ cm) \ (1.00 \times 10^{-5} \ mol/l)$$

$$(\epsilon_A)_{340} = 1.87 \times 10^4 \ l/mol \ cm$$

In a similar way for p-hydroxybenzaldehyde

$$(\epsilon_I)_{355} = 1.59 \times 10^4 \ l/mol \ cm$$

$$(\epsilon_I)_{340} = 2.83 \times 10^4 \ l/mol \ cm$$

If these values are substituted into Equations 3–7 and 3–8, we obtain

$$A_{355} = (2.69 \times 10^4)(1.00)(C_A) + (1.59 \times 10^4)(1.00)(C_I)$$

and

$$A_{340} = (1.87 \times 10^4)(1.00)(C_A) + (2.83 \times 10^4)(1.00)(C_I)$$

These two equations can be used to calculate the vanillin concentration in any sample. One must measure A_{355} and A_{340} for the sample and then calculate. For example, a patient's urine was carried through the procedure and was subjected to spectroscopic evaluation.

A_{355} was found to be 0.173
A_{340} was found to be 0.121

$$0.173 = (2.69 \times 10^4)(1.00)C_A + (1.59 \times 10^4)(1.00)(C_I)$$

Equation 3–9

$0.121 = (1.87 \times 10^4)(1.00)C_A + (2.83 \times 10^4)(1.00)(C_I)$

$$\text{Equation } 3\text{--}10$$

Multiplying Equation 3–10 by $-1.59/2.83$

$$\frac{-1.59}{2.83}(.121) = \frac{-1.59}{2.83}(1.87 \times 10^4)(1.00)C_A$$
$$+ \frac{(-1.59)}{2.83}(2.83 \times 10^4)(cm)C_I$$

we obtain

$$-.0680 = (-1.05 \times 10^4)(1 \text{ cm})C_A + (-1.59 \times 10^4)(1 \text{ cm})C_I$$

If the preceding equation is added to Equation 3–9, one obtains

$$0.173 - .068 = (6.69 \times 10^4 - 1.05 \times 10^4) \, 1 \text{ cm } C_A + 0$$

and

$$C_A = \frac{.105}{(5.64 \times 10^4) \, 1 \text{ cm}}$$
$$= 1.86 \times 10^{-6} \text{ mol/l vanillin in the cell}$$

If one wants to know C_I, C_A can be substituted into Equation 3–9 or 3–10 and C_I can be obtained.

Problems

1. Convert the following transmittance values to absorbance values.
   - (a) 0.10
   - (b) 0.235
   - (c) 0.583
   - (d) 0.010
   - (e) 0.001
   - (f) 0.862

2. Convert the following per cent transmittance values to absorbance values.
   - (a) 10.0%
   - (b) 23.5%
   - (c) 58.3%
   - (d) 1.00%
   - (e) 0.10%
   - (f) 8.67%

3. Convert the following absorbance values to per cent transmittance values.
   - (a) 0.100
   - (b) 0.55
   - (c) 1.00
   - (d) 2.00
   - (e) 1.20
   - (f) 0.051

4. A solution produces a per cent transmittance value of 33 per cent when held in a 1.16-cm cell. What will be the transmittance of the same solution when measured in a 2.00-cm cell? Assume that all measurements are made at the same wavelength.

5. A 10.0-ppm solution of an organic dye produces a transmittance of 0.831 when measured in a 0.80-cm cell. What is the absorbance of a 23.0-ppm solution of the same dye when measured in a 2.20-cm cell? Assume that all the measurements are made at the same wavelength.

6. The molar absorptivity of quinoline at 227 nm is 37,000 l/mol cm. If the solution is held in a 2.00-cm cell and the absorbance is 0.850:
   - (a) What is the molar concentration of quinoline in the solution?
   - (b) What is the quinoline concentration in parts per million? The molecular weight of quinoline is 129.2 g/mol. Assume that water is the solvent. The density of water is 1.00 g/ml.
   - (c) What is the quinoline concentration in milligrams per deciliter?

7. Epsilon (ϵ) for anthracene is 199,000 l/mol cm. A solution held in a 1-cm cell produced an absorbance of 0.35. The molecular weight of anthracene is 178.2 g/mol. What is the concentration:
   - (a) in mol/l?
   - (b) in g/ml?
   - (c) in mg/dl?

8. Consider the two cells illustrated in the next figure. If the two cells contain the same solution, what is the absorbance of the second cell?

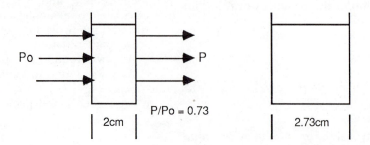

9. A 5.0-ppm solution of acetaminophen produces an absorbance of 0.529 when measured at 250 nm. The absorbance was measured using a 1.16-cm cell. Assume that the solvent is water and that the molecular weight of acetaminophen is 151.2. What is the molar absorptivity of acetaminophen at 250 nm? Acetaminophen is an analgesic.

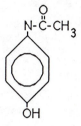

10. A 0.277-ppm solution of amitriptyline produces an absorbance of 0.014 when measured at 240 nm. The solution was held in a 1-cm cell; water was the solvent. The molecular weight of amitriptyline is 277.4 g/mol. What is the molar absorptivity of amitriptyline at 240 nm? The compound is an antidepressant.

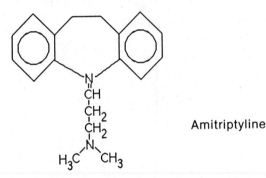

Amitriptyline

11. A 2.38 mg/dl solution of secobarbital produces an absorbance of 0.516 when measured at 260 nm. The solution was held in a 1.20-cm cell; dilute sodium hydroxide was the solvent. The molecular weight of secobarbital is 238.3 g/mol. What is the molar absorptivity of secobarbital at 260 nm? Secobarbital is a short-acting sedative.

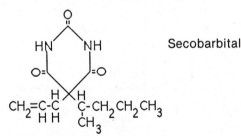

Secobarbital

12. $KMnO_4$ has a λ_{max} at 524 nm; the molar absorptivity at 524 nm is 2500 l/mol cm. A solution of $KMnO_4$ produces a transmittance of 0.95 when measured in a 1-cm cell.
    (a) What is the molar concentration of the $KMnO_4$ solution?
    (b) What is the concentration in mg/dl? The molecular weight is 158 g/mol.
    (c) If the solvent is water, what is the concentration of the solution expressed in per cent weight/weight?

13. Consider the following data:

Solution	Analyte Concentration	Absorbance
Standard 1	1 mg/100 ml	0.17
Standard 2	3 mg/100 ml	0.51
Standard 3	6 mg/100 ml	1.03
Blank	0	0.00
Sample 1	?	0.89

What is the concentration of the analyte in sample 1?

14. An acid organic compound is a pollutant in the river. If 10.0 ml of river water is extracted with 100 ml of $CHCl_3$ and 90 ml of the resulting chloroform solution is extracted with 5.0 ml of 0.1 molar NaOH, what is the concentration of the organic acid in the river water? The cell produced an absorbance of 0.37. The cell path was 1.00 cm and the compound's molar absorptivity is 1.0×10^4. Assume 100 per cent extraction in each step.

15. Three ml of blood serum is extracted with 50 ml of chloroform. 40 ml of the chloroform solution is extracted with 4.0 ml of base. The absorbance of the base solution is 0.40 when held in a 1-cm cell. Epsilon (ϵ) for compound Y is 4000 l/mol cm; its molecular weight is 150 g/mol. Assume 100 per cent extraction in each step. What is the concentration of compound Y in mg/100 ml?

16. If a two-component mixture has an absorbance of 0.70 at 350 nm and 0.79 at 400 nm, what is the concentration of compound A? What is the concentration of compound B? Assume a 1-cm cell.

Compound A
$\lambda = 350$ nm, $\epsilon = 10^4$ 1/mol cm

$\lambda = 400$ nm, $\epsilon = 5 \times 10^3$ 1/mol cm

Compound B
$\lambda = 350$ nm, $\epsilon = 7.5 \times 10^3$ 1/mol cm

$\lambda = 400$ nm, $\epsilon = 1.3 \times 10^4$ 1/mol cm

17. Consider the following figure.

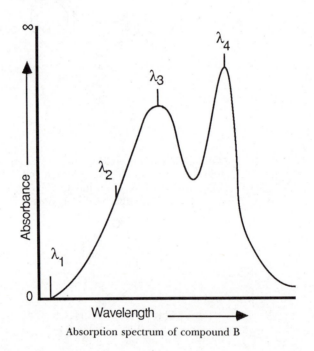

Absorption spectrum of compound B

(a) Which wavelength would be the best for trace analysis of compound B?
(b) Which wavelength would be the best for the analysis of compound B if it were present at greater than trace concentrations?
(c) Which wavelength would be useless for the analysis of samples for the concentration of compound B?

18. Consider the following data:

	A_{455}	A_{490}
10 ppm badger red	0.280	0.550
10 ppm wolverine yellow	0.800	0.150
Distilled H_2O	0.000	0.000
An unknown	0.655	0.400

Assume that all measurements were made in 1.20-cm cells. What is the concentration of badger red and wolverine yellow in the unknown? Give the values in parts per million.

19.   Consider the following data:

Uric acid concentration	Absorbance
1.00 mg/dl	0.261
0.75 mg/dl	0.176
0.50 mg/dl	0.114
Blank	0.000
1:10 protein-free dilution of a serum sample	0.134

What is the uric acid concentration in the serum sample? Assume that each solution was held in a 1.20-cm cell.

20.   Consider the following data:

Uric Acid Concentration	Absorbance
0.50 mg/dl	0.114
1:10 protein-free dilution of a serum sample	0.153
Blank	0.000

What is the uric acid concentration in the serum sample? Assume that each solution was held in a 1.20-cm cell.

4

Visible Absorption Spectroscopy

Visible spectroscopy is spectroscopy that is done using radiant energy that is between 380 and 800 nm wavelength. It is called visible spectroscopy because this wavelength range can be detected by the human eye. If one were to examine the color of various wavelengths of visible radiation, the results listed in Table 4–1 would be found.

TABLE 4–1. Colors of Various Wavelengths of Radiation

Radiation Viewed	Color Perceived
380–435 nm	Violet
435–480 nm	Blue
480–490 nm	Greenish blue
490–500 nm	Bluish green
500–560 nm	Green
560–580 nm	Yellowish green
580–595 nm	Yellow
595–650 nm	Orange
650–800 nm	Red

It is important to remember that sunlight and electric lights are polychromatic sources of radiant energy, that is, sources of white light. The perceived color of objects is generally caused by the interaction between polychromatic light and the object. This interaction results in the unabsorbed wavelengths being reflected to our eyes. For example, a blue car is absorbing yellow radiant energy from the polychromatic sunlight; our eyes then see the unabsorbed radiant energy that is reflected. As a second example, the purple solutions produced in

the determination of protein by the biuret method are really absorbing green photons of 540 nm wavelength (Table 4–2).

VISIBLE INSTRUMENTATION

In Chapter 3 the work of Beer and Lambert is presented, and it has been shown that all absorption spectroscopy requires "monochromatic" radiation.* It has also been shown that the radiation intensities P and Po must be measured in order for

---

* "Monochromatic" means approaching monochromatic. True monochromaticity is a goal that is seldom realized, whereas "monochromatic" (that is, approaching monochromatic) is generally attainable.

TABLE 4–2. The Relationship between Absorbed Wavelength and Observed Color When Viewed in White Light

Color of Object in White Light	Wavelength Absorbed	Color of the Absorbed Photons
Yellowish green	380–435 nm	Violet
Yellow	435–480 nm	Blue
Orange	480–490 nm	Greenish blue
Red	490–500 nm	Bluish green
Purple	500–560 nm	Green
Violet	560–580 nm	Yellowish green
Blue	580–595 nm	Yellow
Greenish blue	595–650 nm	Orange
Bluish green	650–800 nm	Red

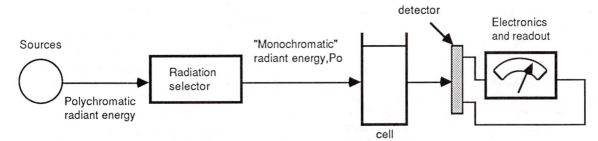

Figure 4–1. Simplified optical path of a spectrophotometric instrument.

the analyst to do qualitative and quantitative analysis. With these facts in mind, then, let us proceed with our study of visible spectroscopic instrumentation.

Consider the simplified optical path of a spectrophotometric instrument (Fig. 4–1). The source provides photons for Po; the radiation selector accepts polychromatic radiation and disperses it into "monochromatic" radiation. The cell holds the sample while the radiation intensity detector measures P and Po. The exact nature of these components depends on the nature of the measurements to be made, but many generalities concerning each part can be presented. Let's consider the parts of the instrument one by one.

Sources of Visible Radiation

Two closely related sources of visible radiation are common in instruments. The first is a tungsten lamp, which is made by carefully preparing a tungsten filament and enclosing it in an evacuated glass envelope (see Fig. 4–2). Lamps of this type are

Figure 4–2. Tungsten lamp. (Courtesy of Oriel Corporation, Stratford, Conn. Reprinted with permission.)

powered by well-regulated AC power supplies. The lamp output is a function of the temperature of the filament, which in turn is a function of the current through the filament. The current that is allowed to flow is carefully controlled so that the lamp output is constant with time. The lamp output, even at constant filament temperature, is a function of wavelength (Fig. 4–3). Note that as the lamp temperature increases, the output is greater and the wavelength of maximum emission shifts to shorter

wavelength. Although instrument designers have always desired higher intensity, the shortened lamp life has been a limiting factor.

A few years ago quartz halogen sources were introduced. Quartz halogen lamps are tungsten

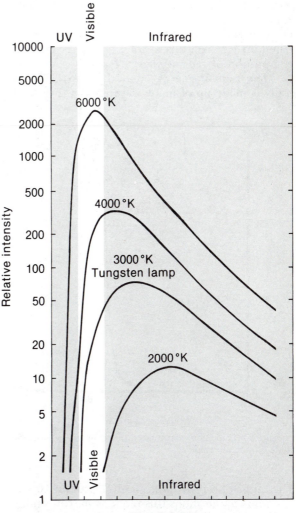

Figure 4–3. Adapted from: A.J. Diefenderfer, Principles of Electronic Instrumentation, 2nd Ed., Philadelphia, Saunders College Publishing, 1979.

lamps that have quartz envelopes. The lamps also contain a low pressure of iodine. These lamps are operated at higher temperatures than ordinary tungsten lamps. The quartz envelope is needed to withstand the higher temperature. The iodine is included in the lamp to keep the tungsten filament in good shape. Any tungsten atoms that boil off the filament will react with the iodine to form gaseous tungsten iodide. When tungsten iodide contacts the hot filament, it decomposes to form tungsten and iodine. This process tends to redeposit tungsten on the filament. Both quartz halogen and tungsten lamps are popular sources of visible radiation.

A less common source of visible radiation is the light-emitting diode (LED). These sources are manufactured from one of the following semiconductor materials: gallium arsenide (GaAs), gallium phosphide (GaP), or gallium arsenide phosphide (GaAsP). The semiconductors may be doped to form N and P type materials. (See Appendix C for greater detail.) When a suitable voltage is applied to an LED, it drives electrons into the N material. The electrons migrate to the junction and fill holes. When an electron fills a hole, energy is given off as a photon of radiant energy.

The radiant energy emitted by an LED generally is made up of many wavelengths (Fig. 4–4).

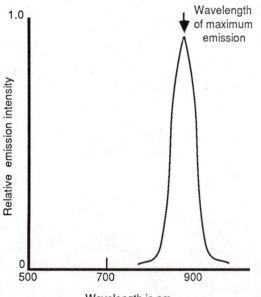

Figure 4–4. Emission characteristic of a gallium arsenide phosphide light-emitting diode.

The range of wavelengths emitted is about 100 nm. For some applications this range is narrow enough that no filter or monochromator is needed.

The orginal GaAs LED emitted its peak intensity in the near infrared range at about 900 nm. More recently, diodes that emit their peak intensity

at about 670 nm (red), 630 nm (orange), 590 nm (yellow), and 570 nm (green) have been manufactured. These diodes are easily pulsed by powering them with an AC power supply. Light-emitting diodes are popular in coagulation timers and analyzers. They are also used in some reflectance photometers.

Filters and Filter Photometers

The next component in the optical path illustrated in Figure 4–1 is the radiation isolator. Several types of devices are used to isolate "monochromatic" radiation from the polychromatic input. Optical filters may be used when one needs to work at only a few wavelengths. When a filter is used to isolate "monochromatic" radiation, the instrument is called a filter photometer. For other applications it is necessary to change wavelength frequently and in a continuous way. Instruments used for such applications are called spectrophotometers or spectrometers. They have a device called a monochromator that isolates "monochromatic" radiation from the polychromatic input. A monochromator allows one to work at any wavelength by just turning the wavelength selector. Monochromators are somewhat expensive—hence the interest in filters.

When a user of absorption spectroscopy realizes that he or she has many samples of the same type to analyze, it is wise to get a filter photometer. With the proper filter, the photometer is dedicated to the single-analysis method. Filter photometers are relatively inexpensive, so the purchase is easy to justify. Let's redraw Figure 4–1 and replace the radiation isolator with a filter (Fig. 4–5).

There are two common types of filters. The first is the so-called glass filter, which is prepared from two pieces of special glass. For example, the first piece of glass might transmit no radiation below 450 nm but 100 per cent of the radiation above 550 nm. The second piece of glass might transmit 100 per cent of the radiation below 450 nm and none of the radiation above 550 nm. Consider Figure 4–6. The combination of glass 1 and glass 2 produces a filter that transmits radiation ranging from 450 nm to 550 nm. Although a range of 100 nm is not monochromatic, for some applications it is "monochromatic" enough. The range of wavelengths transmitted is called the spectral bandwidth and is a measure of monochromaticity. The range of wavelengths transmitted at one-half the peak transmittance is called the effective bandwidth and is a measure of monochromaticity favored by some. The nominal wavelength is the wavelength at peak transmittance and is the value written on the filter. If a filter with a smaller bandwidth is desired, it can be produced by selecting glasses that cut off at, say, 470 nm and 530 nm respectively, but lower peak transmittance through the filter will always come

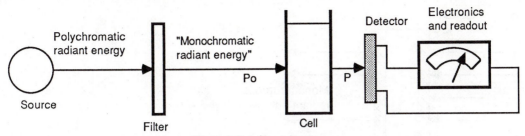

Figure 4–5. A filter photometer.

with this approach. The best of the glass filters have spectral bandwidths of about 60 nm. Glass filters are inexpensive, but they do get hot, and the monochromaticity of radiation that they isolate is poor.

When the "monochromaticity" of the radiant energy provided by a glass filter is insufficient for an application, an interference filter can be used. Interference filters are made from magnesium fluo-

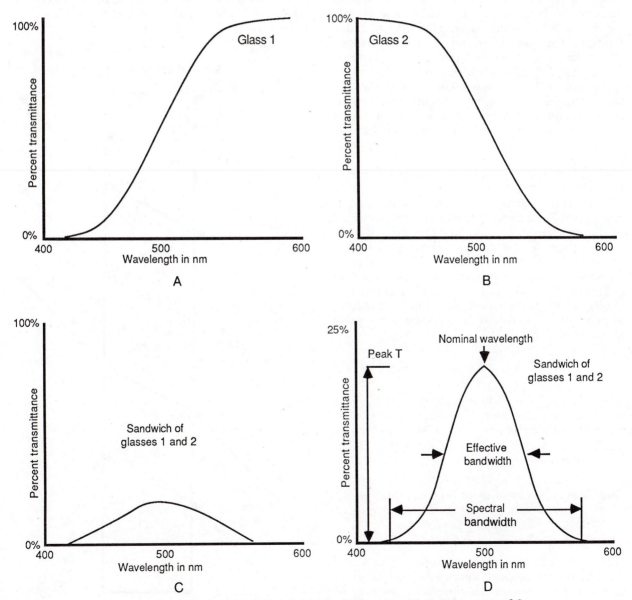

Figure 4–6. *A*, Glass 1. *B*, Glass 2. *C*, Sandwich of glasses 1 and 2. *D*, Enlargement of *C*.

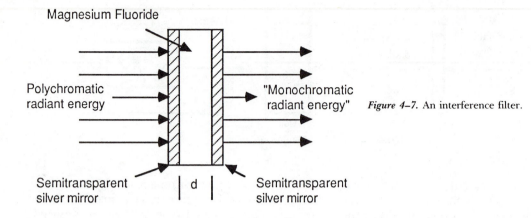

Magnesium Fluoride

Polychromatic radiant energy

"Monochromatic radiant energy"

Semitransparent silver mirror

Semitransparent silver mirror

d

Figure 4–7. An interference filter.

ride (MgF_2) and silver (Ag) (Fig. 4–7). Polychromatic radiation comes to the filter and most passes through the semitransparent silver mirror on into the magnesium fluoride layer. The incident radiation reflected from the outside of mirror a is lost. The radiation that enters the magnesium fluoride reflects back and forth off the mirrored surfaces many times before it finds small holes in mirror b and escapes the filter. While the radiation is reflecting back and forth many times, it is subjected to the process of interference (discussed in detail in Chapter 2). Drawing the incident radiation at an angle other than 90° helps one gain a better understanding of the process of interference. Consider Figure 4–8.

If radiation arrives in phase at points a, c, e, and g, and if the paths abc, cde, and efg are equal to the wavelength or some multiple thereof, then the condition for constructive interference is met. The radiation entering at point a will arrive at point c just in time to add to the incoming radiation at c. This process will continue until the radiation that entered at points a, c, e, and g combines and escapes at point h. The equation that predicts this type of behavior is

$$M\lambda = 2dN \sin \theta \qquad \text{Equation } 4\text{–}1$$

where M can be any whole number 1, 2, 3 . . .
    d is the thickness of the MgF_2 layer
    N is the refractive index of MgF_2 and is 1.38
    θ is the angle of incidence and is 90° in instrument applications

Substituting N and sin θ into the equation gives

$$M\lambda = 2.76d$$

Thus a filter for any wavelength can be produced by selecting the proper d value in the manufacturing process. The equation also predicts that only selected wavelengths can pass through the device. For all other wavelengths the distances abc, cde, and efg are incorrect, and destructive interference will take place. These distances are, of course, controlled by the thickness, d.

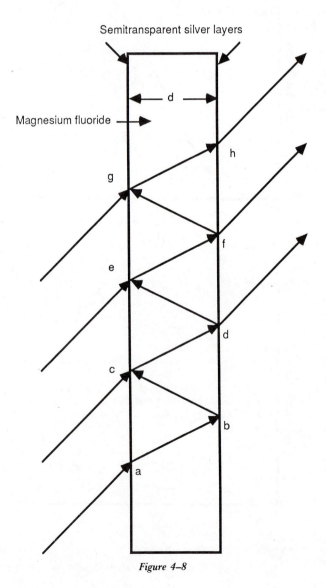

Semitransparent silver layers

Magnesium fluoride

d

Figure 4–8

Example: Consider an interference filter of d = 290 nm that is used at θ = 90°. What wavelength will pass through the filter?

For M = 1 1λ = (290 nm) 2.76
                           λ = 800 nm

For M = 2 2λ = (290 nm) 2.76
                           λ = 400 nm

For M = 3 3λ = (290 nm) 2.76
                           λ = 266.7 nm

Plotting the data from the example calculation, we see that the filter will pass several wavelength ranges. The radiation that enters the filter is made much more "monochromatic" than before, but it is still not as good as it could be. If our interference filter were combined with glass 1 from Figure 4–6, then the radiation leaving the combination would be much more monochromatic because glass 1 would not transmit the 400 nm band or any other band below 400 nm (Fig. 4–9). Glass 1 used in this application would be called a blocking filter.

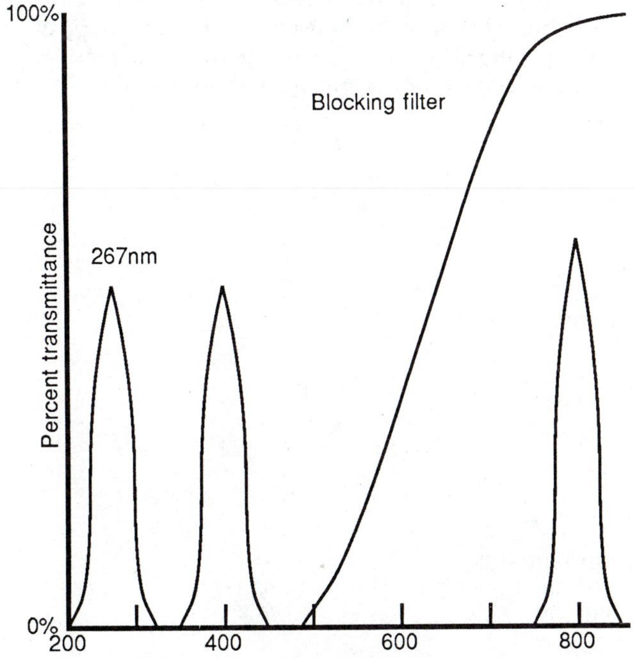

Figure 4–9

Interference filters, when combined with suitable blocking filters, produce radiation of fairly good monochromaticity; they are widely used in instruments in the clinical laboratory.

Diffraction Gratings and Monochromators

In some laboratory situations a filter photometer is not well suited to the analysis work that must be done. For example, much of my drug-screening work has required the rapid determination of ab-

sorption spectra of compounds. In other labs, the work is varied and would require the availability of many filters. In either case, the radiation isolator in Figure 4–1 should have a wavelength control that allows for continuous and easy variation of wavelengths. There are several such devices in existence. The most common is the diffraction grating monochromator. Before we can really appreciate the nature of a monochromator, we must carefully study the diffraction grating.

DIFFRACTION GRATINGS

Diffraction gratings are prepared by grooving a glass or aluminum plate with a diamond stylus. The grooving is done on a machine that carefully spaces the grooves and cuts them in a specific shape. A more modern technique for making gratings involves using a laser to create a holograph on a photographic plate. The photographic plate is then developed and enhanced to create a so-called holographic grating.

The specifications for an instrument may speak of the blaze wavelength of a grating. The blaze wavelength is the wavelength that is diffracted most efficiently. A plot of efficiency versus wavelength is shown in Figures 4–10 and 4–11. Figure 4–10 is

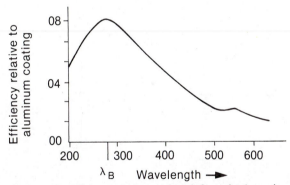

Figure 4–10. Efficiency versus wavelength for a classic grating. (Courtesy of Oriel Corporation, Stratford, Conn. Reprinted with permission.)

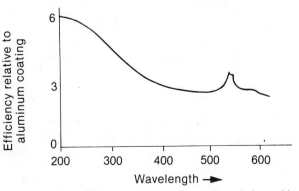

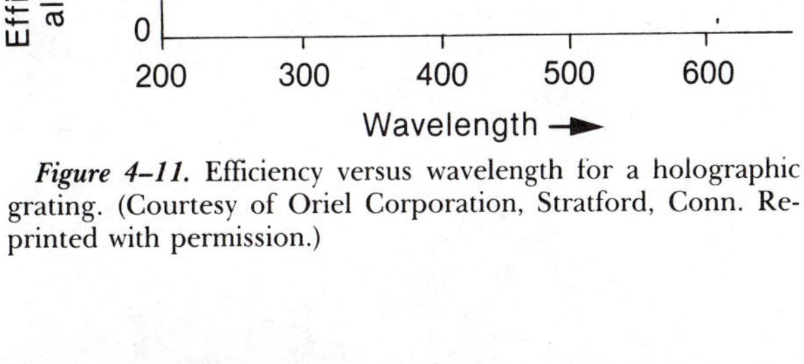

Figure 4–11. Efficiency versus wavelength for a holographic grating. (Courtesy of Oriel Corporation, Stratford, Conn. Reprinted with permission.)

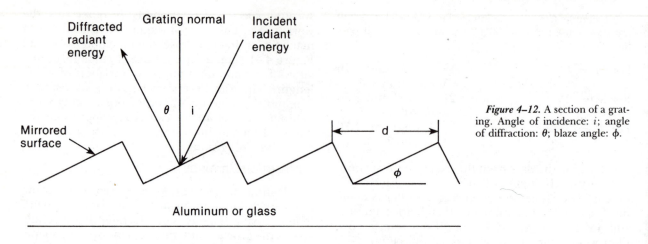

Figure 4-12. A section of a grating. Angle of incidence: i; angle of diffraction: θ; blaze angle: ϕ.

for a classic grating. It shows the blaze wavelength λ_B. Figure 4-11 is for a holographic grating. Note that a holographic grating is not quite as efficient as a classic grating. The holographic grating tends to have a more consistent efficiency. The holographic gratings have more stray radiation.

Gratings that are used in the visible region are generally grooved with a saw-toothed pattern. They generally have either 600, 1200, or 1800 grooves per millimeter and are between 2 and 10 cm² in area. Although most gratings used in instruments have reflective surfaces, transmission gratings also exist and are useful for teaching purposes. Consider Figures 4-12 and 4-13. The blaze angle, ϕ, is a design consideration related to a subsequent discussion. The distance between grooves is the grating spacing, d, which can be neither smaller than the wavelength nor larger than about five times the

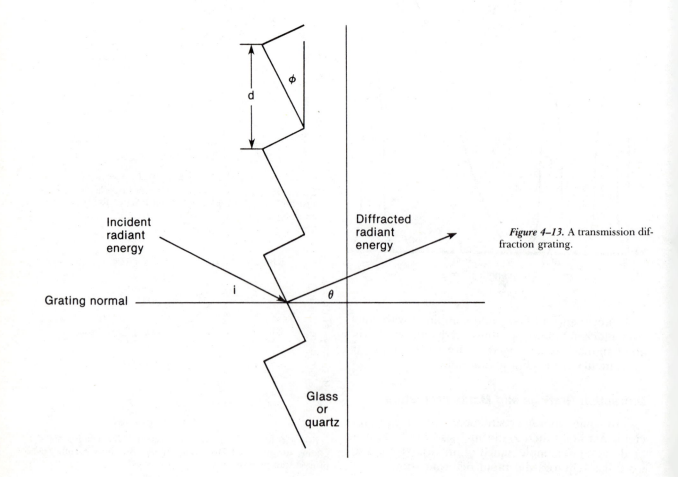

Figure 4-13. A transmission diffraction grating.

wavelength. Grating spacings in this range cause the incident radiation to scatter in all directions. It is the scattering that allows the device to disperse radiation.

In order to gain insight into the dispersion of radiation by the diffraction grating, consider a transmission grating with two grooves (rulings) (Fig. 4–14). When incident radiation strikes the rulings, the

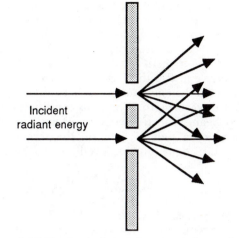

Figure 4–15. Simpler depiction of a transmission grating with two rulings.

I have provided two small slits from which the incident radiation can be scattered. The distance between them is equal to d just as in Figure 4–14. The radiation scatters from these slits is just like the radiation being scattered in Figure 4–14. In Figure 4–16, I have provided Figure 4–15 enlarged but with a few stipulations and the addition of a screen for viewing the scattered radiation.

Let's discuss the radiation that would be seen on the screen as point P is moved up and down the screen. Initially we must consider the case when the two-ruling grating is illuminated with monochromatic radiation. If a line ac is drawn such that the distance aP is equal to the distance cP, it can be

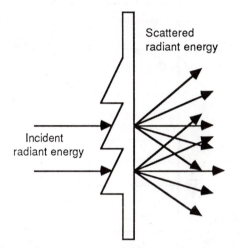

Figure 4–14. A transmission grating with two rulings.

radiation from each ruling is scattered in all directions down the optical path. The depiction of this process is a bit cluttered, so, with your indulgence, I would like to illustrate the same grating in a simpler way. Consider Figure 4–15. In this illustration,

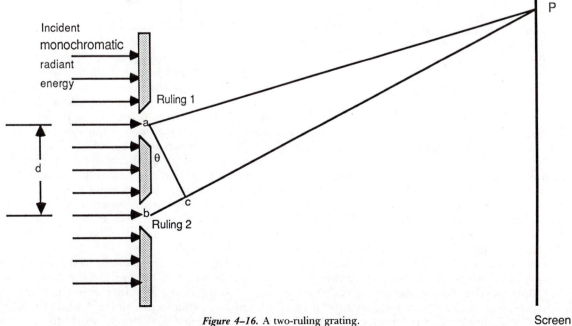

Figure 4–16. A two-ruling grating.

seen that the length of the line bc must vary as the location of the point P is changed. The monochromatic radiation arriving at the two slits is phase coherent, and the distances aP and cP are equal. The only requirement for the two beams of radiation to arrive at P with phase coherence is for the distance bc to be equal to the wavelength or a multiple of the wavelength. When this condition is met, the two beams constructively interfere (combine) and produce a spot of light on the screen. The mathematical equation that predicts this behavior is

$$M\lambda = d\sin\theta \qquad \text{Equation } 4-2$$

where $M = 1, 2, 3, \ldots$ or $-1, -2, -3, \ldots$
Thus even with monochromatic incident radiation, there are many angles of θ at which the scattered radiation constructively interferes to produce points of illumination on the screen.

Destructive interference would occur when the distance bc in Figure 4–16 is equal to the wavelength divided by 2 or some multiple thereof. Thus

$$\frac{(2M+1)\lambda}{2} = d\sin\theta \qquad \text{Equation } 4-3$$

predicts the angles of θ that would correspond to points on the screen where destructive interference would be occurring.

In summary, the screen, even for monochromatic radiation, would have regions of darkness where the scattered radiation would be destructively interfering and regions of illumination where the scattered radiation would be constructively interfering (Fig. 4–17). If one had a bigger screen, other values of M might be visible. However, with the grating rulings shaped like saw teeth and with the proper value of the blaze angle, most of the incident radiation would go into $M = -2, -1, 0, 1,$ and 2.

Remember Equation 4–2:

$$M\lambda = d\sin\theta$$

If we select a value for M, such as $M = 1$, then the equation becomes

$$\lambda = d\sin\theta \qquad \text{Equation } 4-4$$

where d is a constant. If we were to illuminate the slits in Figure 4–16 with polychromatic radiation, say from 400 to 800 nm, then Equation 4–4 predicts that every wavelength in the range would be found on the screen in a slightly different position. Remember that the distance bc must equal the wavelength for the $M = 1$ case, and, hence, every different wavelength would require a different distance bc. The only way for this to happen is for every different wavelength to have its own position P in the $M = 1$ area. In all other positions in the $M = 1$ region the process of destructive interference

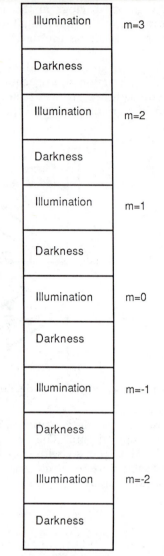

Figure 4–17

would destroy the particular wavelength. In this way the grating disperses polychromatic radiation into an array of monochromatic radiation not unlike a rainbow. The same process occurs in $M = 2, -1,$ and -2. In the $M = 0$ region, no dispersion takes place. For M equals zero, θ must equal zero.

Before we probe some of the fine points, let us consider a grating with three rulings (Fig. 4–18). It can be seen that the new ruling at point d is to point b what point b is to point a. The same proof that showed constructive interference would occur in Figure 4–16 would also show that constructive interference would occur in Figure 4–18. Now consider Figure 4–19.

In Figure 4–19, the previous three-ruling grating has been given an additional ruling. It can be seen that the ruling at point f would also be capable of sending radiation to point P while maintaining phase coherence. If the proof works for two, three, and four rulings, it will also work for larger num-

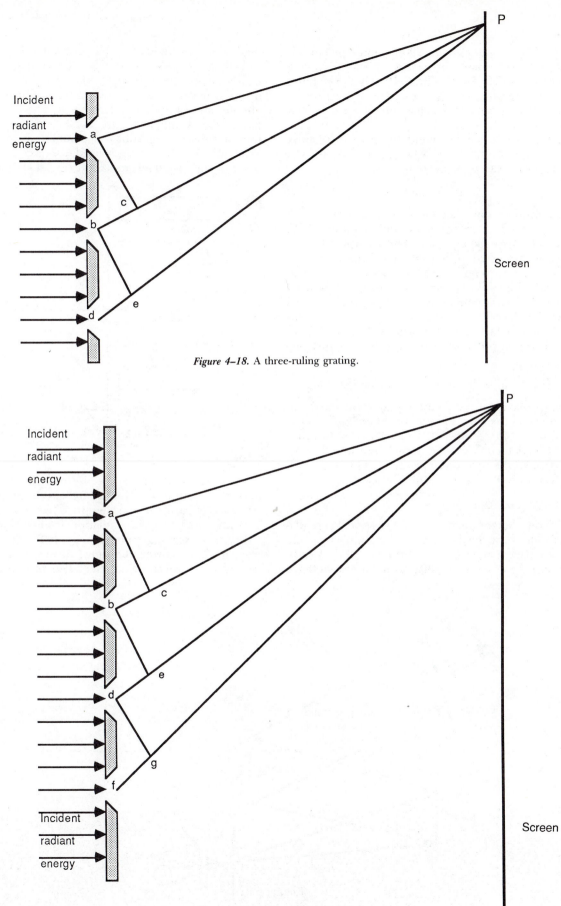

Figure 4–18. A three-ruling grating.

Figure 4–19. A four-ruling grating.

bers. Thus, a grating with 2000 rulings per mm is undergoing the same processes we saw occurring in Figures 4–16, 4–18, and 4–19. The large number of rulings on a grating tends to bring the process of constructive and destructive interference to near perfection.

With these facts in mind we must reconsider Equation 4–2; the news is not good, but at least the problem can be managed.

$$M\lambda = d\sin\theta$$

Equation 4–2 says that for a given grating and a given angle θ, there are many combinations of $M\lambda$ that will satisfy the equation. For example, if a scientist is interested in first-order 800 nm radiation, he or she must keep in mind that second-order 400 nm radiation will also be mixed with the desired radiation. As a second example, if the scientist needs second-order 600 nm radiation, he or she must contend with the fact that third-order 400 nm radiation as well as first-order 1200 nm radiation will be present.

$$1(1200 \text{ nm}) = 2(600 \text{ nm}) = 3(400 \text{ nm}) = d\sin\theta$$

This problem can be handled in several ways, depending on the case. Frequently a blocking filter is used to reject the unwanted wavelengths. In other cases the detector cannot detect the unwanted wavelengths, so they are not a problem. This topic will be considered in more detail a little later.

MONOCHROMATORS

Now that diffraction gratings have been discussed, let's use the device in conjunction with some slits to make a grating monochromator (Fig. 4–20). Slits are used to isolate radiation and are sometimes

made from razor blades. The entrance slit is used to isolate the small area of the source that is used to illuminate the grating. A blocking filter can be inserted when needed to avoid second-order contamination of the first-order region. The front surface of the grating in Figure 4–20 is mirrored. The use of reflection gratings is common in modern instruments. The exit slit is used to isolate a narrow band of the dispersed radiation (Fig. 4–21).

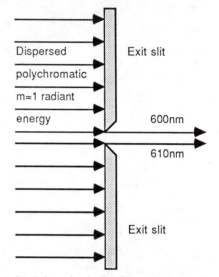

Figure 4–21. Enlarged exit slit of a grating monochromator.

In Figure 4–21 we see the exit slit enlarged and the radiation that is "falling out" through the slit. In this case the wavelength selector is set on 605 nm, but, even so, radiation from 600 nm to 610 nm is escaping from the slit. If the slit width were made narrower, the range of wavelengths escaping would become smaller. The profile of the

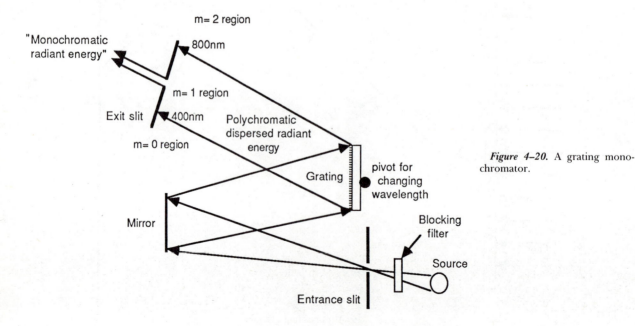

Figure 4–20. A grating monochromator.

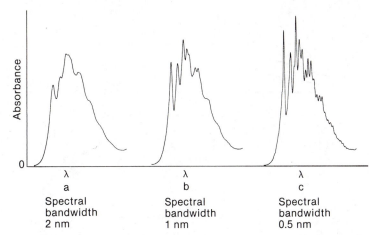

Figure 4–22. Absorption spectra of toluene at three different spectral bandwidths.

radiation leaving the slit is much like that in Figure 4–6D except that the spectral bandwidth would be much narrower. Middle-quality visible spectrometers have spectral bandwidths of about 1 nm. On most middle-quality visible spectrometers, the slit width and hence the bandwidth can be selected. For example, the instrument I frequently use can be set to produce a spectral bandwidth of 0.5 nm, 1 nm, or 2 nm. The wider the slit, the lower the resolution of the instrument. The effect of lower resolution is seen in Figure 4–22. The proper value of slit width depends on what you are doing with the instrument. The narrow slit width provides the best resolution between close absorptions, but narrow settings also cut down the amount of radiation that the detector has to work with. This condition may lead to less reliable P and Po measurements. So, although narrow settings allow more of spectral fine structure to be detected, there is a practical limit on how narrow the slit width can be made. If resolution of fine structure is not a big problem, and no other materials are present that would interfere, the use of slit width settings that are in the middle of the range on the instrument is indicated. When working in the visible region, there is seldom a need for bandwidths of less than 1 nm.

Summarizing, then, diffraction gratings provide radiation dispersion. Entrance and exit slits are used in conjunction with gratings to isolate radiation for spectroscopic measurements. Blocking filters can be used to absorb the unwanted wavelengths that contaminate the desired wavelength. The only thing that has not been discussed is how the wavelength is changed.

The wavelength is changed by moving a wavelength control that pivots the grating around a pivot point. The angle of the grating relative to the exit slit is changed. This causes different wavelengths of radiant energy to fall on the gap in the exit slit and to pass on through. On some instruments, the wavelength control is driven by a motor so that wavelength can be systematically changed during the experiment. When this is done, the blocking filter is automatically inserted at the proper time.

Cells and Lenses

Throughout this chapter, we have been considering the components of the optical path in Figure 4–1. We have considered sources of visible radiation and the most common radiation isolators. It is now time to consider the optical materials that are used in cells, lenses, and windows.

In instrumentation, it is sometimes necessary to isolate the monochromator from the rest of the instrument. This is frequently done in order to protect the grating from the corrosive atmosphere common to laboratories. When this is done, the monochromator is built into a black box that has plain, flat windows through which the radiation enters and leaves the device. Many detectors also have a protective window through which P and Po must pass. These windows must be highly transmissive to the radiation.

In most instruments it is necessary to have an occasional lens. Lenses are sometimes provided to collect and condense radiation. Sometimes lenses are used to collect divergent radiation and make it parallel. In either case, the material used to make the lens must have a high transmittance for the wavelengths of interest. See Figure 4–23.

In most absorption spectrophotometric instruments, it is necessary to contain liquid samples in some sort of cell. There are two common types of cells, round and square. The round cells look a lot like test tubes and are carefully manufactured so as to avoid imperfections in thickness and diameter. Regular test tubes should not be substituted for round cells. Square cells are used in some instruments; their main advantage is that the cell path, b, can be measured and used. In either case, the material that is used to make the cell must be highly transmissive to the radiation of interest.

Borosilicate glass is a good material for use in

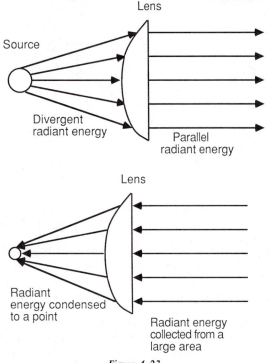

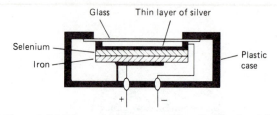

Figure 4–24. Schematic of a photovoltaic cell. (Reprinted from: D.A. Skoog, Principles of Instrumental Analysis, 3rd Ed., Philadelphia, Saunders College Publishing, 1985.)

by sandwiching a thin layer of the semiconductor selenium between iron and silver layers. The iron layer is thick enough to give the device some mechanical strength. The silver layer is semitransparent and is provided for electrical contact with the selenium; the iron layer is the other electrical contact.

The three-layer array has two junctions between dissimilar conductors—the iron selenium junction and the silver selenium junction. Junctions that are formed between two dissimilar conductors always have an electrical potential energy; thus, the photovoltaic cell is a cell that produces its own voltage. No power supply is needed; the sum of the two junction potentials provides all the necessary voltage.

The cell is able to detect visible radiation because the photons of radiation have enough energy to break the covalent selenium bonds. The electron-hole pairs that are produced provide current carriers that allow conduction across the selenium layer. The number of electron-hole pairs produced is directly proportional to the intensity of the radiation striking the selenium. Thus there is a linear relationship between the detector photocurrent and the radiation intensity (Fig. 4–25). The device that is used to measure the photocurrent is generally a

Figure 4–23

the visible region of the electromagnetic spectrum. It transmits well from about 350 nm to about 2000 nm. Quartz is also a good material, as it transmits well from 200 nm to about 3300 nm. Quartz is more expensive and should be used only when necessary. Some plastics transmit well enough to be used in the visible region.

Cells must be kept very clean and free of scratches. The cleaning instructions supplied with the cells should be used. Cells should not be handled on their optical faces. If they have been handled on their optical faces, they should be wiped free of fingerprints with Chem-wipes or a similar product. Do not use paper towels or handkerchiefs, as they can leave scratches, which cause error in the measurement of P and Po. The topic of error will be covered in detail in Chapter 5.

Radiation Intensity Detectors

The last item in the optical path in Figure 4–1 is the radiation intensity detector that is used to measure P and Po. There are several types of detectors, and each has its good and bad points.

PHOTOVOLTAIC CELL

The first example of a radiation intensity detector is the photocell. The photocell has been called a photovoltaic cell or a barrier layer cell. I prefer the use of the term photovoltaic cell; it is more descriptive and less ambiguous than the other names.

The photovoltaic cell (Fig. 4–24) is prepared

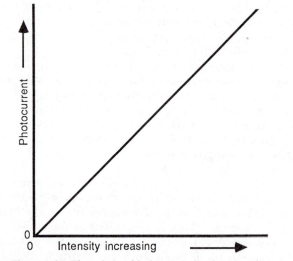

Figure 4–25. The relationship between the detector photocurrent and the radiation intensity is linear. (Assumption for purposes of illustration: external circuit resistance is less than 400Ω.)

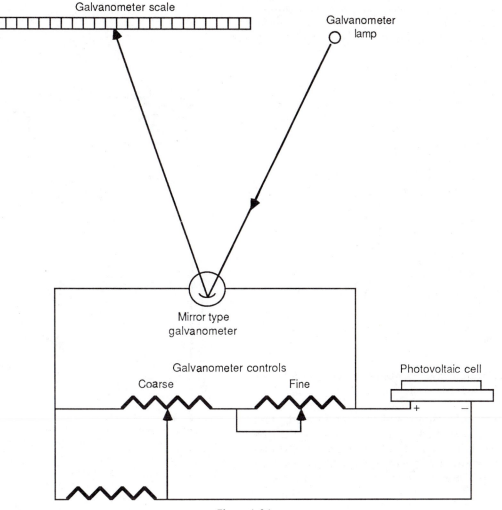

Figure 4-26

d'Arsonval meter of some type. It is not unusual to find the device connected to a mirror-type d'Arsonval galvanometer (Fig. 4–26). In such a device, the pointer is replaced with a mirror.

The photovoltaic cell does not produce photocurrent from all wavelengths of radiation with equal efficiency. The relationship between relative response and wavelength is shown in Figure 4–27.

In summary, the photovoltaic cell is an inexpensive, linear radiation intensity detector. When used with a suitable low-resistance d'Arsonval meter or galvanometer, the photovoltaic cell provides enough power to drive its readout device without amplification. The cell has variable response to radiation of differing wavelengths. When illuminated, the cell does not produce an instantaneous current but requires perhaps a second to develop its equilibrium current output. The photovoltaic cell is found in many common laboratory instruments. The ubiquitous Coleman Junior IIs and the older Technicon Autoanalyzers are prime examples.

PHOTOTUBE

The second visible radiation intensity detector that I would like to discuss is the vacuum photodiode, which is more frequently called a phototube. The phototube uses the photoelectric effect to con-

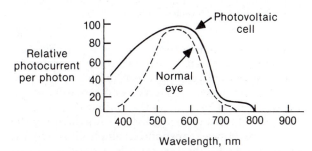

Figure 4–27. Spectral response of a selenium photovoltaic cell with glass cover. (From Instrumental Methods of Analysis, 6th Ed., by H. Willard, L. Merritt, J. Dean, & F. Settle, Copyright © 1981 by Litton Educational Publishing, Inc. Reprinted by permission of Wadsworth, Inc.)

vert radiant energy to electrical current. A photo-tube is built from a wire anode and a concave photocathode (Fig. 4–28). The photocathode is

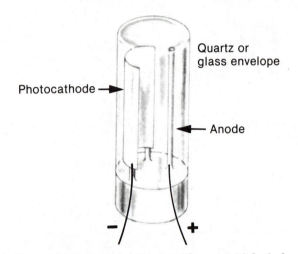

Figure 4–28. A phototube. (Reprinted from: A.J. Diefenderfer, Principles of Electronic Instrumentation, 2nd Ed., Philadelphia, Saunders College Publishing, 1979.)

manufactured by silver plating a concave nickel electrode. The silver layer is then oxidized to produce some silver oxide. The electrode is then covered with a layer of cesium and is heat treated. The resulting photocathode has a surface that absorbs photons well. The energy of the photons is transferred to the surface electrons, which, in turn, are ejected from the electrode surface. This is best seen by looking at the top view of a phototube (Fig. 4–29). The electrons that are produced from the pho-

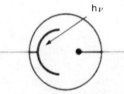

Figure 4–29. Top view of a phototube. (Reprinted from: A.J. Diefenderfer, Principles of Electronic Instrumentation, 2nd Ed., Philadelphia, Saunders College Publishing, 1979.)

tons are attracted to the anode by the application of about 90 volts DC relative to the photocathode. The applicable circuit is shown in Figure 4–30.

The phototube is a very good radiation intensity detector. It has a linear response with increasing radiation intensity (Fig. 4–31). Unfortunately, the phototube does not respond to all photons equally. In Figure 4–32, the response of a phototube that has been exposed to the same number of photons at each wavelength has been plotted against wavelength. Note that the photocurrent production dif-

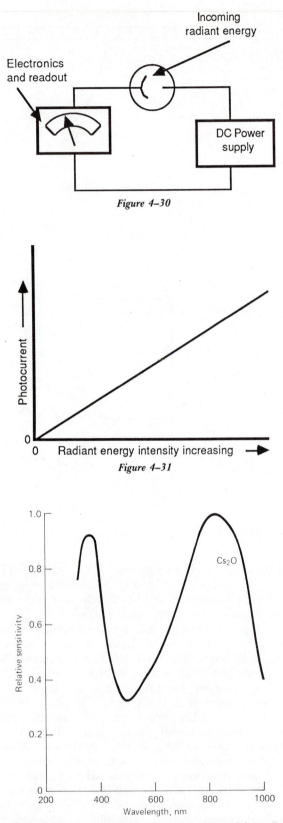

Figure 4–30

Figure 4–31

Figure 4–32. Phototube response curve. (Adapted from: D.A. Skoog and D.M. West, Principles of Instrumental Analysis, 2nd Ed., Philadelphia, Saunders College Publishing, 1980.)

fers at various wavelengths. It is more than twice as effective in the production of photocurrent at the ends of the visible region as it is at 580 nm. Extremely efficient photocathodes can be made that work well from 180 nm to about 1100 nm. No single cathode will cover the whole range, however.

Phototubes, unfortunately, produce a small amount of current even in the dark. This so-called dark current is caused by an occasional thermal electron that escapes the cathode surface. All instruments make some provision for the management of the dark current. It is frequently done by an electrical adjustment of the amplifier. The adjustment makes the readout say "zero transmittance" when no radiation is reaching the detector, even though a small current is flowing.

In summary, the phototube is a good radiation intensity detector; it has linear response. Its function is based on the photoelectric effect. The detector, by virtue of the ease with which its output can be amplified, is very sensitive and can function with much smaller radiation intensities than can a photovoltaic cell. The phototube is almost instantaneous in its response to radiation. These desirable properties make it a more common detector than the photovoltaic cell.

PHOTOMULTIPLIER TUBE

A third important visible radiation intensity detector is a close relative of the phototube and is called a photomultiplier tube. The device, along with its electrical block diagram, is illustrated in Figure 4–33. Look at the diagram carefully, then return to the following description.

Like the photocathode in the phototube, the photocathode in the photomultiplier tube is a photoemissive surface that produces electrons from photons. Electrodes 2 through 9 are called dynodes and are prepared from materials that hold their electrons loosely. The dynodes are given voltage from the power supply; each dynode is about 90 volts more positive than the one before it. This voltage is high enough to accelerate an incoming electron to the point of being able to knock two to five secondary electrons out of a dynode. In this way, at least two electrons leave a dynode for each electron that arrives. By the time this process has gone on at each of the dynodes, the anode receives about 10^7 electrons per photon that hit the photocathode. In this way a lot of amplification takes place in the photomultiplier tube.

Because of its internal amplification, the photomultiplier tube is easily the most sensitive visible radiation detector. Photomultiplier tubes and their associated circuits are expensive; they are used only when they offer considerable advantage.

The facts concerning the linearity and response characteristics of the phototube are applicable to the photomultiplier tube as well (see Figs. 4–31 and 4–32).

The photomultiplier tube is a fast detector. The photocurrent changes with changes in radiation intensity almost instantaneously. They can be manufactured to cover the wavelength range of 180 to about 1200 nm. As with phototubes, however, no single photomultiplier tube covers the entire range.

SILICON PHOTODIODE

A fourth important visible radiation detector is the silicon photodiode. A PIN silicon photodiode

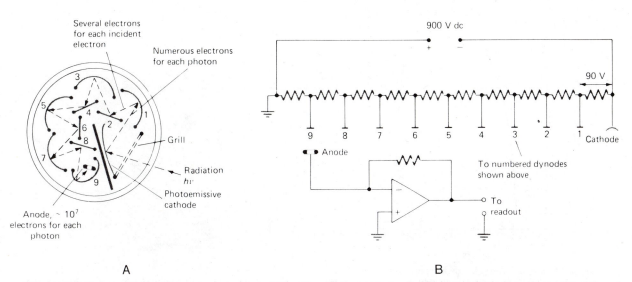

Figure 4–33. Photomultiplier tube: *A*, cross-section of the tube; *B*, electrical circuit. (Reprinted from: D.A. Skoog, Principles of Instrumental Analysis, 3rd Ed., Philadelphia, Saunders College Publishing, 1985.)

Figure 4–34. S1723 photodiode. (Courtesy of Hamamatsu Corporation, Middlesex, New Jersey. Reprinted with permission.)

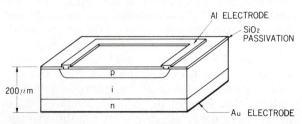

Figure 4–35. Schematic of the S1723 photodiode. (Courtesy of Hamamatsu Corporation, Middlesex, New Jersey. Reprinted with permission.)

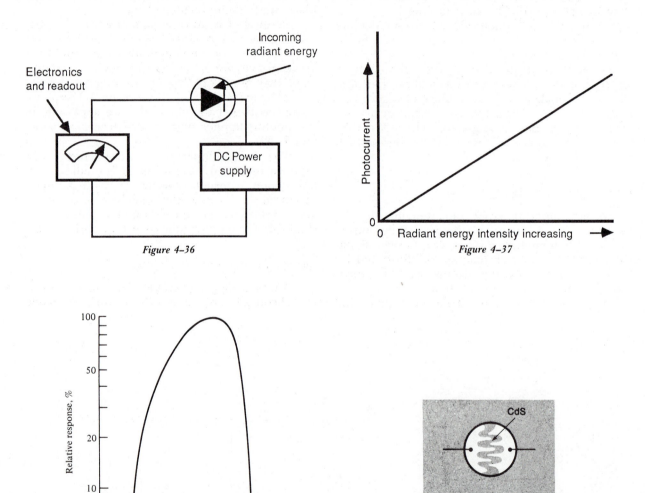

Figure 4–36

Figure 4–37

Figure 4–38. Spectral response of a typical *p–n* photodiode. (From Instrumental Methods of Analysis, 6th Ed., by H. Willard, L. Merritt, J. Dean, & F. Settle, Copyright © 1981 by Litton Educational Publishing, Inc. Reprinted by permission of Wadsworth, Inc.)

Figure 4–39. Photoconductivity cell. Bottom, symbol; top, physical layout. (Reprinted from: A.J. Diefenderfer, Principles of Electronic Instrumentation, 2nd Ed., Philadelphia, Saunders College Publishing, 1979.)

is illustrated in Figures 4–34 and 4–35. The detector is called a PIN photodiode because of the layer of P-type silicon that is connected to the intrinsic silicon, which is in turn connected to the N-type silicon. This is illustrated in Figure 4–35. The detector is connected to a voltage source that tries to pump electrons into the P-type material (Fig. 4–36).

In the absence of radiant energy the circuit has almost no current flow because of this reversed bias situation. When radiant energy falls on the thin film of p-material, hole-electron pairs are formed. The holes ultimately accept electrons from the aluminum electrode and the external circuit. The electrons cross the $i-n$ junction with ease and exit the diode through the gold electrode, causing a significant current flow. The increase in current is linearly related to the incident radiation intensity. This is shown in Figure 4–37.

The silicon photodiode does not have the same ability to convert photons of all wavelengths to current with equal efficiency. This is shown in Figure 4–38.

CADMIUM SULFIDE PHOTORESISTORS

A fifth detector is the cadmium sulfide photoresistor, also known as the cadmium sulfide photoconductivity cell. It is made by depositing a thin film of polycrystalline cadmium sulfide between two metallic electrodes. The film is generally deposited in a serpentine pattern and is doped with a trace of copper. A cadmium sulfide photoresistor is illustrated in Figure 4–39. If the device is connected to a source of voltage, the current that flows is proportional to the number of hole-electron pairs that form per second. The number of hole-electron pairs that form is proportional to the intensity of radiant energy that falls on the device (Fig. 4–40).

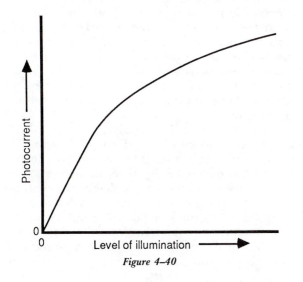

Figure 4–40

It is unfortunate that the relationship is not strictly linear. This tends to limit the use of the device.

The wavelength has a considerable effect on the amount of photocurrent that can be generated from a given number of photons of radiant energy (Fig. 4–41). The cadmium sulfide cell can convert

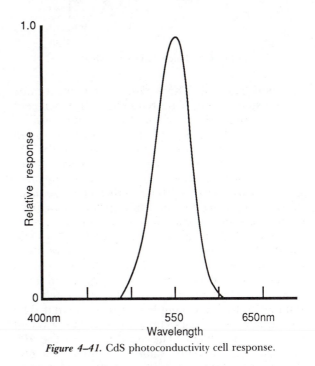

Figure 4–41. CdS photoconductivity cell response.

photons to hole-electron pairs only if the photons are between 500 and 600 nm. The detector is frequently used in conjunction with a green LED.

The main clinical application of the cadmium sulfide photoresistor is in the detection of clot formation in coagulation timers or analyzers. These instruments are the topic of Chapter 24.

CATEGORIES OF VISIBLE INSTRUMENTS

Single-Beam Photometers

Now that the characteristics of optical components have been discussed, let us consider the major types of visible photometric and spectrophotometric instruments. These instruments fall neatly into five categories, even though the various examples within a category have their unique qualities.

First let's consider a single-beam photometer. The optical path of this category of instrument is shown in Figure 4–42. It is called single beam because it has only one radiation path from source to detector. In order to use the single-beam photometer, three important things must be done. (1) The proper filter for the analysis must be inserted into

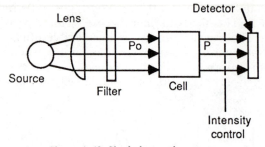

Figure 4–42. Single-beam photometer.

Double-Beam Photometers

A second category of photometer is the double-beam photometer (Fig. 4–44). This instrument is

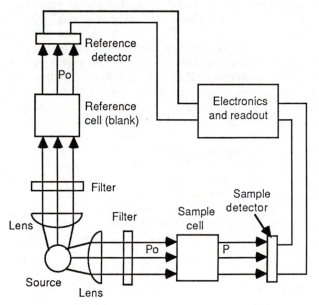

Figure 4–44. Double-beam photometer (ratio-indicating).

the instrument. (2) The dark-current control must be adjusted to make the readout read 0 per cent transmittance when the beam of radiation is blocked. The beam is generally blocked with a solid black "cell" that is inserted into the cell holder. (3) The instrument must be adjusted to read 100 per cent transmittance when a cell containing the analysis solvent is placed in the beam. Once these three things have been done, the photometer can be used to measure the transmittance of a solution.

If the photometer is stable, several samples and standard solutions can be measured before steps two and three must be repeated. On instruments that are not so stable, the 100 per cent T and 0 per cent T adjustments must be made every time a new standard solution or sample is to be evaluated. The adjustment of the dark current is almost always an electrical adjustment. The adjustment of the 100 per cent transmittance value is frequently a mechanical adjustment of a radiation intensity control. On some other instruments it is an electrical adjustment. When the control can be adjusted mechanically, it frequently takes the form of a metal variable occluder (Fig. 4–43). As the knob on the

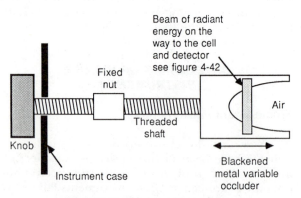

Figure 4–43. Metal variable occluder.

front of the instrument is turned in one direction, the black variable occluder is moved to the right and cuts off more of the radiation beam. However, if more radiation is needed in order to make the readout read 100 per cent T, then the knob is turned the other way. This makes the occluder move to the left, which lets more radiation pass.

called double beam because the value of Po is determined in a second radiation path. Instruments of this category are frequently more accurate than single-beam instruments because slow variations in the source intensity affect both the sample detector and the reference detector in the same way. In order to use instruments of this category, three important things must be done. (1) Identical filters must be inserted. (2) With the sample beam blocked, the instrument must be adjusted to 0 per cent transmittance. (3) The solvent that will be used to dissolve the analyte must be placed in both the reference and the sample cells. With this done, the electronics can be adjusted to make the readout read 100 per cent transmittance. So adjusted, the instrument can be used to measure the transmittance of standards and unknowns. Depending on the stability of the instrument, the settings of 100 and 0 per cent transmittance will require periodic adjustment.

The electronics and readout of double-beam instruments can be of two different designs. The first design is that shown in Figure 4–44. In double-beam photometers of this type, the reference detector current and the sample detector current are electrically ratioed. As the reference detector current is directly proportional to Po, and the sample detector current is directly proportional to P, then the ratio of the two currents is the transmittance. The transmittance so calculated is read out on the

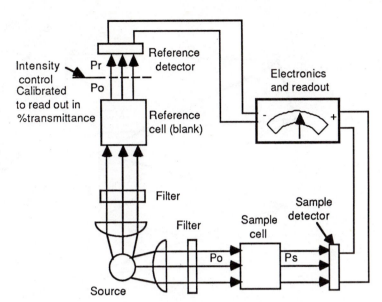

Figure 4–45. Optical null double-beam photometer.

read-out device. This type of instrument is a ratio-recording, double-beam photometer.

The second design used for finding the transmittance of a sample on a double-beam photometer is shown in Figure 4–45. This type of double-beam photometer is called an optical null double-beam photometer. In this type of instrument the reference detector current is compared electrically with the sample detector current. If the current from the sample detector is smaller than the current from the reference detector, then the meter indicates a current difference on the minus side. This means that the per cent T control must be changed to a lower value. When the sample detector current is larger than the reference detector current, then the

meter indicates a positive current difference. This means that the per cent T control must be changed to a higher value. When the meter is brought to a zero reading, then Ps and Pr are equal, and the sample transmittance can be read off the per cent T adjusting knob.

Single-Beam Spectrophotometers (Spectrometers)

A third category of visible instrumentation is a single-beam spectrophotometer. This spectrophotometer is much like a single-beam filter photometer except that a grating monochromator is provided in place of the filter. The single-beam spectrophotometer is illustrated in Figure 4–46. Ex-

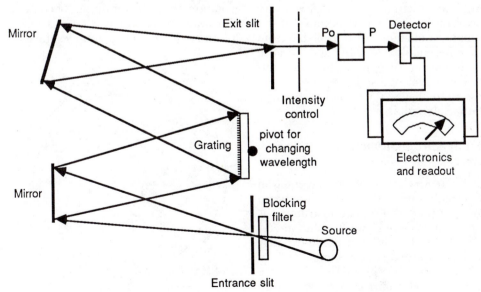

Figure 4–46. Single-beam grating spectrophotometer.

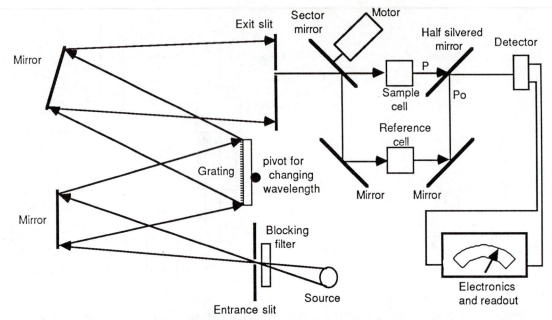

Figure 4–47. Double-beam, ratio-recording, grating spectrophotometer.

cept for the need to set the wavelength of the monochromator, the instrument shown in Figure 4–46 is used like a single-beam photometer. It is still necessary to adjust the instrument to 0 per cent and 100 per cent transmittance. So calibrated, the instrument can be used to measure the transmittance of standards and unknowns. Remember that the 0 per cent and 100 per cent transmittance adjustments must be repeated as often as instability requires.

The spectrophotometer can measure the absorption spectrum of a compound, but remember to adjust the 0 per cent and 100 per cent transmittance settings at each new wavelength. The 100 per cent transmittance setting in particular will vary a lot with wavelength. This is true because of the differing amount of radiation that is available from the source at different wavelengths, and because the detector efficiency in current production is different at different wavelengths. These two factors make if necessary to change the radiation intensity control as the wavelength is changed.

Double-Beam Spectrophotometers (Spectrometers)

The fourth category of visible instrumentation is the highly versatile ratio-recording double-beam spectrophotometer (Fig. 4–47). The instrument is similar to the single-beam spectrophotometer shown in Figure 4–46. The differences are two new optical components located on the sample side of the exit slit. These must be considered before we go on. The first is a sector mirror; consider Figure 4–48.

The sector mirror has a round metal frame that is mounted with a mirrored section and a section with no mirror at all. It is spun in the optical path by a constant-speed motor. The sector mirror serves several purposes. First when the hole is in the radiation beam, the radiation goes to the sample. When the mirrored section is in the radiation beam, the visible radiation is reflected 90 degrees to the fixed mirror and then through the reference cell. In this way the sector mirror is a radiation director. Second, the sector mirror interrupts the radiation twice per rotation. The metal frame between the mirror and hole is what interrupts the beam. By interrupting the radiation to the detector twice per rotation, the detector output becomes pulsed direct current. The mirror is rotated a few hundred revolutions per minute. This gives the electronics nice,

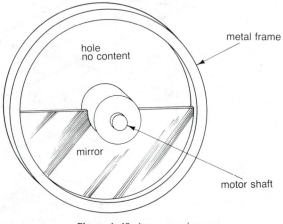

Figure 4–48. A sector mirror.

pulsating DC to work with. The electronics designed to work with pulsating DC are particularly low-noise in character. The act of interrupting the visible radiation beam is called chopping the beam; the sector mirror is sometimes called a chopper. There are choppers, however, that are not sector mirrors.

The second new optical component is a half silvered mirror (Fig. 4–49). The half silvered mirror

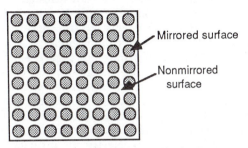

Figure 4–49. Half silvered mirror.

reflects half of what it receives. The other half is transmitted through the nonmirrored surface.

Now let's go back and consider the double-beam recording instrument. The entire instrument is illustrated in Figure 4–47, but let's consider only the double-beam portion that is redrawn in Figure 4–50.

I have illustrated this system with the mirrored portion of the sector mirror down in the visible radiation beam. The radiation is reflected off the sector mirror to fixed mirror M_1. From M_1 the radiation is reflected through the reference cell onto fixed mirror M_2. From M_2 the radiation goes to the half silvered mirror, where half of it is reflected to the detector. This half is $\frac{Po}{2}$. The other half passes through the half silvered mirror and is lost. In this way the detector receives $\frac{Po}{2}$ and produces a current that is directly proportional to $\frac{Po}{2}$. This information is stored.

In the next instant the sector mirror rotates so that the hole is down in the radiation beam. When this happens, the visible radiation goes through the sample cell onto the half silvered mirror. At the mirror, half the radiation falls through the unsilvered portions and becomes $\frac{P}{2}$. The detector receives $\frac{P}{2}$ and produces a current from it that is directly proportional to $\frac{P}{2}$. This current is electrically ratioed with the $\frac{Po}{2}$ current that was just measured. This ratio, of course, is the transmittance.

There is a second way that this optical path can be used to find the transmittance of a sample. This second approach cannot be done with any detector other than a photomultiplier tube. In this approach, the value of $\frac{Po}{2}$ is not used to calculate the transmittance. The value is routed to a differential amplifier, where it is compared with a reference value of 100 arbitrary units.

If the $\frac{Po}{2}$ value is different from the reference value, the photomultiplier tube's power supply is electrically adjusted in order to modify the photomultiplier tube's amplification factor. This is done many times per second and in effect makes the detected value of Po 100 at all wavelengths. With the photomultiplier tube adjusted in such a way,

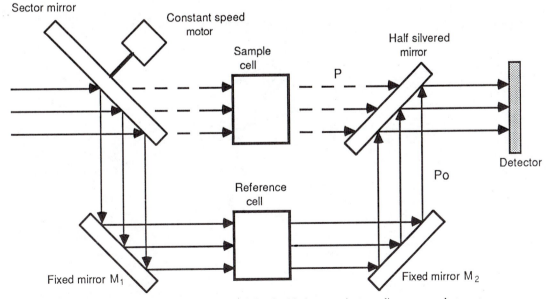

Figure 4–50. The double-beam portion of the double-beam ratio-recording spectrophotometer.

the value of $\frac{P}{2}$ that is detected is equal to the transmittance for the sample.

In both approaches to the calculation of the transmittance, the double-beam approach minimizes some errors. For example, if the sector mirror is spun at 600 revolutions per minute, then Po and P are measured ten times each second. Transmittances measured in this way are less error-prone than those measured on a single-beam instrument. This is true because anything that affects Po and P more slowly than in .01 second will affect them both about the same, and the resulting error will be canceled.

The primary reason for making a double-beam instrument is to make automatic wavelength-scanning possible. If a motor and suitable mechanical components are connected to the grating, it can be moved about its pivot point automatically. Continuing our example from the preceding paragraph, if it were set up to move fast enough to sweep 60 nm of wavelength past the exit slit per minute, then the instrument could measure the transmittance every 0.1 nm. Remember that 60 nm per minute is 1 nm per second and the transmittance is being measured ten times per second. If the transmittance were passed through a computer that would make the −log transformation, and then to a linear recorder, the absorption spectrum could be graphed automatically. There is no need to adjust the radiation intensity control even though the radiation output and detectability change with wavelength. In fact, there is no radiation intensity control. The ratioing of P and Po takes all of these variations into account.

Diode-Array Spectrophotometers

The previous discussion concerns photometers and spectrophotometers that are designed to measure the absorbance of an analyte at a specific wavelength. Although it is true that the single- and double-beam spectrophotometers discussed could determine the absorbance of an analyte at many wavelengths, the absorbance values are determined one wavelength at a time. The fastest common instruments of this type could scan through 100 nm of wavelength in about 1 minute. In direct contrast to these classic instruments are the recently introduced diode-array spectrophotometers. A simplified optical path is illustrated in Figure 4–51.

The instruments use classic sources, cells, and gratings. The detector and associated circuitry differ from those of the classic instruments. The detector is made up of about 200 narrow silicon photodiodes. Each of the diodes is typically about 50 μm in width. They are electrically insulated from one another. When the visible spectrum is dispersed by the grating and is allowed to fall on the array, every diode receives a band of radiant energy that is about 2 nm wide. The photocurrent that is developed in each of the diodes is proportional to the radiant energy for that 2 nm segment of the spectrum. The measurements of radiant energy can be made very rapidly; a computer can measure the intensity of the radiant energy at all 200 diodes in less than a second.

At the time of writing, I know of two diode-array instruments manufactured in the United States. Perkin-Elmer Corporation is introducing the

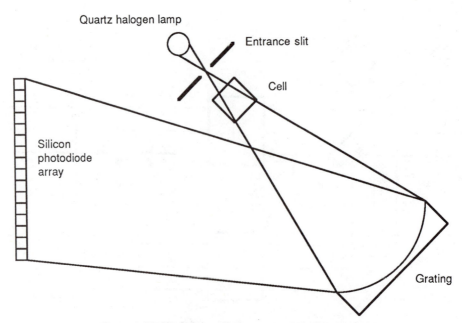

Figure 4–51. Single-beam diode-array spectrophotometer.

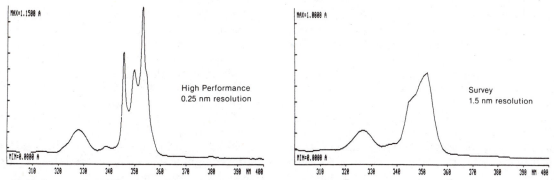

Figure 4–52. Spectra of a NdCl$_3$ solution illustrating the difference between a 0.25-nm/diode dispersion and 1.5-nm/diode dispersion. (Reprinted with permission from: J. Cahill and M. Retzik, American Laboratory, volume 15, number 11, page 47, 1984. Copyright 1984 by International Scientific Communications Inc.)

single-beam Lambda Array 3840. Perkin-Elmer has not as yet published the optical path of the instrument. Nor have they indicated the manner in which the dark current and the value of Po are determined. The determination of these values must, of course, be made along with the determination of P at each wavelength. The instrument is known to have two gratings. One is a low-dispersion grating for obtaining the spectrum of a sample over a broad wavelength range. The second grating is a high-dispersion grating that is used for higher resolution measurements. The instrument is reported to be capable of 0.25 nm resolution when the better grating is used. The resolution is reported to be 1.5 nm using the coarser grating. The spectra in Figure 4–52 illustrate the effect of this variation in resolution.

The Hewlett-Packard HP 8450 is a five-beam, diode-array instrument. The five beam positions allow for the measurement of Po for a blank, the dark current, P for a sample, and P for two standards. This can be done in less than 5 seconds. The optical path of the HP 8450 is shown in Figure 4–53. Only two of the five beams are illustrated.

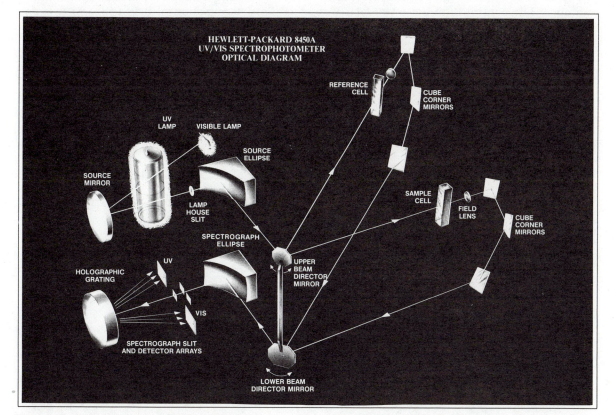

Figure 4–53. Optical layout of the HP 8450A UV-visible spectrophotometer. (Courtesy of Hewlett-Packard Co., Scientific Instruments Div., Palo Alto, Calif. Reprinted with permission.)

The instrument is built to operate simultaneously in the ultraviolet as well as the visible regions. The instrument has a pair of beam directors. The first director sends the polychromatic radiation down each of the five paths. The polychromatic radiation is then reflected back under the original path back to the second beam director. From the second beam director, it is sent to the grating and the diode arrays. The computer then determines the dark current for each diode. It then evaluates Po for the blank at each diode, and then P for each sample at each diode. This information makes it possible to determine the absorbance at all wavelengths of operation.

Diode-array instruments are the result of the union of computer technology, semiconductor technology, and refinements in optical technology. They have impressive capabilities.

DETERMINATION OF COLORLESS COMPOUNDS BY VISIBLE SPECTROSCOPY

In the early years of spectroscopy almost every laboratory had a visible-radiation photometer or spectrophotometer. As medical science matured through this same period, there was a lot of interest in the analysis of colorless compounds of medical significance. The outgrowth of these two facts was that clinical scientists developed a large number of new procedures. These procedures involved the use of visible instruments to determine the concentration of colorless analytes. These procedures, in

many cases, are still in use and are still very good methods. These methods always involve the addition of a compound called a chromogen. The chromogen is caused to react with the colorless analyte. The product of the reaction is a colored product, the concentration of which is proportional to the concentration of the original colorless analyte. I would like to provide you with two examples of such reactions.

First, serum proteins are colorless analytes of extreme significance. These compounds undergo reaction in basic solution to produce compounds similar in structure to biuret.

Biuret

These biuret-like compounds react with copper ion in the same way that biuret would. Using biuret as a model compound, we obtain the first reaction shown at the bottom of page. The tetra coordinated Cu(II) complex anion that forms absorbs strongly at 540 nm, and its absorbance is directly proportional to the concentration of protein in the original sample. The structure of the product is fairly well understood.

A second example of the use of a chromogen is the determination of uric acid, which is an important compound in clinical science. Uric acid is commonly determined by its reaction with phosphotungstic acid (see the second reaction shown at the bottom of the page). As is so common in these

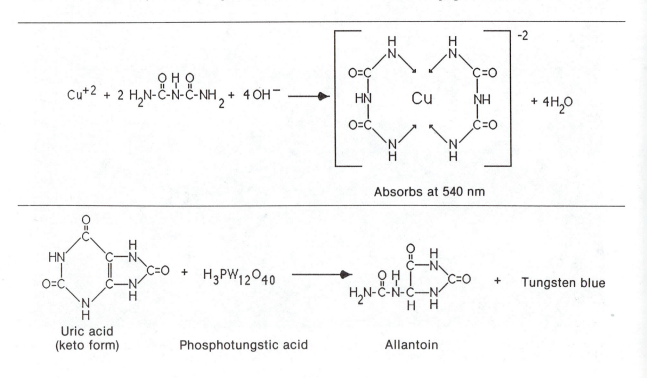

chromogenic reactions, the structure of the colored product, tungsten blue, is not known. All that is known is that it absorbs well at 710 nm and that the absorbance is directly proportional to the uric acid concentration in the blood serum.

Reactions of this type have dramatically increased the use of visible spectroscopy. It is important to remember, however, that some of these reactions are troublesome. When developing a procedure of this type, be sure to evaluate the method carefully with respect to the following:

1. How specific is the chromogen in its reaction with the analyte? Many chromogens are not as specific as they should be. For example, it has been estimated that 11 per cent of the tungsten blue produced in the uric acid method outlined previously does not come from the reaction of uric acid with phosphotungstic acid. Fortunately the 11 per cent systematic error is about constant, so the procedure is still useful.

2. How stable is the chromogen and the colored product? Some chromogens are stable for only 2 or 3 hours. Some of the colored products are stable for only a few minutes. Poor stability is an inconvenience at the least. In some cases the timing of the measurement of the absorbance is critical.

3. What are the effects of various operating variables? Does the order of reagent addition affect the outcome of the results? Does pH have an effect? For example, the uric acid procedure is one in which the order of addition of reagents is significant.

4. Does the method produce linear Beer's law plots?

5. Does the method have sufficient sensitivity to detect the analyte in your samples? In my work in toxicology, I found that many methods could find the drug in an overdose patient (postmortem), but the same method was not sensitive enough to measure the drug at therapeutic levels.

A careful evaluation of these matters can prevent untimely and embarrassing surprises.

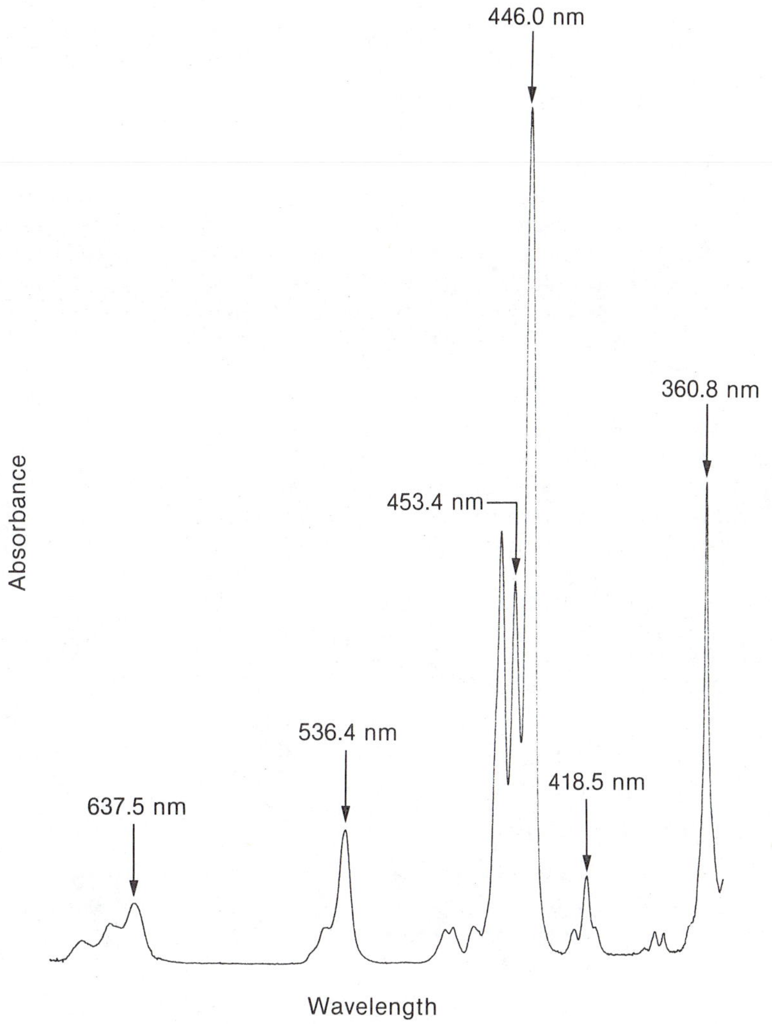

Figure 4–54. Absorption spectrum of holmium oxide. Wavelength assignments are from Beckman publication 015-081506C (August 1977). The values were obtained on a Beckman Model 25 spectrometer. The spectrum was run on a Coleman 124 spectrometer with a spectral bandwidth of 0.5 nm.

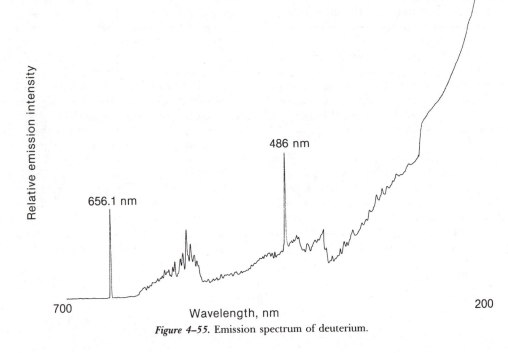

Figure 4–55. Emission spectrum of deuterium.

WAVELENGTH CALIBRATION

A further matter of importance is the periodic wavelength calibration for spectrophotometric instruments. Most instruments these days hold their calibration well, but, as part of the quality assurance program, they should be checked on a regular basis. Instruments used daily should be checked weekly. There are several approaches to calibration; the best one for your instrument depends on the instrument.

One way that this can be done is to scan the absorbance of a holmium oxide filter between 700 nm and 350 nm. Holmium oxide has a number of useful absorbances that can be used for calibration. The spectrum of holmium oxide is shown in Figure 4–54. The spectrum in Figure 4–54 was determined on an instrument with a spectral bandwidth of 0.5 nm. Thus the wavelength assignment to the nearest 0.1 nm is an estimate in the last place.

Didymium filters have been suggested for use in wavelength calibration. I do not recommend it. The absorptions of didymium are broader than those of holmium oxide, and hence there is more uncertainty in the wavelength of maximum absorption.

On instruments that have a deuterium lamp as an ultraviolet source, provisions are frequently made to allow the analyst to calibrate the instrument with the deuterium lamp. A deuterium lamp emits a sharp line at 656.1 nm (Fig. 4–55).

OPTICAL FIBERS

Some instruments use optical fibers (sometimes called light pipes) to transfer radiant energy from one location in an instrument to a second location. The core of the fibers is made from glass for visible wavelengths and quartz for ultraviolet wavelengths. The fibers are coated with a material of low refractive index. The coating material is said to be the cladding material. Radiant energy passes through an optical fiber because it undergoes multiple total internal reflections as it passes down the fiber (Fig. 4–56). The cladding material prevents

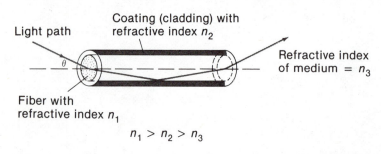

Figure 4–56. Schematic drawing showing the light path through an optical fiber. (Adapted from: D.A. Skoog, Principles of Instrumental Analysis, 3rd Ed., Philadelphia, Saunders College Publishing, 1985.)

the loss of much of the radiant energy through the side walls of the fiber.

Bundles of optical fibers are sometimes used to transfer images. This can be done if the fibers are the same length, and if the individual fibers have the same position in the bundle at both ends.

For a more detailed discussion of optical fibers please see Willard, Merritt, Dean, and Settle's book *Instrumental Methods of Analysis*, 6th Edition, Wadsworth Publishing Company, 1981.

ODDS AND ENDS

Some instruments use either an analog or a digital computer to convert transmittance to absorbance. On such instruments, count your lucky stars and use the absorbances that the instrument provides. On other instruments, the readout is supplied linearly in transmittance and nonlinearly in absorbance. It is more accurate to measure the transmittance on such instruments and then to calculate the absorbance mathematically.

When purchasing instruments, carefully consider your needs. Double-beam recording instruments are relatively expensive. If you never need to record an absorption spectrum, why purchase one? If you need to record spectra, don't buy a filter photometer or single-beam spectrophotometer. If you analyze only one thing, maybe a filter photometer would be your best choice; it surely costs less. There is certainly room for both spectrophotometers and photometers in a large clinical laboratory.

Problems

1. Consider the transmission spectra of the six pieces of colored glass shown in the illustration. If glasses D and E were combined to make a filter, what would be the filter's nominal wavelength, its peak transmittance, and its spectral bandwidth?

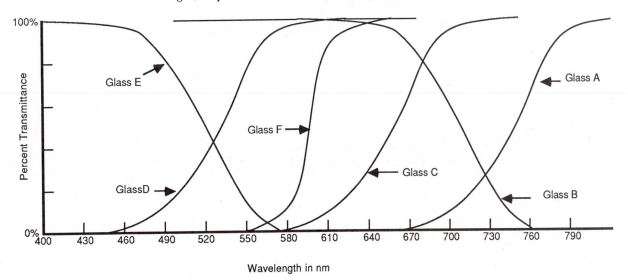

Wavelength in nm

2. If glasses A and B are combined to make a filter, what will be its nominal wavelength, its peak transmittance, and its spectral bandwidth?

3. What color is glass B?

4. For an interference filter with a MgF_2 thickness of 5.5×10^{-5} cm, at what wavelengths will transmission peaks be observed? Assume that $\theta = 90$ degrees and calculate for M = 1,2,3,4.

5. Assume that a source of radiant energy produces energy from 400 to 1000 nm. Further assume that the interference filter from problem 4 is to be used in combination with the source to produce "monochromatic" radiant energy at a nominal wavelength of 760 nm. What glass from problem 1 could be used as a blocking filter?

6. Repeat problem 5, but select a proper glass from problem 1 that would allow the isolation of 507 nm radiant energy.

7. Consider the following figure. Assume that the angle of incidence is equal to the angle of diffraction. Calculate the wavelengths that could pass through a grating monochromator at 37 degrees. Consider only M = 1,2,3; assume that a 1200 line/mm grating is used. For this calculation, the grating equation becomes $M\lambda = d(\sin i + \sin \theta)$.

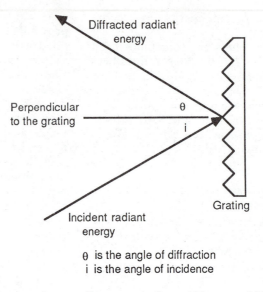

θ is the angle of diffraction
i is the angle of incidence

8. Assume that a source produces radiant energy from 400 nm to 1500 nm. Further assume that the grating in problem 7 is to be used to produce "monochromatic" 500 nm light from the grating's second-order diffraction at 37 degrees. Which glass from problem 1 could be used for the blocking filter?

9. A diffraction grating has 600 grooves per millimeter. The incident radiant energy contains all wavelengths between 400 and 1000 nm. The incident radiant energy is incident at 0 degrees relative to a perpendicular to the grating. What is the angle of diffraction for 800 nm and 400 nm radiant energy? Make calculations for $M = 1$ and 2.

10. Continuing problem 9, calculate the angle of diffraction for $M = 3$, 400 nm radiant energy.

11. What glass from problem 1 would make a good blocking filter if one wanted to isolate the 400 nm energy of the second-order in problem 9?

12. Gratings are generally used in their $M = 1$ region. On the other hand, gratings are used in some instruments in the $M = 2$ region. This is because the angular dispersion is two times greater for $M = 2$ than for $M = 1$. With the use of $M = 2$ comes an engineering problem. Use the data in problems 9 and 10 to identify the problem.

13. If the grating in problem 9 were used in the $M = 2$ region, what glass from problem 1 would be useful as a blocking filter in order to avoid $M = 3$ contamination problems?

Deviations from Beer's Law and Errors in Absorption Spectroscopy

Many aspects of absorption spectroscopy are discussed in Chapters 3 and 4. In this chapter we will continue the discussion by considering the non-ideal behavior and the errors that are encountered when performing spectrophotometric measurements. In many cases, the easiest way to study these deviations and errors is to think of the optical path of the instrument (Fig. 5–1).

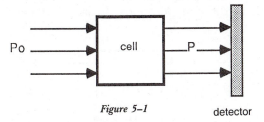

Figure 5–1   detector

In our discussion we will be thinking a lot about Po and P. Remember that the transmittance is the ratio of P to Po.

$$t = P/Po$$

Also remember that

$$A = -\log t = -\log P/Po = \epsilon bC$$

It will be necessary for you to interrelate transmittance and absorbance rapidly. Of particular interest will be the extremes of transmittance $t = 0$ and $t = 1$. Remember that when the transmittance is 0, then P must be 0; in order for P to be zero, the absorbance must be infinite. When the transmittance is 1, P must equal Po, and the absorbance must be 0. For quick reference, this information is presented in Table 5–1 along with 10 per cent and 1 per cent transmittance.

DEVIATIONS FROM BEER'S LAW

Beer's law is the basic law of all of absorption spectroscopy. The equation

$$A = -\log P/Po = \epsilon bC$$

predicts that a linear relationship exists between absorbance and concentration. The limits of the equation are absorbance equals zero and absorbance equals infinity. Thus, when Beer's law is followed, the plot of A versus concentration starts at zero concentration and goes to infinite concentration, while absorbance goes from zero to infinity

TABLE 5–1. Selected Values of Transmittance and the Corresponding Absorbance Values

Transmittance	% Transmittance	Value of P	Absorbance	Concentration
1.0	100	Po	0	0
0.1	10	0.1 Po	1.0	Depends on ϵb
0.01	1	0.01 Po	2.0	Depends on ϵb
0	0	0	∞	∞

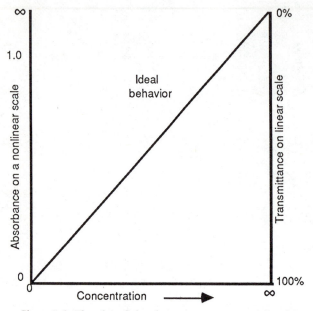

Figure 5–2. The plot of absorbance versus concentration that conforms to Beer's law.

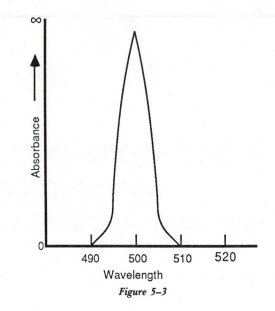

Figure 5–3

(Fig. 5–2). Fortuitously, Beer's law is frequently followed. We prepare standards over the range of our interest, and we find absorbances. The values plot out linearly. Occasionally, however, we do find that our experimentally determined plots are non-linear. When this occurs, we have a deviant Beer's law plot. Deviant Beer's law plots can be useful, but they are inconvenient. They require that many more standards be run in order to define the line well. When a deviant plot is found for a method, the procedure must make use of graphical concentra-tion determination or have a well-programmed computer for data reduction. Equations like

$$C = \frac{A_u}{A_s} C_s$$

and

$$C = \frac{A_u}{\epsilon b}$$

from Chapter 3 fail under these circumstances. Because analysis under these conditions is so in-convenient, scientists will do much to avoid the situation. The avoidance always involves an under-standing of the source of the deviation.

One of the most common deviations from Beer's law occurs when Po is insufficiently mono-chromatic. Insufficiently monochromatic radiant energy contains wavelengths of radiation that the analyte in the cell cannot absorb. This happens, for example, when the absorption spectrum of the an-alyte is narrow (Fig. 5–3), but the radiant energy supplied to the cell is a broad band of wavelengths ranging from, say, 480 to 520 nm. With this example

in mind, let's symbolize the intensity of two types of radiant energy. Let's call the intensity of the radiation that is between 490 and 510 nm, $(Po)_a$. This symbol stands for Po absorbable. Let's call the intensity of the radiation that is between 480 and 490 nm as well as the radiation that is between 510 and 520 nm, $(Po)_u$. The symbol stands for Po un-absorbable. As far as the instrument is concerned, Po is $(Po)_a$ plus $(Po)_u$.

In Figure 5–4 I have illustrated $(Po)_a$ and $(Po)_u$ as two separate beams that are passing through the cell in different locations. In an instrument, this

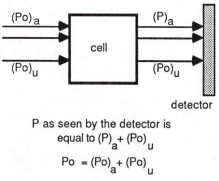

P as seen by the detector is
equal to $(P)_a + (Po)_u$

$Po = (P)_a + (Po)_u$

Figure 5–4

separation does not exist, but the effect would be the same even if it did. When the cell contains only nonabsorbing solvent, both $(Po)_a$ and $(Po)_u$ pass through the cell so that $(P)_a$ plus $(Po)_u$ is equal to Po. In this case, the transmittance will be 1.0 and the absorbance will be 0. Now let's let the concen-tration of the analyte increase without limit. As the concentration increases, the value of $(P)_a$ will start to drop, but $(Po)_u$ will still equal $(Po)_u$. As the concentration approaches infinity, $(P)_a$ will approach zero, but $(Po)_u$ will still equal $(Po)_u$. The detector will detect $(Po)_u$ and confuse it for P. This causes the transmittance and absorbance to reach a limiting value. The outcome of this situation is illustrated in Figure 5–5. The higher the

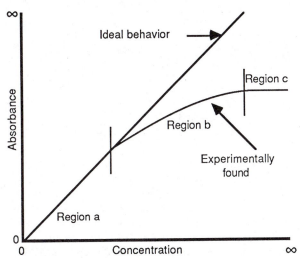

Figure 5–5. Region a is the linear dynamic range. Region b is the nonlinear dynamic range. Region c is the nondynamic range.

ratio $\dfrac{(Po)_u}{(Po)_u + (Po)_a}$, the more deviation one can expect. The higher the ratio, the smaller the linear dynamic range, and the larger the nondynamic range. If the ratio is small, that is, the radiant energy is almost monochromatic enough, the nondynamic range will be small and the dynamic range large. In this case, the Beer's law plot is usable if the concentrations are low enough to stay in the linear range.

The solution to the problem outlined here is to get an instrument with a smaller spectral bandwidth. Using our example, if an instrument with a spectral bandwidth of about 5 nm were used instead of the one with a 40 nm spectral bandwidth, no nonlinearity would be found.

A second situation that can lead to deviations in Beer's law is the one that occurs when the compound being measured fluoresces. The resulting deviation is like the deviation illustrated in Figure 5–5. Molecular fluorescence is a phenomenon that occurs when a compound re-emits absorbed radiant energy rather than converting it to heat. Generally the fluorescent emission is of longer wavelength than Po and P, but the detector still confuses the fluorescent intensity (F) for P (Fig. 5–6). In a situation like this, the higher the concentration of the absorbing compound, the lower P becomes, but the higher F becomes. Because of F, the detector never can realize that P has approached 0. Thus the deviation develops. The better a compound fluoresces, the greater the deviation will be. The prob-

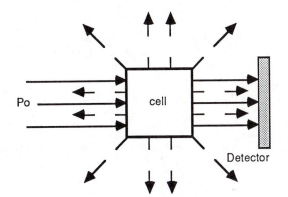

Figure 5–6. Solid arrows represent Po or P. Broken arrows illustrate the fluorescent intensity (F).

lem can be minimized by placing a sharp cutoff filter between the cell and the detector. The filter must pass P but cut off F.

A third source of deviation of the type depicted in Figure 5–5 is the deviation caused by stray radiant energy, which can be of two types. First, stray radiant energy can be of the same wavelength as Po but somehow miss the cell on its way to the detector. Second, it can be radiation of any nonabsorbable wavelength that, because of poor design, reaches the detector. This stray radiant energy problem is illustrated in Figure 5–7. The source of

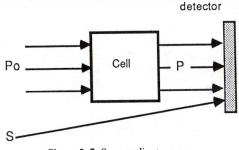

Figure 5–7. Stray radiant energy.

stray radiant energy could always be found if one were to look hard enough. All instruments have some stray radiation; however, for most instruments built after 1975, the stray radiant energy level is generally acceptable. Consider the problem from the mathematical point of view. P, as far as the detector is concerned, is really P + S. Thus the transmittance would be

$$t = \frac{P + S}{Po + S}$$

With this equation in mind, let us examine what would happen if the concentration of the compound in the cell were increased without limit. As C approaches ∞, P approaches 0, but t as calculated by the instrument approaches $\dfrac{S}{Po+S}$. For example, if

TABLE 5–2. The Effect of Concentration on the Per Cent Ionization

Concentration of HA Added (mol/l)	$K_i = 10^{-3}$	For an acid, HA, with $K_i = 10^{-5}$ % Ionization	$K_i = 10^{-7}$
1.0 mol/l	3.2	0.3	0.03
10^{-1} mol/l	11.0	1.0	0.18
10^{-2} mol/l	27.0	3.1	0.32
10^{-3} mol/l	62.0	9.5	2.00
5×10^{-4} mol/l	73.0	—	—
10^{-4} mol/l	90.0	27.0	3.16
Infinite dilution	100.0	100.0	100.00

an instrument has a stray radiation level of 1 per cent of Po, then

$$t \text{ approaches } \frac{0.01\ Po}{Po + 0.01\ Po} \text{ or } 0.01$$

Thus the instrument with a stray radiation level of 1 per cent of Po could never read an absorbance higher than A = 2.0. There would be error in absorbance well below the A = 2.0 level also. The way to avoid this error is to buy or build instruments that have lower stray radiation levels.

A fourth source of deviation from Beer's law that results in a plot like the one in Figure 5–5 is the deviation that is caused by the use of an improper wavelength for the analysis. Consider the absorption spectrum shown in Figure 3–9 (p. 28). Lambda$_2$ and lambda$_4$ are wavelengths where ϵ is varying rapidly with wavelength. When this is the case, the Beer's law plots tend to be deviant. The more rapidly ϵ changes with wavelength, the more likely it is that this will be a problem. Also, the larger the spectral bandwidth of the instrument, the more likely it is that this will be a problem. Using an instrument with a 1 nm bandwidth, I have never encountered this deviation, but most reference works discuss it. The reason for this deviation is beyond the scope of this book; however, be aware that it could be a problem.

A fifth source of deviation of the type depicted in Figure 5–5 is the deviation caused by the variation in per cent ionization of a slightly ionized analyte. In order to understand the origin of this deviation, it is necessary for you to recall some of your prior chemistry. To refresh your memory, I will hit the high points using a hypothetical, red, weak acid (HA). Weak acids are weak because they do not ionize well; they undergo the following reaction only slightly, but how slightly is dependent on the concentration.

$$HA + H_2O \rightleftharpoons H_3O^+ + A^{-1}$$

Also recall the concept of per cent ionization. For our acid, the per cent ionization is

$$\% \text{ ionization} = \frac{[A^{-1}]\ 10^2\%}{[HA] + [A^{-1}]}$$

The brackets mean molar concentration at equilibrium. Consider the data in Table 5–2. You can see from the table that the per cent ionization varies a great deal with concentration. You can also see that the K_i of the compound has a considerable effect on the relationship between per cent ionization and concentration. For our example of the red acid, let's assume that the K_i is 10^{-3} and plot the data from Table 5–2 (Fig. 5–8). Next, consider the spectroscopic data on the acid (Table 5–3).

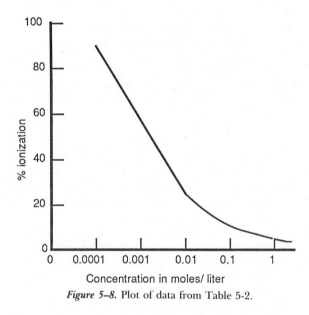

Figure 5–8. Plot of data from Table 5-2.

Now let's put together the information in Tables 5–2 and 5–3. Let's think about a Beer's law plot prepared at $\lambda = 400$ nm that covers the concentration range 0 to 10^{-3} mol/l in added HA. Let's have one solution that contains no HA (a blank) and three standards. The blank contains no A^{-1}, so it will have 0 absorbance. The solution prepared from 10^{-4} moles of HA and enough water to make a liter of solution will contain 9×10^{-5} mol/l of

TABLE 5-3. Spectroscopic Data on a Hypothetical Acid

	HA	A⁻	H₃O⁺	H₂O
$\epsilon_{495\ nm}$	10^3	0	0	0
$\epsilon_{400\ nm}$	0	10^3	0	0

A^{-1} and will have an absorbance of 0.09. The solution prepared from 5×10^{-4} moles of HA and enough water to make a liter of solution will contain 3.65×10^{-4} mol/l of A^{-1} and will have an absorbance of 0.365. The solution prepared from 10^{-3} moles of HA and enough water to make a liter of solution will have 6.2×10^{-4} mol/l of A^{-1} and an absorbance of 0.62. If we plot this data along with a line that assumes 100 per cent ionization at all concentrations, the deviation becomes apparent (Fig. 5-9). The calibration curve is close to the ideal

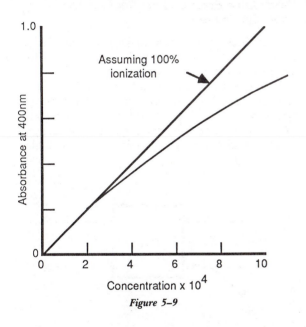

Figure 5-9

line at concentrations between 0 and 10^{-4} mol/l, but as the concentration increases above 10^{-4} mol/l, the per cent ionization decreases. This decrease causes the absorbance to be lower than expected.

The extent of this type of deviation is affected by the concentration range of interest and the Ki of the analyte. This deviation could be avoided by making all the solutions very basic. The high pH would make the per cent ionization 100 per cent at all concentrations. Under these conditions, the ideal plot would be obtained. If the proper pH is selected, the management of the problem is simple.

Not all deviations from Beer's law fall below the line of ideal behavior. One of the reasons for deviations above the line of ideal behavior is the variation of per cent ionization with increasing concentration. This case is the opposite of the case

shown in Figure 5-9. Using our previous example, if there is less A^{-1} at higher concentrations than is ideal, there must be more HA. Thus, a plot of absorbance at 495 nm versus concentration must yield results like those shown in Figure 5-10. This prob-

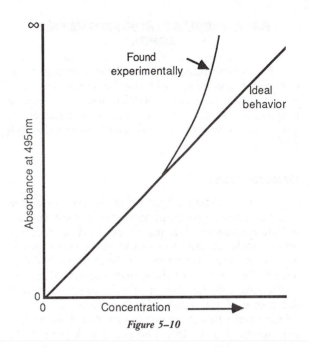

Figure 5-10

lem with deviation can be solved by making all your standards very acid. The common ion effect will make the weak acid have almost a 0 per cent ionization and, hence, will remove the deviation. The control of pH is very important in dealing with slightly ionized acids and bases.

A second source of deviation above the line of ideal behavior is the deviation caused by the reflection and scattering of radiant energy. Consider the case of a compound that is used for making standards and that contains a small percentage of particles of insoluble material. These particles serve as sites for the reflection or scattering of radiant energy. Consider Figure 5-11. The radiation that

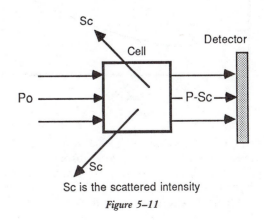

Sc is the scattered intensity

Figure 5-11

is lost as the result of reflection and scattering causes the detector to receive less radiant energy than it should. The detector sees P-Sc instead of Po. The Beer's law plot would be like the one shown in Figure 5–10.

RELATIVE SPECTROPHOTOMETRIC ERROR

In any chemical analysis there is always some uncertainty associated with the measurements. The uncertainties have many origins, and their relative importance is not even a general phenomenon. There are, however, several trends worth discussing.

Detector Noise

One of the major errors encountered in visible and ultraviolet spectrophotometric measurements is that associated with the detector. The detector noise produces variation in the absorbance that is measured. This noise tends to limit precision and, hence, the accuracy of these instruments. Although a discussion of the origin of this noise is beyond the scope of this work, it is beneficial to see the relative error that it creates in our absorbance measurements (Fig. 5–12). The curve in Figure 5–12

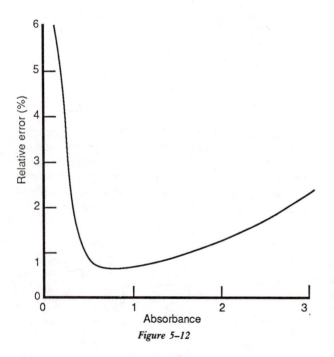

Figure 5–12

tells us that the value of the measured absorbance is more subject to error if the absorbance is very high or very low. It also tells us that great error is likely in measurements of absorbance below 0.1.

Generally, procedures for the analysis of bioan-

alytes are designed to avoid excessively high or low absorbance values. Do, however, remain mindful of this error problem.

Beer's Law and High Concentrations

Beer's law, as discussed previously, is a law that may not be followed at concentrations above about 10^{-2} molar. At concentrations higher than about 10^{-2} molar, the molecules of the absorbing compound may get close enough to start interacting. The result will be that ϵ will no longer be a constant. This will lead to deviations from Beer's law. The solution to this problem is dilution. Once the compound is in more dilute surroundings, the nonideal behavior should stop. At times, dilution is not possible because ϵ is small. When this is the case, the nonlinearity must be tolerated.

Predictable Errors, Their Causes, and Their Elimination

An inattentive analyst can do many things to spoil an analysis, even after a good, straight Beer's law plot has been obtained.

Fingerprints, dirt, or any chemicals on the outside of the cell will cause the measured transmittance to be low. The reason is that the materials on the outside of the cell scatter radiant energy (Fig. 5–13). The low transmittance causes a high ab-

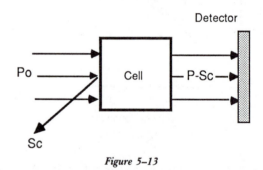

Figure 5–13

sorbance and a high result for the analysis. Keep cells clean! Follow the cleaning procedures suggested by the cell manufacturer. When cells develop scratches, discard them and get new ones. It does not take much to make a 2 or 3 per cent transmittance error.

A second source of error that leads to low results is cell misalignment. When the radiation energy comes to a cell, if the angle of incidence is not 90°, there are greater reflective losses (Fig. 5–14). Here the detector sees P-Sc instead of Po, which leads the instrument to read out a low transmittance or a falsely elevated absorbance. The solution to the problem is to insert the cell properly.

A fluorescing contaminant excited to fluoresce

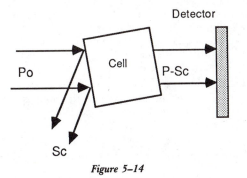

Figure 5-14

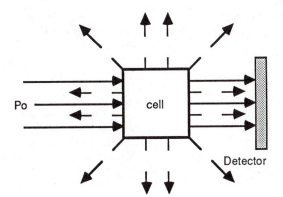

Figure 5-15. Solid arrows represent Po or P. Broken arrows represent fluorescent intensity.

by Po can re-emit radiation energy that the detector receives (Fig. 5–15). In this way, the analyte in the cell produces a higher transmittance than it really should. This produces a lower absorbance and, hence, a low result for the analysis. This is a classic example of a so-called matrix effect. The matrix is everything in the cell except the analyte. In this case, the fluorescing contaminant is part of the matrix. It causes an error, so the error is called a matrix effect. The way to avoid this matrix effect is to use a sharp cutoff filter between the cell and the detector. The filter should transmit P completely but should cut off F.

A fourth phenomenon that can introduce an error in an analysis is the effect of temperature on Ki values. The value of Ki is a function of temperature. Sometimes the variation of Ki is great enough to be a problem. The way to control this problem is to have your standards and samples at about the same temperature.

Sometimes it is necessary to work with solutions that have high concentrations of dissolved material in addition to the analyte compound. When this is the case, standard solutions may have quite a different refractive index than do the unknowns. This can lead to error because refractive index does have an effect on the value of ϵ. Consider the equation

$$\epsilon = \epsilon_0 \frac{N}{(N^2 + 2)^2}$$

ϵ_0 can be thought of as the molar absorptivity in very dilute solutions— say 10^{-3} mol/l or less

ϵ is the molar absorptivity in more concentrated solutions

N   is refractive index of the solution

This is an example of a matrix effect.

An insidious error that can mislead you on a double-beam instrument occurs when the solvent you have selected absorbs radiation strongly at the wavelength of interest. When this is the case, the analyte can be overlooked because the solvent absorbs almost all of Po. The ironic part of this is that the reference beam and the sample beam do have the same intensity. The intensity is near zero. The instrument forms the ratio of the two and calls it 1.0. As far as the instrument is concerned

$$t = \frac{\text{sample beam current}}{\text{reference beam current}} = \frac{\text{almost 0}}{\text{almost 0}} = 1.0$$

Thus, the solvent absorbs in both beams; the instrument says that the transmittance is 100 per cent, and, in the meantime, the analyte is overlooked for want of photons to absorb. The same thing happens on optical null instruments. Pick your solvents well! (See Table 5–4.)

TABLE 5–4. Solvents for the Ultraviolet and the Visible Regions

Solvent	Approximate transparency minimum (nm)
Water	180
Ethanol	220
Hexane	200
Cyclohexane	200
Benzene	280
Carbon tetrachloride	260
Diethyl ether	210
Acetone	330
Dioxane	320
Cellosolve	320

Reprinted from: D. A. Skoog and D. M. West, Principles of Instrumental Analysis, 2nd Ed., Philadelphia, Saunders College Publishing, 1980.

Be informed that lipemic samples tend to produce lower transmittance values than they should. This occurs because radiation is scattered off the fat globules in the serum. The resulting losses make P lower than it should be. The lower transmittance value creates a higher absorbance value and a falsely elevated result of the analysis.

The presence of large amounts of any abnormal constituent in a sample should be considered with respect to its effect on other analyses. The material might constitute a source of a matrix effect.

Problems   In each of the following problems assume that a perfect Beer's law plot has been obtained by measuring the absorbance of a set of standard solutions. Assume that the standard solutions were prepared from perfect solutes and solvents. Assume that the cells containing the standard solutions and the blank were perfect. The question to be answered in each of the next 16 problems is: Will the concentration that is found experimentally be high, low or correct?

1. The solution of unknown concentration is cloudy.
2. The cells that were used to make the standard curve were dropped on the floor and broken. An old scratched cell was used in order to finish the analysis. The solution of unknown concentration was analyzed in the scratched cell.
3. The standard solutions of the weak acid HX were prepared using 0.1 molar sodium hydroxide, (NaOH) as the solvent. The solution of unknown concentration was prepared with distilled water.
4. Repeat question 3 except assume that 0.1 molar HCl is the solvent for the unknown.
5. The solution of unknown concentration contained a fluorescent compound that is excited to fluoresce by Po.
6. The standard solutions of methylamine (CH_3NH_2) were prepared by using 0.1 molar sodium hydroxide as a solvent. The unknown solution was prepared using the same solvent.
7. The cell that contained the solution of unknown concentration has a white streak in the cell face.
8. The cell path of the cell that contained the solution of unknown concentration is larger than the other cells that were used in the analysis.
9. The cell that contained the solution of unknown concentration was placed in the instrument carelessly. The result was that the radiant energy was not incident at 90° to the cell face.
10. The cell that contained the unknown was handled carelessly. This resulted in many fingerprints on the optical faces.
11. The HX solution of unknown concentration was 50°C hotter than were the standards. The K_a of HX is 1×10^{-3}. The analysis was done at the λ_{max} of HX.
12. The cell that contained the solution of unknown concentration was very cold. The cell was placed in the instrument. Water vapor condensed on the outside of the cell.
13. Very small bubbles of gas formed and stuck to the inside of the cell that contained the solution of unknown concentration.
14. A double-beam instrument was used for the determination. Small bubbles of gas formed and coated the inside of the reference cell. The sample cell formed no bubbles.
15. There was not enough solution of unknown concentration to fill the cell properly. This left the optical path partially unexposed to the solution.
16. The standard solutions were made from 90 per cent pure analyte. The analyst assumed that the compound was 100 per cent pure. The solution of unknown concentration was analyzed.
17. The absorption spectrum of the analyte is shown in the next figure.

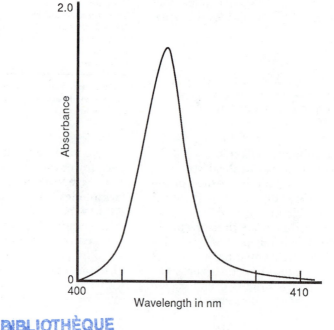

The procedure for the use of an analysis kit says that the results may be calculated from the following equation:

$$C = 124 \text{ A mg/dl}$$

The equation assumes that an instrument with 1 nm spectral bandwidth is used. If you use an instrument with 10 nm spectral bandwidth, and you use the equation, will your results be high, low, or correct?

18. Consider the spectrum of the analyte shown in the next figure.

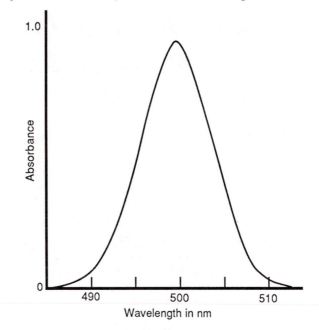

The Beer's law plot was made using 500 nm as the analytical wavelength. The instrument has a 1 nm bandwidth. The solution of unknown concentration was analyzed by error at 499 nm. Would the resulting value be high, low, or correct?

19. What is the highest absorbance that an instrument can detect if the stray radiation level is 0.1 per cent of Po?

20. Repeat question 18 for an analyte with spectrum shown in the next figure.

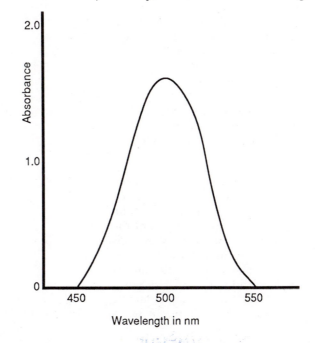

6

Ultraviolet Absorption Spectroscopy

The classic ultraviolet region of the electromagnetic spectrum is the wavelength range from 200 to 380 nm. It is called the ultraviolet region because it is composed of radiation that is shorter in wavelength than is violet visible radiation. Ultraviolet radiation is more energetic than visible radiant energy. This narrow wavelength range is an important region to clinical scientists because many compounds of biomedical interest absorb strongly in this range. The wavelength region is widely used for both qualitative and quantitative analysis. Fortunately, the information about quantitative analysis that was presented in Chapter 3 is applicable in the ultraviolet region. Unfortunately, the errors and inconveniences discussed in Chapter 5 are also potential problems in the ultraviolet region.

QUALITATIVE ANALYSIS

The topic of qualitative analysis has not been fully developed in this text. There were some introductory remarks about it in Chapter 3, but let us consider the matter in greater detail. In the ultraviolet region of the spectrum, many compounds absorb radiation in a unique way. This uniqueness has its origin in the fact that the molar absorptivity of a compound varies over quite a range, and it does so in a concentration-independent, characteristic manner. Consider the spectra in Figures 6–1 through 6–5. These are spectra of 10 mg/l solutions of various drugs that have been scanned between 200 and 340 nm. The solvent is either 0.1 molar hydrochloric acid (HCl) or 0.1 molar sodium hydroxide (NaOH) depending on the needs of the sample. The solvents are nonabsorbing in this range. The spectra were determined on a double-beam, recording, ultraviolet spectrophotometer with a 1 nm spectral bandwidth.

First consider the spectrum of amphetamine (Fig. 6–1). The spectrum is very simple; it has a

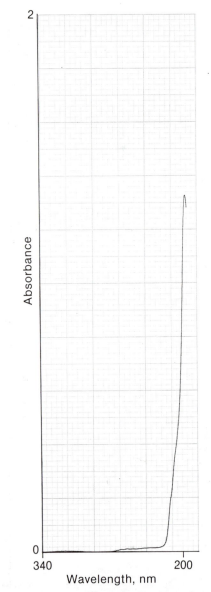

Figure 6–1. Spectrum of amphetamine; 10 mg/l amphetamine in 0.1 molar HCl.

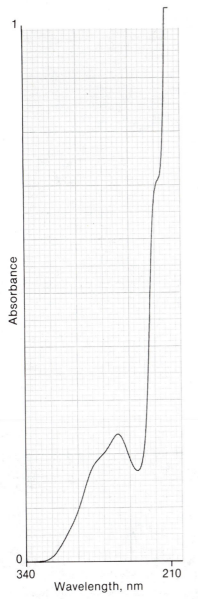

Figure 6–2. Spectrum of imipramine; 10 mg/l imipramine in 0.1 molar HCl.

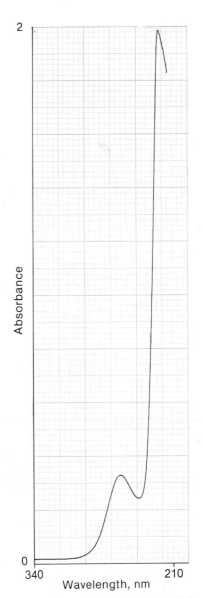

Figure 6–3. Spectrum of phenobarbital; 10 mg/l phenobarbitol in 0.1 molar NaOH.

wavelength of maximum absorption (λ_{max}) at 197 nm. Next consider the spectrum of imipramine (Fig. 6–2). This spectrum is different from the spectrum of amphetamine in that it has an absorbance peak at 253 nm. Compare these two spectra with the spectrum of phenobarbital (Fig. 6–3). Phenobarbital has a λ_{max} at 218 nm and a secondary λ_{max} at 256 nm. Compare the last two spectra with the rest (Figs. 6–4 and 6–5), and note that each spectrum is at least slightly different from each of the other four. In light of this spectral uniqueness, the spectra of unknown compounds and of bona fide pure compounds can be compared and, in many cases, qual-

itative identification is "possible." I place the word possible in quotation marks because the identification does contain an element of uncertainty. If the spectrum of a known bona fide compound is different from that of an unknown compound, they cannot be the same. If, on the other hand, the spectra match well, the two spectra were probably produced by the same compound. I say probably because it is possible for two closely related but slightly different compounds to have spectra that are very similar. Although it frequently is not a problem, one must be aware of this insidious potential for misidentification.

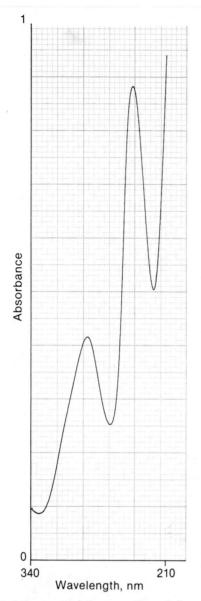

Figure 6–4. Spectrum of diazepam; 10 mg/l diazepam in 0.1 molar HCl.

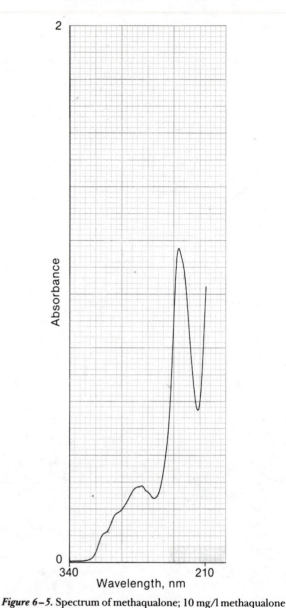

Figure 6–5. Spectrum of methaqualone; 10 mg/l methaqualone in 0.1 molar HCl.

ULTRAVIOLET INSTRUMENTATION

Because qualitative analysis by spectral comparison is so common and so important, the vast majority of ultraviolet instruments are of the recording, double-beam variety. This is because recording instruments make the determination of an absorption spectrum so convenient. In addition to the recording spectrophotometers, both single-beam spectrophotometers and double-beam filter photometers are in existence. The filter photometers are used as liquid chromatograph detectors and will be discussed in Chapter 17. For now let's concern ourselves with the more common double-beam recording instruments.

Modern ultraviolet instruments are almost always grating spectrophotometers much like the one that is illustrated in Figure 4–47 (p. 54). The instruments generally have a visible as well as an ultraviolet source, so that they can be used in the visible as well as in the ultraviolet regions. Consider the optical path of the instrument illustrated in Figure 6–6.

The source mirror can be moved through enough of an arc to reflect either the radiation from the ultraviolet source or that from the visible source into the entrance slit. Otherwise, this instrument has much in common with the instrument shown in Figure 4–47 (p. 54). The only new things are the ultraviolet source and the quartz lens, cells, and

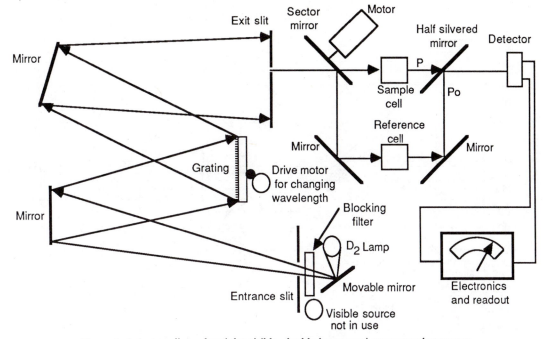

Figure 6-6. A recording, ultraviolet-visible, double-beam grating spectrophotometer.

detector envelope. The quartz optical parts are not a problem in the visible region, but glass optics would not be transparent in the ultraviolet region. The detector may be a phototube, photomultiplier, or a silicon photodiode. Photomultipliers are probably the most common of the instruments in this class.

Many diode-array instruments have the capacity to be used in the ultraviolet and visible regions (see Fig. 4-53, p. 57).

ULTRAVIOLET RADIATION SOURCES

Over the years there have been several important sources of ultraviolet radiation. All of these sources produce the radiation by causing an electrical discharge to take place in a sealed tube that contains a gas. The electrical energy causes the gas atoms or molecules to become excited and hence to emit ultraviolet radiation. In older instruments the filling gas was hydrogen, and the resulting source was called a hydrogen lamp. In recent years deuterium gas has been used in place of hydrogen because the deuterium lamp produces a greater intensity of ultraviolet emission and has a longer lifetime. The lamp is hard to illustrate, so I have included the photograph in Figure 6-7. The geometry of the lamp causes the radiation to be produced in an intense ball about 1.5 mm in diameter. These lamps emit continuous radiation from approximately 160 to 375 nm. The emission

spectrum of a typical lamp is illustrated in Figure 6-8.

It is important to remember that these sources emit a high flux of ultraviolet radiation. The ultraviolet radiation can blind you. Do not look into an operating ultraviolet lamp. High intensities or long exposures to ultraviolet radiation can cause skin burns as well. Avoid exposure to ultraviolet radiant energy.

SOME APPLICATIONS

Many drugs have a unique ultraviolet absorption spectrum. This has led to the widespread use of ultraviolet spectroscopy in the field of analytical toxicology.

If the toxicant is available to the analyst in the form of a fairly pure compound, it is generally brought into solution, and its spectrum is deter-

Figure 6-7. A deuterium discharge lamp. (Courtesy of Oriel Corporation, Stratford, Conn. Reprinted with permission.)

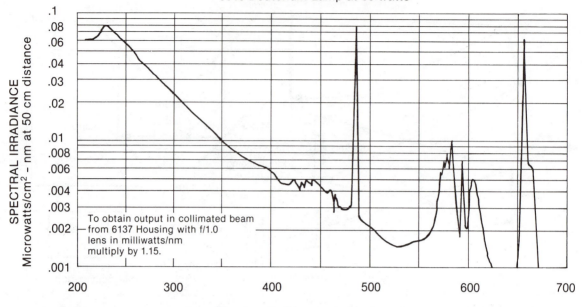

TYPICAL SPECTRAL OUTPUT
6316 Deuterium Lamp at 30 watts

Figure 6–8. Emission spectrum of a dueterium lamp. (Courtesy of Oriel Corporation, Stratford, Conn. Reprinted with permission.)

mined. If, on the other hand, the toxicant must be identified in a blood or urine sample, it is generally necessary to carry the sample through a solvent extraction procedure. This is done in order to separate the toxicant from interfering compounds that are frequently a part of the sample matrix. Once the toxicant has been isolated from the sample matrix, its spectrum can be determined. The spectrum can then be compared, either visually or by computer, to the spectra of known toxicants. Once a match is made between the spectrum of a known compound and the spectrum of the unknown toxicant, the toxicologist can say that the two may well be the same compounds.

Quantitative analysis is frequently done by ultraviolet spectroscopy. Toxicants that are ultraviolet-absorbing compounds can be measured in this way.

In addition to the analysis of toxicants, ultraviolet spectroscopy is widely used for the analysis of enzyme concentration. The following is an example of enzyme analysis.

The determination of serum aspartate transaminase (AST), formerly serum glutamate-oxaloacetate transaminase (GOT), is a common laboratory test for the diagnosis of some liver diseases. In diseases that involve liver cell destruction, the AST enzyme "leaks" from the damaged cells and is picked up by the blood stream. This elevates the serum AST level. This situation is easily detected because normal blood does not contain much AST enzyme.

The analysis involves the incubation of the patient's serum with aspartic acid and α-ketoglutaric acid. The AST enzyme in the serum catalyzes the reaction shown at the bottom of the page. The AST enzyme is not destroyed in the reaction, but it behaves like a typical catalyst. The oxaloacetic acid that is produced then undergoes a second reaction (top of p. 77).

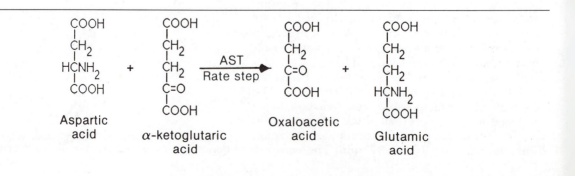

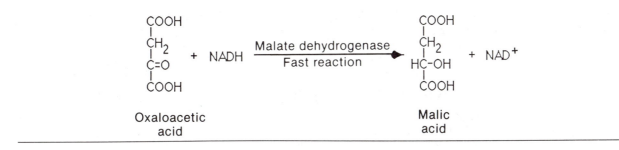

The second reaction converts oxaloacetic acid to malic acid. It also converts NADH to NAD$^+$. NADH is the reduced form of nicotinamide adenine dinucleotide; it absorbs strongly at 340 nm. NAD$^+$ is nicotinamide adenine dinucleotide; it does not absorb at 340 nm.

Under suitable experimental conditions, the rate of production of malic acid and NAD$^+$ is directly proportional to the concentration of AST in the reaction mixture. The concentration of AST in the reaction mixture is directly proportional to the AST concentration in the patient's serum. Thus, the rate of disappearance of NADH as measured by the rate of change in absorbance at 340 nm is directly proportional to the concentration of AST in the patient's serum. The concentration of AST in the patient's serum may be calculated by an equation of the form

$$\text{Concentration of AST} = \left(\frac{\Delta A_{340\ nm}}{\text{sec}}\right)\left(\text{constant}\right)$$

where the constant is a known value. It contains the molar absorptivity of NADH and other terms that are dependent on the sample size, the procedure, and the units that are to be used to calculate the results.

This approach to the analysis of an enzyme is called a kinetic or rate method. These methods are very common in the clinical laboratory. Most of the enzymes that are determined are done by rate methods. Most of these procedures make use of the NAD$^+$, NADH system.

It is also possible to use an enzyme to catalyze a reaction that is run in order to determine the concentration of the *enzyme's substrate*. The determination of serum ethanol (ethyl alcohol) is such an example.

The ethyl alcohol concentration of a blood sample can be measured by determining the concentration of the reaction product that forms when the enzyme dehydrogenates the alcohol. Consider the reaction below. Ethyl alcohol, NAD$^+$, and ace-taldehyde do not absorb ultraviolet radiation at 340 nm. On the other hand, the NADH, which forms on an equal molar basis, relative to the alcohol, absorbs very well at 340 nm. The molar absorptivity of NADH at 340 nm is 6220 l/mol cm. The molarity of NADH that forms is equal to

$$C = A/\epsilon b = \frac{A}{(6.22 \times 10^3\ \text{l/mol cm})(1.00\ \text{cm})}$$

This equation assumes that the measurement is made in a 1-cm cell. As the alcohol that was present produced NADH in equal molar amounts, the value C is also the alcohol concentration in the reaction mixture.

Many methods of analysis use the biochemical hydrogen acceptor NAD$^+$ in conjunction with an enzyme in order to analyze for a compound of interest. These methods all measure the NADH at 340 nm.

ODDS AND ENDS

On instruments used for qualitative analysis it is necessary to check the instrument's resolution periodically. This can be done by scanning the absorption spectrum of toluene vapor. The reference beam contains air. This is done by placing one drop of toluene in the bottom of a standard quartz cell and putting the cell cap in place. After a few minutes, the toluene vapor comes to equilibrium with the liquid. The cell can be placed in the sample beam of the instrument, and the spectrum can be determined. The spectrum of toluene determined at three different bandwidths is shown in Figure 4–22 (p. 45). Depending on your instrument, you should get something close to these spectra. How much resolution you need depends on your application. The spectral resolution of an instrument can decrease with time; it should be checked now and then. Toluene is not to be touched or inhaled.

The 360.8 nm value for holmium oxide absorption can be used to check wavelength calibration in the ultraviolet region (see Fig. 4–54, p. 59).

$$\text{CH}_3\text{CH}_2\text{OH} + \text{NAD}^+ \xrightarrow[\substack{\text{dehydrogenase} \\ \text{(enzyme)}}]{\text{alcohol}} \overset{H}{\text{CH}_3\text{C}}=O + \text{NADH}$$

ethyl alcohol · acetaldehyde

Problems

1. Ethyl alcohol is determined in serum by deproteinizing the serum and diluting it 1 to 5. Of the resulting protein-free solution, 0.10 ml is added to a 2.90-ml solution containing alcohol dehydrogenase, NAD^+, buffer, and semicarbazide. The alcohol undergoes the following reaction:

$$CH_3CH_2OH + NAD^+ \xrightarrow[\text{dehydrogenase}]{\text{alcohol}} CH_3 - \overset{H}{\underset{}{C}} = O + NADH + H^+$$

NADH absorbs strongly at 340 nm; its molar absorptivity is 6220 l/mol cm. None of the other compounds absorbs at 340 nm. The equation for calculating the results is

$$(A_{max} - A_{initial})_{340\ nm}\ b(111) = \text{ethanol in the serum in mg/dl}$$

If $A_{initial}$ is 0.11 and A_{max} is 2.13, what is the alcohol concentration for the serum sample? Assume that b is 1 cm.

2. The molecular weight of ethyl alcohol is 46 g/mol. Using the value and information in problem 1, derive the equation that was used to calculate the results in problem 1.

3. Serum aspartate aminotransferase (AST) can be determined in the ultraviolet region by the reactions outlined in this chapter in the section entitled *Some Applications*. The analysis is done by mixing 3.00 ml of solution that contains aspartic acid, α-ketoglutaric acid, NADH, and buffer with 0.20 ml of serum. The reaction is given 2 minutes to gain a steady reaction rate. The absorbance at 340 nm is then read and recorded. Exactly 5 minutes later the absorbance is read again. The results are calculated by the following equations:

$$\frac{\Delta A}{5\ min} = \text{initial A} - \text{final A}$$

$$\text{AST expressed units per liter} = (\Delta A/5\ min)\ 515$$

Based on the following data, what is the serum AST level?

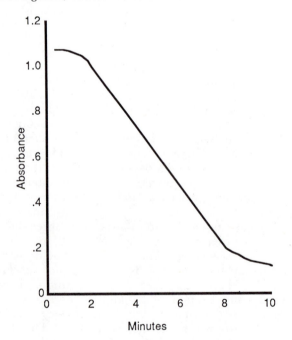

4. Serum alanine aminotransferase (ALT) can be determined by the following reactions:

$$\text{L-alanine} + \text{2-oxoglutarate} \xrightarrow[\text{rate step}]{\text{ALT}} \text{pyruvate} + \text{L-glutamate}$$

$$\text{Pyruvate} + \text{NADH} + \text{H}^+ \xrightarrow[\text{dehydrogenase}]{\text{lactate}} \text{L-lactate} + \text{NAD}^+ + \text{H}_2\text{O}$$

Serum (0.20 ml) is mixed with 3.00 ml of solution containing L-alanine, 2-oxoglutarate, NADH, lactate dehydrogenase, and buffer. The reaction is given 2 minutes to gain a constant reaction rate. The absorbance at 340 nm is measured and recorded. Exactly 5 minutes later the absorbance is measured again. The results are calculated by the following equations:

$$\frac{\Delta A}{5 \text{ min.}} = \text{initial A} - \text{final A}$$

$$\text{ALT expressed in units per liter} = \left(\frac{\Delta A}{5 \text{ min}}\right)(515)$$

Based on the following data, what is the serum ALT level?

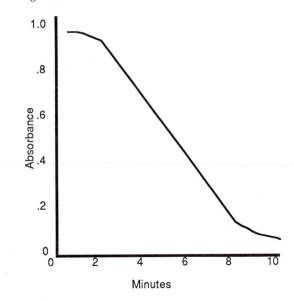

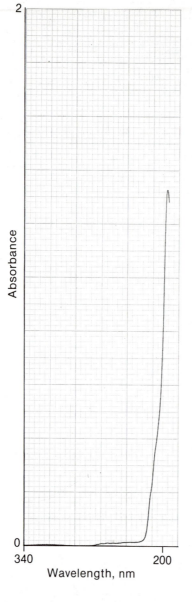

5. Limiting yourself to the data in this chapter, what com-
    pound is this likely to be?

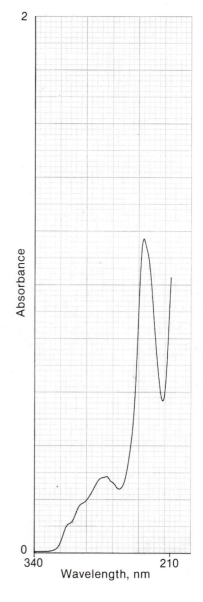

6. Limiting yourself to the data in this chapter, what com-
    pound is this likely to be?

7

Molecular Fluorescent Spectroscopy

Fluorescent spectroscopy is closely related to absorption spectroscopy. The nature of this relationship will be made clear in the next few pages, but, by way of introduction, let me say that fluorescent spectroscopy is a technique used mainly for quantitative analysis. The technique makes use of radiant energy that is emitted by the analyte. The limit of detection for fluorescence is frequently 1000-fold lower than that for ultraviolet spectroscopy. Many compounds of biomedical interest can be determined by fluorescent spectroscopy. There are also a few qualitative analysis applications.

In order to get a better understanding of the nature of fluorescence, consider gaseous sodium and its ability to absorb and re-emit 589 nm photons. This process can be represented symbolically.

$$Na_{(g)} + h\upsilon \rightarrow Na_{(g)}* \rightarrow Na_{(g)} + h\upsilon$$

| Ground state atom | 589 nm photon | Excited state atom | Ground state atom | 589 nm photon |

It is also shown diagrammatically in Figure 2–9 (p. 20). The ground state atom is so called because the valence electron in the atom is in its lowest energy level. The 3s orbital is where the electron should be; thus the ground state atom is stable. The absorption of a 589 nm photon forces the valence electron from the 3s orbital to the 3p orbital. The atom that results is said to be excited and is in an unstable configuration. After about 10^{-13} seconds, the excited atom relaxes and re-emits the 589 nm photon. In so doing, the sodium atom returns to the ground state. The process of re-emitting the photon is said to be an atomic fluorescent process. Atomic fluorescence is a universal property of monatomic atoms in the gas phase; it is the process whereby these atoms get rid of absorbed energy. In summary, atomic fluorescence is a two-part process. First, excitation energy is absorbed, then a flu-

orescent photon is emitted. Think about this example carefully. These concepts will be used as a model for molecular fluorescence.

Molecular orbitals have fixed energies just like the atomic orbitals. The electron pairs that form the bonds in molecules are thought to occupy these molecular orbitals. Molecules absorb energy by forcing electrons from ground-state molecular orbitals into higher-energy empty orbitals. Molecules that have undergone this transition are in an excited state and must relax through some mechanism. Molecules generally do this by converting the absorbed energy to heat. This process is called collision deactivation and is the most common relaxation mechanism. Some types of molecules can give off a photon of radiant energy and, in so doing, revert to the ground state. This relaxation mechanism is called molecular fluorescence. Molecular fluorescence is like atomic fluorescence except that the emitted photon is less energetic than the absorbed photon. This means that the fluorescent emission is at longer wavelength than the fluorescent excitation. Although most molecules convert absorbed photons to heat, many compounds of biomedical interest do fluoresce.

Molecular fluorescence is frequently activated by the absorption of ultraviolet radiation. The fluorescent emission frequently occurs in the visible region. There are, however, many examples of fluorescence induced by short-wavelength visible radiation with fluorescent emission occurring at longer visible wavelengths. The reason that the fluorescent emission takes place at longer wavelengths than the excitation is beyond the scope of this book; however, a discussion of this topic can be found in any good instrumental analysis text that has been written for chemists. Note again that fluorescence is a two-step process; first, there is excitation and then emission.

THE MATHEMATICAL BASIS OF FLUORESCENT SPECTROSCOPY

In this section we will work out the equations that relate concentration to the experimentally measured fluorescent intensity (F).

It is helpful to think about the cell in which fluorescence is being excited (Fig. 7–1). Po is

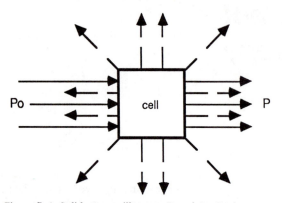

Figure 7–1. Solid arrows illustrate P and P_o. Broken arrows illustrate fluorescent emission.

illustrated as three solid arrows; P is illustrated as three solid arrows. The total fluorescent output (F′) is shown as 12 arrows with dashed shafts. The fluorescent output leaves the cell in all directions at equal intensity. With these facts in mind, let us consider the mathematics of fluorescence.

In order to relate the fluorescent intensity to the concentration of an analyte, we must start by considering the fraction of the incident radiant energy that is absorbed. The fraction of radiation absorbed is

$$\frac{Po - P}{Po}$$

This may be written

$$\frac{Po - P}{Po} = \frac{Po}{Po} - \frac{P}{Po} \qquad \text{Equation 7–1}$$

If we write Beer's law in its exponential form, we obtain

$$P/Po = e^{-\epsilon bC} \qquad \text{Equation 7–2}$$

where e = is 2.718
If we substitute Equation 7–2 into Equation 7–1, we get a useful equation that relates the fraction of absorbed radiation to ϵbC.

$$\frac{Po - P}{Po} = \frac{Po}{Po} - e^{-\epsilon bC} \qquad \text{Equation 7–3}$$

$$\frac{Po - P}{Po} = 1 - e^{-\epsilon bC}$$

If we multiply both sides of Equation 7–3 by Po, it leads us to an equation that relates the absorbed intensity to the concentration of the fluorescing compound.

$$Po - P = Po(1 - e^{-\epsilon bC}) \qquad \text{Equation 7–4}$$

Scientists have found experimentally that not all of the absorbed photons are re-emitted as fluorescence. Generally, part of the molecules in the excited state release their energy as heat. In order to relate the total fluorescent emission intensity to the absorbed photons, one must introduce a constant called the quantum efficiency (ϕ). The quantum efficiency is the ratio of excited molecules that fluoresce to the total number of excited molecules. Its value varies between 0 and 1 and is greatly affected by temperature. Introducing the quantum efficiency into Equation 7–4, we obtain

$$F' = \phi(Po - P) = \phi Po(1 - e^{-\epsilon bC}) \qquad \text{Equation 7–5}$$

This equation relates the total fluorescent emission intensity to the analyte concentration. Equation 7–5 is a useful equation and is of particular interest to physical chemists, but it needs modification for practical use. Fluorescent spectroscopy is done on instruments that can observe only a portion of one side of the cell. This means that many of the photons in F′ go where the detector is not. In order to relate Equation 7–5 to an equation that predicts the fluorescent intensity (F) that is seen by the detector, we must realize that

$$F = kF' \qquad \text{Equation 7–6}$$

where k is a geometrical constant that is the ratio of the detected photons to the total number of emitted photons. Thus, if Equation 7–5 is modified by Equation 7–6, we get

$$F = kF' = k\phi Po(1 - e^{-\epsilon bC}) \qquad \text{Equation 7–7}$$

Equation 7–7 is the most general law of fluorescent spectroscopy. It is to fluorescent spectroscopy what Beer's law is to absorption spectroscopy. Equation 7–7 clearly predicts that the measured fluorescent intensity is not a linear function of analyte concentration. This nonlinearity is seen in experimental work in some cases. On the other hand, when ϵbC is less than approximately 0.02, Equation 7–7 simplifies to

$$F = k\phi Po \, (\epsilon bC) \qquad \text{Equation 7–8}$$

This simplification is possible because

$$(1 - 10^{-\epsilon bC}) \cong (\epsilon bC)$$

for small values of ϵbC. Thus, in a given experiment, on a given instrument, when ϵbC is small, Equation 7–8 becomes

$$F = KC \qquad \text{Equation } 7\text{–}9$$

where $K = k\phi P_o (\epsilon b)$

Equation 7–9 predicts that under favorable conditions, linear calibration curves can be expected.

FLUORESCENT SPECTROPHOTOMETRIC INSTRUMENTS

There are three basic types of fluorescent instruments: filter fluorometers, single-beam fluorescent spectrophotometers, and double-beam fluorescent spectrophotometers. These instruments contain many of the optical components discussed in Chapters 4 and 6.

Filter Fluorometers

Filter fluorometers should be called compensated fluorescent photometers, but this more descriptive term is not in common usage. The optical path of a typical filter fluorometer is illustrated in Figure 7–2. Let's consider this instrument part by part.

The source in fluorescent instruments is a powerful ultraviolet source. I have never seen one that was less than 120 watts. Equation 7–8 predicts that a large value for P_o will make for greater instrument sensitivity, so the powerful sources are not

surprising. There are two common sources in fluorescent instruments. The mercury vapor lamp (Fig. 7–3) is the first.

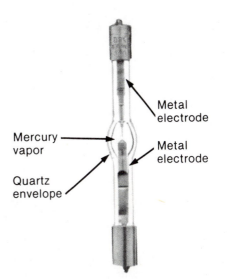

Figure 7–3. Mercury lamp. (Courtesy of Oriel Corporation, Stratford, Conn. Reprinted with permission.)

The mercury vapor lamp is a simple device, powered by a constant-voltage power supply. It provides a number of intense emission lines that are superimposed on a weak continuum. The mercury vapor emission spectrum is illustrated in Figure 7–4. The mercury lamp is a good source if your sample can be excited by one of the lines. If your sample needs radiation at 210 nm, though, the analysis is not possible. The mercury source does not produce

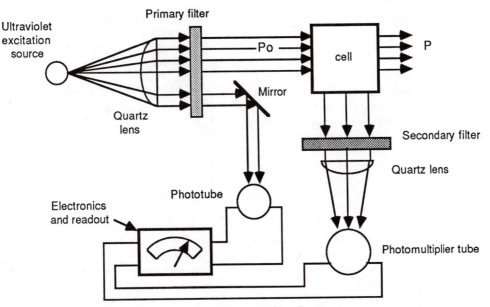

Figure 7–2. A "compensated" filter fluorometer.

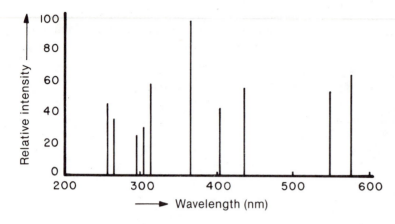

Figure 7–4. Mercury lamp spectrum plot. (Courtesy of Perkin-Elmer Corporation, Norwalk, Conn. Reprinted with permission.)

much intensity at 210 nm. The big advantage of the mercury source is that it is not very expensive and is very durable. I have seen a 120-watt mercury source last for more than 10 years in our academic laboratory.

The second popular source for fluorescent spectroscopy is the xenon arc source (Fig. 7–5).

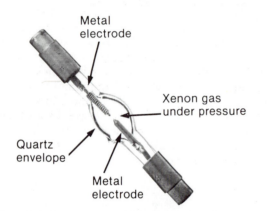

Figure 7–5. Xenon arc source. (Courtesy of Oriel Corporation, Stratford, Conn. Reprinted with permission.)

The source is powered by a constant-voltage power supply. The source provides an intense output that is useful throughout the visible and ultraviolet regions. The emission spectrum of a typical xenon lamp is illustrated in Figure 7–6. The source is more versatile than the mercury lamp, but it is more expensive and less durable. Both the mercury and xenon sources produce high intensities of ultraviolet radiation. These intense ultraviolet sources will burn your eyes badly; don't look at them without proper eye protection.

Many of the compounds that are determined by fluorescent measurements must be excited by ultraviolet radiation. Hence, the envelopes on the sources and detectors are made from quartz. So are the lenses and cells. These cells are expensive; don't use them on the visible absorption instrument.

Depending on the situation, filters of several types can be used. The common approach is to use a glass or quartz filter like the ones discussed in Chapter 4 for the primary filter (see the section entitled *Filters and Filter Photometers*). This is done because interference filters produce more stray radiation. Stray radiation is more of a problem in fluorescence than in ultraviolet and visible work. Even so, interference filters are used in some applications. The primary filter is used to supply the sample with reasonable monochromatic excitation energy. This is done for three reasons. If there were an interfering second compound in the cell, one might be able to avoid exciting the interfering compound. This could be done if it were excited at a different wavelength relative to the analyte. The second reason for exciting the analyte with monochromatic radiation is that a certain amount of excitation radiation is scattered into the secondary optical system. This stray radiation would, in part, reach the detector and would lead to error. The exact extent of the error would depend on the situation. The third reason for using monochromatic excitation is that there is less difficulty with photodecomposition of the analyte. If

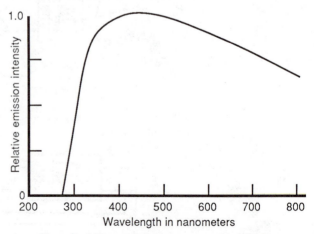
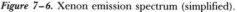

Figure 7–6. Xenon emission spectrum (simplified).

there were no primary filter, the entire source output would reach the sample cell. If the analyte were photosensitive, the source would decompose the analyte in short order.

Sometimes the secondary filter is just a sharp, cut-off glass filter similar to glass 1 in Figure 4–6a (p. 37). This approach is a little crude, but in the absence of an excited interfering compound in the sample, it can work out satisfactorily. Otherwise, a proper filter—either glass or interference—can be used.

The most popular detector for fluorescent measurements is the photomultiplier tube, which was discussed in the section entitled *Radiation Intensity Detectors* in Chapter 4.

The small fixed mirror and phototube in the filter fluorometer are provided in order to measure the intensity of the source. A knowledge of the source intensity allows for the compensation of source intensity variations. If the mirror and phototube were not used, the measured fluorescence of a sample would be very susceptible to source intensity variations. After all, Equation 7–8 predicts that Po has quite an effect on the fluorescent intensity. The instrument corrects for the source variations by providing the instrument's electronics with a photocurrent that is directly proportional to the source intensity. The electronics can then compensate for the variation in the source intensity by adjusting the photomultiplier voltage to offset source intensity changes. The change in photo-

multiplier amplification factor makes the correction. A second way that this compensation can be made is by having the readout respond to the ratio of the photomultiplier output to the phototube output. The ratio is determined electronically. This type of optical configuration makes the instrument much more stable.

In summary, the ultraviolet excitation source provides polychromatic radiation; the primary filter refines the polychromatic radiation so that the cell is exposed to a narrow band of proper wavelength. The secondary filter selects which wavelength of the emission will be seen by the detector. The detectors detect Po and F. The electronics calculates and presents the data.

The main advantage of a filter fluorometer is that it is inexpensive. The main disadvantage is that it has little versatility. It may require different primary and secondary filters for each different analyte that is to be measured. A second disadvantage is that qualitative analysis is impossible on these instruments.

Single-Beam Fluorescent Spectrophotometers

The next type of instrument is the single-beam fluorescent spectrophotometer. Consider the optical path shown in Figure 7–7. The optical components are all familiar to you; there are no new components. The primary monochromator selects

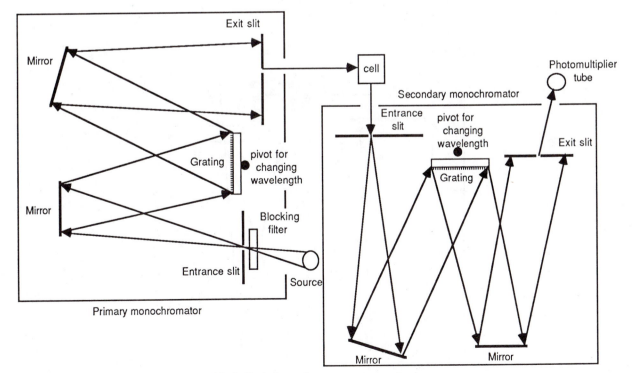

Figure 7–7. Single-beam fluorescent spectrophotometer.

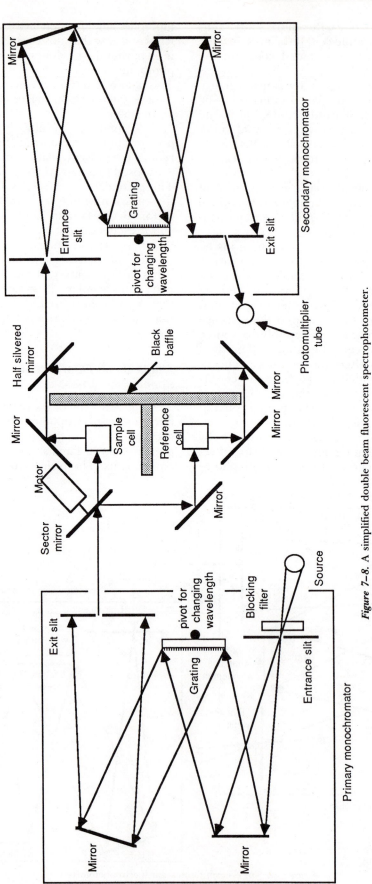

Figure 7–8. A simplified double beam fluorescent spectrophotometer.

the wavelength of excitation that will be supplied to the sample, and the secondary monochromator selects the wavelength of emission to be detected. The monochromators give the instrument great versatility and selectivity. The proper choice of primary and secondary wavelength can avoid many problems with interfering compounds in the sample.

Single-beam fluorescent spectrometers are built with very stable source power supplies. The instruments are, however, susceptible to drifts in source intensity. The problem can be minimized by frequent calibration. Sometimes a compensation system is included. This system was discussed in the previous section.

Double-Beam Fluorescent Spectrophotometers

The third type of fluorescent instrument is the double-beam spectrophotometer (Fig. 7–8). The instrument contains no new optical components and is much like the ratio-recording instrument discussed in Chapter 4, Figures 4–47 through 4–50. The source and primary monochromator supply Po to a sector mirror that sends it to the sample and reference cells alternately. The fluorescent output of each cell is detected by the photomultiplier—first the sample output, then the reference output. This is possible because the two fluorescent outputs are separated in time after recombination at the half silvered mirror. The advantage of this type of instrument is that a reagent blank can be placed in the reference cell. The difference in fluorescent output between the two cells is then read out.

QUANTITATIVE ANALYSIS

If an analyst knows which wavelengths to use for the primary and secondary monochromators, or knows which filters to use in the filter fluorometer, quantitative analysis becomes a straightforward matter. Several standard solutions are prepared from pure analyte and a suitable nonfluorescing solvent. The fluorescent output of these standards is determined and plotted against their concentrations (Fig. 7–9). The measurements are made by adjusting the readout so that the fluorescent intensity (F) reads 0 when the pure solvent is in the cell. The instrument is then adjusted so that it reads 100 when the most concentrated standard is placed in the cell. The less concentrated standards then produce readings between 0 and 100. The fluorescent outputs of the samples are then determined on the instrument and the results are graphically interpreted to find the analyte concentrations. If you are lucky, you will get linear calibration as shown in Figure 7–9, but don't be surprised if you don't.

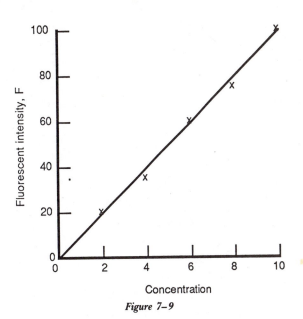

Figure 7–9

Remember that linear calibration cannot be expected if your standards have absorbances much in excess of 0.02 A.

If a double-beam instrument is used, the reference cell is filled with either pure solvent or the reagent blank, and the various standards and samples are alternately placed in the sample cell. Beyond this difference, the procedure for the analysis is the same as that outlined previously.

If the proper wavelengths for the analysis are not known, it is necessary to study the analyte's fluorescent properties systematically. Even if you know which wavelengths to use, it is a good idea to verify the information by going through the procedure outlined in the next paragraph.

First, look at the ultraviolet and visible spectra of the compound. For an example, look at the ultraviolet spectrum of imipramine in 0.1 sodium hydroxide (Fig. 7–10). If the instrument has a xenon source, adjust the primary monochromator to 216 nm. The reason for this is that imipramine absorbs 216 nm photons better than any other wavelength, and the xenon source emits intensely at 216 nm. If the instrument has a mercury source, adjust the primary monochromator to 254 nm. The reason for this choice is that the 254 nm emission line of mercury is the most intense line that the compound will absorb. With the primary monochromator so adjusted, obtain the apparent fluorescent emission spectrum of the imipramine. This spectrum is illustrated in Figure 7–11 and was determined on an instrument with a mercury source.

The spectrum in Figure 7–11 is a plot of the fluorescent emission intensity (F) versus the wavelength of the secondary monochromator. It tells the analyst that 405 nm is the wavelength at which imipramine emits the greatest fluorescent intensity.

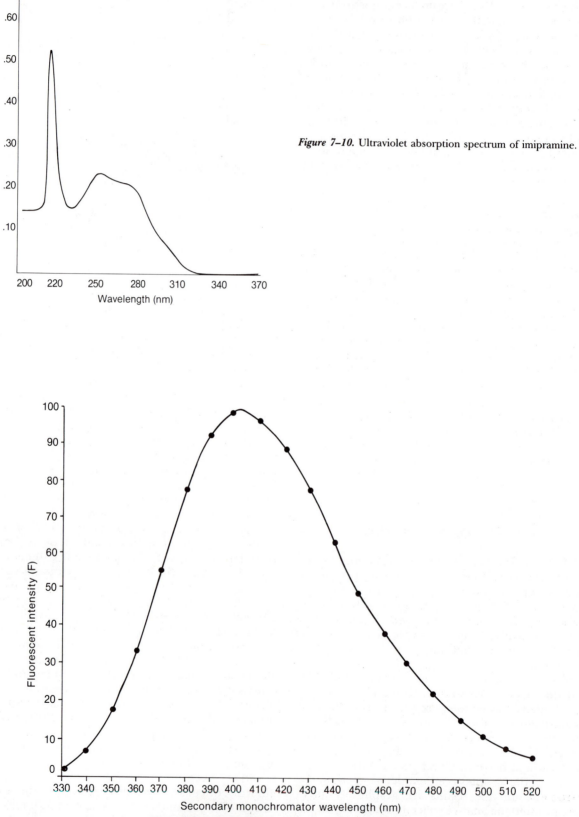

Figure 7-10. Ultraviolet absorption spectrum of imipramine.

Figure 7-11. Apparent emission spectrum of imipramine.

Hence, it is the best wavelength for the secondary monochromator. This spectrum is called the apparent emission spectrum because the detector's characteristics, as well as the drug's fluorescence, have contributed to its shape. For practical use, the apparent emission spectrum does allow the analyst to select the best secondary monochromator wavelength for the drug, but only when measured on that specific instrument.

Once the secondary monochromator wavelength has been selected, the primary monochromator can be varied to find its optimum wavelength. In our example, the secondary wavelength would be adjusted to 405 nm, and the primary monochromator would be varied between 200 nm and 320 nm. The range was selected on the basis of the absorption spectrum that is shown in Figure 7–10. If the compound does not absorb, it cannot fluoresce, so there is no point in looking outside this range. The resulting plot of F versus the primary monochromator wavelength is called the apparent excitation spectrum. The apparent excitation spectrum of imipramine determined on an instrument with a mercury source is shown in Figure 7–12. This spectrum tells the analyst which wavelength on the primary monochromator will produce the greatest fluorescence for a given amount of the drug. This is the wavelength, on this instrument, that should be used for imipramine analysis. Frequently, the accuracy of our original guess will be confirmed by these experimental measurements. The actual determination of the primary spectrum is recommended just in case a better wavelength exists. This spectrum is an apparent spectrum because its shape is a combination of the source's output and the analyte's fluorescent characteristics.

QUALITATIVE ANALYSIS BY FLUORESCENT SPECTROSCOPY

In some cases, such as when the analyte is in very dilute solution, it is desirable to do qualitative analysis by fluorescent spectroscopy. The great sensitivity of fluorescence is exploited in these cases; often the analysis is not possible by any other means.

In order to do qualitative analysis by fluorescent spectroscopy, one must compare the excitation and emission spectra of bona fide compounds that are suspected of being in the unknown with the spectra of the unknown. The spectra of the suspected pure compounds and the unknown should be determined on the same instrument because the apparent excitation and apparent emission spectra are significantly affected by the instrument on which they are determined.

If I were in the business of doing a lot of qualitative analysis, I would buy an instrument that could automatically remove the effects of the source and detector on the spectra. The resulting spectra would be called the true excitation spectrum and the true emission spectrum, and they would be instrument independent. Such instruments do exist, but they are expensive. They are not common and will not be discussed further.

THE FLUORESCENT DETECTION OF NONFLUORESCING COMPOUNDS

In much the same way that colorless compounds are determined by visible spectroscopy, nonfluorescing compounds can be reacted with suitable compounds to produce fluorescent products. The resulting fluorescent products are then investigated to find their concentrations and, hence, the concentrations of the analyte. For example, phenylalanine can be determined by its reaction with ninhydrin. The product that forms fluoresces well. Another example is the determination of magnesium. Magnesium ion is reacted with 8-hydroxyquinoline to form bis-8-hydroxyquinoline magnesium (II) ion, a very fluorescent material. The 8-hy-

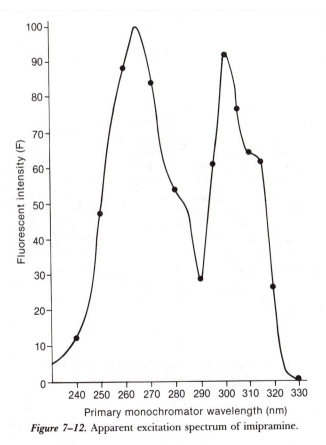

Figure 7–12. Apparent excitation spectrum of imipramine.

droxyquinoline itself is fluorescent, so a reagent blank must be used to subtract the small amount of fluorescent intensity of the starting material. The use of "fluorogens" has extended the scope of usefulness of fluorescent spectroscopy.

SOME PRACTICAL CONSIDERATIONS

Fluorescent spectroscopy is a very sensitive technique that, in favorable cases, can detect concentrations down to about 1 μg/l. When working with concentrations in this range, an analyst could have a troublesome analysis. First of all, microorganisms in the solutions can eat the analyte. The microorganisms that live in distilled water systems are hardy bugs and can eat many organic compounds. A second problem is that glassware can absorb the analyte onto its surface and, in so doing, can leave a solution that is devoid of the analyte. A third problem is the occurrence of quenching, which is a phenomenon that takes place when a trace of some contaminant prevents the analyte from fluorescing. For example, the fluorescence of thiamine can be quenched by traces of iodine vapors. Each of these problems is hard to detect; the effects are really insidious.

In published methods of analysis using fluorescent spectroscopy, one often finds very specific instructions for the preparation of glassware. These instructions may seem foolish in some cases, but they really must be followed. In some cases, they recommend not getting the glass too clean, so that surface absorption can be avoided. In other cases, the instructions suggest removing quenching agents and interfering compounds. If one does not follow the glassware preparation instructions, bad results can be expected.

Even when an analyst is making measurements on more concentrated solutions, a number of experimental factors can affect fluorescent emission intensity. A good place to start this discussion is with Equation 7–8,

$$F = k\phi Po \, (\epsilon bC)$$

Anything that affects any of the terms of this equation can affect fluorescent intensity. For example, temperature can affect ϕ, the quantum efficiency. A 10-ppm solution of imipramine hydrochloride at 0°C produces a fluorescent intensity of 100 units. The same solution at 21°C produces a reading of 92, whereas the same solution at 100°C produces a reading of 69 units. Thus, the temperature of standards and unknowns should be about the same.

The pH of a solution can have a considerable effect on the fluorescent output of a sample. Imipramine hydrochloride fluoresces 5 times better in a pH 13 solution than it does in a pH 1 solution. The variation of fluorescent intensity is generally related to molecular structure. Imipramine in a pH 13 solution has the following structure:

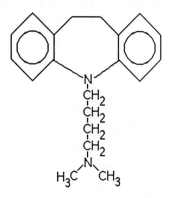

In more acidic solutions, it has the structure

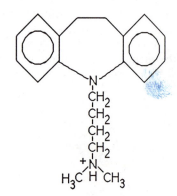

I expect that the fact that ϵ changes with structure is responsible for the difference in fluorescence.

In summary, fluorescent spectroscopy is a very good technique for quantitative analysis. Although experimentally troublesome at times, it is certainly capable of producing good results. The key to good results is close adherence to experimental protocol and, hence, to the proper management of variables. Some qualitative analysis procedures are also known. The very low limit of sensitivity is attractive for trace analysis. The nonlinearity of calibration does limit interest in the technique at concentrations higher than about 10 ppm. At higher concentrations, absorption spectroscopy can generally be used.

FLUORESCENCE MEASUREMENTS USING POLARIZED RADIATION

In some important situations, chemists find that they need to analyze the concentration of very large fluorescent compounds. If the sample matrix is free of fluorescence, then the analysis is trivial. On the other hand, if the sample matrix fluoresces under the same conditions as the analyte, it is sometimes possible to make useful fluorescent measurements with polarized excitation. This reduces the detection of the fluorescence of the matrix while allowing the fluorescence of the analyte to be measured.

The most common fluorescent photometer that provides polarized excitation is the Abbott TD_x instrument. The instrument is a highly specialized device that is designed to measure therapeutic drugs in serum by a competitive-binding immunochemical approach. The technique is discussed in Appendix A (page 303). In the technique, fluorescein or a fluorescein-like compound is used to tag the drug that is to be analyzed. This is done by covalent bonding; the resulting tagged drug is a stable product. The analysis of the patient's drug level is done by adding the tagged drug to the patient's serum and then adding some immunochemical reagents. The solution that results contains a mixture of leftover fluorescein-tagged drug (low molecular weight) and high molecular weight fluorescein-tagged reaction product. The fluorescein in both is excited with 485 nm radiation; they both fluoresce in the 525 to 550 nm range. The instrument is designed to measure the fluorescent output of the high molecular weight reaction product. Please consider the optical path (Fig. 7–13).

The optical path of the TD_x instrument is fairly similar to that of other fluorescent photometers. The source is a quartz–halide tungsten lamp with a reflector. Heat reflectors and absorbers help isolate the heat from the cuvette area. The lenses focus the radiation. The excitation filter has nominal transmission at 485 nm. The half silvered mirror splits off part of the excitation radiation, which goes to a reference detector. The output of current from this detector is used to regulate the lamp intensity to a constant value. The excitation radiation that is transmitted through the beam splitter goes to a device called a polarizer. The polarizer is a selective filter made from a material that exhibits the property of dichroism. A dichroic material is a material that has a much higher molar absorptivity for polarized radiation in one plane than it has for radiation polarized in another plane. The effect of this property is shown in Figure 7–14. The unpolarized radiation enters the filter, and vertically polarized radiation is transmitted. The horizontal component is absorbed.

The polarizer output passes to the liquid crystal. When a voltage is applied to the crystal from top to bottom, it transmits the polarized radiation unaffected. When the voltage is applied from side

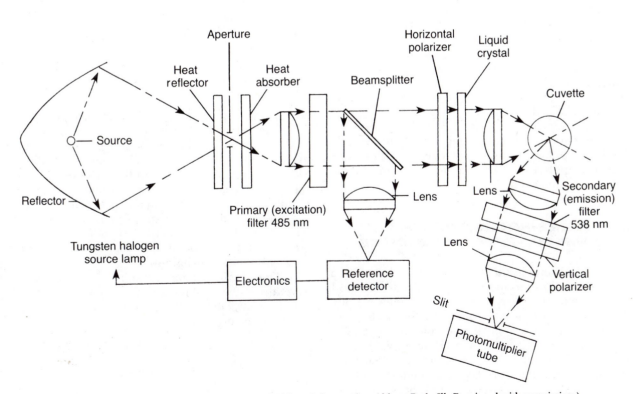

Figure 7–13. TD_x optical system. (Courtesy of Abbott Laboratories, Abbott Park, Ill. Reprinted with permission.)

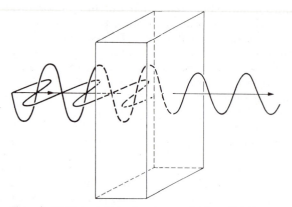

Figure 7–14. Dichroic behavior. As a result of parallel alignment of absorbing functional groups, the horizontal vibration is completely absorbed and the vertical vibration is absorbed only to a slight degree. (Strobel, Chemical Instrumentation, © 1973, Addison-Wesley, Reading, Massachusetts. Pg. 260, Fig. 10–25. Reprinted with permission.)

to side, the vertically polarized radiation is rotated 90 degrees. In this way, the liquid crystal is a polarized radiation rotator; it allows us to excite the sample first with vertically polarized radiation and then with horizontally polarized radiation. These two irradiations can be done many times per second. The output of the liquid crystal falls on the cuvette.

The fluorescence that is produced by the excitation of the sample is emitted in all directions. The emission optical path of the TD_x instrument is at a 75 degree angle to the excitation system. The emitted radiation is passed through a filter that isolates the fluorescein emission from other radiation. The radiation from the emission filter is passed through a second dichroic filter. This filter causes the detector to observe only radiation that is vertically polarized.

In summary, the TD_x instrument *excites* the fluorescein that is used as a tag, first with vertically and then with horizontally polarized radiation. The vertically polarized *emission* is then detected by the photomultiplier tube.

THE MATHEMATICS OF ANALYSIS USING POLARIZED FLUORESCENCE

The cuvette in the TD_x instrument contains a mixture of fluorescein-tagged small molecules and fluorescein-tagged large molecules. When vertically polarized radiation is absorbed by the fluorescein that is bonded to a small molecule, the radiation emitted is polarized in a random way. This is because the small molecules, along with their fluorescein tags, rotate very fast, and, hence, the fluorescein moieties are oriented in all possible directions when fluorescence occurs. If the cuvette contained only these small tagged molecules, the

emitted radiation would be totally unpolarized. On the other hand, very large molecules tagged with fluorescein rotate slowly. When the fluorescein bonded to a large molecule absorbs vertically polarized radiation, it emits vertically polarized radiation. This occurs because the large molecules do not move much during the lifetime of the excited state (about 10^{-9} sec).

If one wants to measure the large tagged molecules, one can do so by measuring the photomultiplier tube output, I_{vv}, when the excitation is vertically polarized, and measuring the output, I_{hv}, when the excitation is horizontally polarized. The difference between these two values should be proportional to the concentration of tagged large molecules. This is true because the value of I_{vv} is a combination of output from the tagged large molecules that have not lost their vertical orientation and the few tagged small molecules that emitted by chance with vertical polarization. The output from tagged small molecules is the same when excited with horizontally polarized excitation as when it is excited with vertically polarized excitation. Because of this fact, I_{hv} is a measure of the small tagged molecules. Thus, the difference

$$I_{vv} - I_{hv}$$

should be directly proportional to the concentration of large tagged molecules.

It was the custom of the early workers in the field to define and use the following equation:

$$p = \frac{I_{vv} - I_{hv}}{I_{vv} + I_{hv}}$$

where p is called the degree of polarization.

The degree of polarization is a value that varies between 0 and 1. When no large tagged molecules are in the cuvette, then $I_{vv} = I_{hv}$ and p is 0. When all of the tagged molecules in the cuvette are large tagged molecules, then I_{hv} approaches 0 and, hence, p approaches 1. In the range between all large tagged molecules and none, there is some nonlinearity in p. This nonlinearity is handled by the instrument's microprocessor by the use of a nonlinear, least squares program.

The TD_x instrument is calibrated by the use of several standards. The p values for the standards are used to generate the constants A, B, C, and D in the following equation:

$$p = A + \frac{B}{C + D(\text{concentration})}$$

The instrument's microprocessor uses this equation to calculate the concentration of drug in the sample.

The TD$_x$ instrument is used to measure a number of drugs in blood serum.

FRONT-FACE FLUORESCENT MEASUREMENTS OF THIN LAYERS

A novel approach to the use of fluorescent photometry for the determination of bioanalytes is the use of the so-called dry reagent format. In this approach, a piece of glass fiber "paper" is impregnated with the reagents needed to do fluorescent analysis. The "slide" or "tab" (Fig. 7–15) is then dried and stored until needed. The sample and necessary additional reagents are added, and the fluorescent product is given time to develop. The developed tab is caused to fluoresce by optical excitation. The excitation energy is projected on the face of the tab at an angle of about 45 degrees relative to the surface of the device. The fluorescent emission is viewed at an angle of 90 degrees relative to the face of the tab. The fluorescent emission that radiates off the face of the tab is then used to find the analyte concentration.

Figure 7–15. A STRATUS immunoassay tab designed for use in the STRATUS immunoassay system. (Courtesy of American DADE, Miami, Fla. Reprinted with permission.)

The exact nature of the calibration depends on the method. Some methods rely on traditional wet chemistry reactions. Others are based on immunoassay approaches. The tabs or slides are generally designed for use on a specific instrument. This approach to clinical analysis has the potential to become very significant.

Problems

1. Consider the following data:

Concentration of Analyte (ppm)	Fluorescent Emission Intensity (arbitrary units)
0	0
2	40
4	69
6	84
8	91
10	97
12	100

Plot a standard curve for the analysis of this analyte. An unknown produces an emission intensity of 47.3. What is the concentration of the unknown?

2. Why is pH an important experimental variable in fluorescence work?

3. Why is temperature an important experimental variable in fluorescence work?

4. Calcium has been determined by its reaction with fluorescein.*

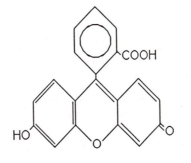

Fluorescein

*B. L. Kepner and D. M. Hercules, Anal. Chem., *35*: 1238, 1963.

The complex that is formed can be excited at 405 nm. It fluoresces at around 520 nm. Consider the following data:

Calcium Concentration (mg/dl)	Fluorescent Emission Intensity (arbitrary units)
0.0	0.0
5.0	45.0
7.5	63.5
10.0	79.0
12.5	93.0
15.0	100.0

What is the calcium concentration of a serum sample if it produces a fluorescent emission of 67.8?

5. Serum magnesium (Mg) has been determined by its reaction with 8-hydroxy-5-quinoline sulfonic acid. At the pH that is used, the reaction is

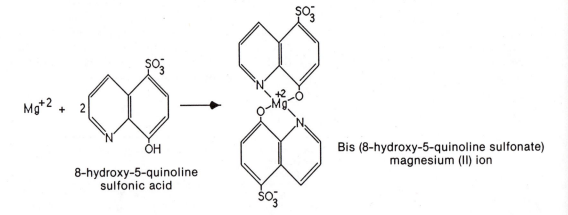

8-hydroxy-5-quinoline sulfonic acid

Bis (8-hydroxy-5-quinoline sulfonate) magnesium (II) ion

The bis (8-hydroxy-5-quinoline sulfonate) magnesium (II) ion can be excited to fluoresce at about 400 nm. The peak of its fluorescent emission is at 510 nm. Unfortunately, the sulfonic acid fluoresces slightly at these wavelengths. Consider the following data:

Tube Number	5 mmol/l Mg Standard (ml)	Distilled Water (ml)	Fluorogen Reagent (ml)	Patient's Serum (ml)	Fluorescent Emission
1	0.5	0.0	4.5	0.0	100.0
2	0.4	0.1	4.5	0.0	82.0
3	0.3	0.2	4.5	0.0	61.0
4	0.2	0.3	4.5	0.0	40.0
5	0.1	0.4	4.5	0.0	20.8
6	0.0	0.5	4.5	0.0	9.0
7	0.0	0.0	4.5	0.5	48.9
8	0.0	4.5	0.0	0.5	0.8

Some data reprinted from: N. W. Tietz, Fundamentals of Clinical Chemistry, 2nd Ed., Philadelphia, W. B. Saunders Company, 1976.

Tube 6 is the reagent blank. Tube 7 contains the sample. Tube 8 is the serum blank. What is the concentration of magnesium in the patient's serum?

6. The concentration of salicylate in serum can be determined by fluorescence. Salicylate is excited at about 310 nm; its fluorescent emission spectrum has a λ_{max} at about 400 nm. The serum may be deproteinized with tungstic acid before analysis. Consider the procedure:

Add 0.5 ml of serum to a centrifuge tube. Slowly add 4.50 ml of tungstic acid reagent. Let this mixture stand for 10 minutes. Centrifuge the tube for 10 minutes at 3000 rpm. Pipet 2 ml of the clear solution into a test tube. Label this tube "sample." Consider the following data:

Tube Number	10 mg/dl Salicylate Solution (ml)	Distilled Water (ml)	Tungstic Acid Solution (ml)	10 Molar NaOH (ml)	Fluorescent Emission
1	0.20	0.00	1.8	3	100
2	0.15	0.05	1.8	3	77
3	0.10	0.10	1.8	3	55
4	0.05	0.15	1.8	3	38
5	0.00	0.20	1.8	3	0
Sample	0.00	0.00	—	3	59

(a) Prepare a standard curve by plotting the fluorescent emission intensity against the concentration of salicylate in each standard.

(b) Calculate the concentration of salicylate in the original serum sample.

Flame Emission Spectroscopy: Flame Photometry

In flame emission spectroscopy (FES), a flame is used to reduce metal ions to free metal atoms. Some of the metal atoms absorb energy from the flame and become excited. When these atoms revert to the ground state, they emit radiant energy. The wavelength of the emitted radiant energy is characteristic of the metal that emitted it. The intensity of emission at the analyte's wavelength can be related to the concentration of the analyte. Flame emission spectroscopy has a broad range of applicability for metals. In favorable cases, the limit of detection is in the range of tenths of parts per billion. In the field of clinical analysis, sodium, potassium, and lithium are the primary elements of interest. With this background in mind, let's consider some of the properties of flames.

FLAMES

When a fuel and an oxidant are brought together and are ignited at the outlet of an orifice, if the flow rate away from the orifice exceeds the velocity of burning, a stable flame results. The flame is the sum of a number of chemical reactions that are occurring simultaneously. Consider the nature of the reactions. First, there is the principal reaction. If propane is the fuel, and oxygen is the oxidant, the main reaction is

$$C_3H_8 + 5O_2 \rightarrow 3CO_2 + 4H_2O + \text{Heat}$$

Consider Figure 8–1. The propane and oxygen move up through the burner head into the pre-

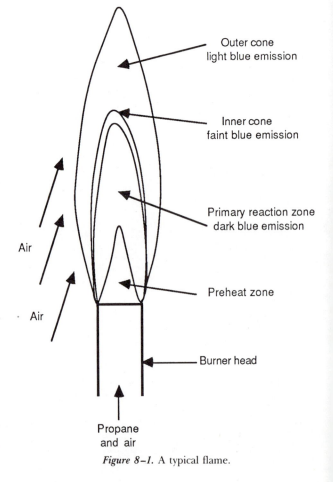

Figure 8–1. A typical flame.

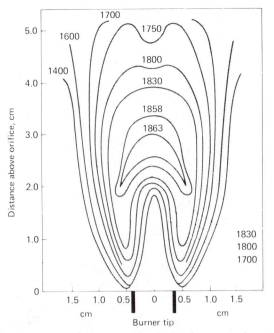

Figure 8–2. Temperature profiles (in °C) for a natural gas/air flame. (Reprinted with permission from: B. Lewis and G. van Elbe, J. Chem. Phys., *11*:75, 1943.)

TABLE 8-1. Flame Temperatures

Fuel	Oxidant	Approximate Temperature (in °C)	Predominant Reducing Species
Propane, C_3H_8	Air	1900	CO, C_2, C
Propane	Oxygen	2700	CO, C_2, C
Acetylene, C_2H_2	Air	2275	CO_2, C_2, C
Acetylene	Oxygen	3100	CO, C_2, C
Acetylene	Nitrous oxide, N_2O	2700	CO, C_2, C
Hydrogen	Air	2025	H, H_2
Hydrogen	Oxygen	2625	H, H_2

Data partly based on: D. A. Skoog and D. M. West, Principles of Instrumental Analysis, 2nd Ed., Philadelphia, Saunders College Publishing, 1980.

heating zone. In this zone, the temperature is increased by the energy coming from the primary reaction zone. In the primary reaction zone, the propane and oxygen react and produce carbon dioxide, water, and heat. Secondary reactions also produce substances such as carbon monoxide, hydrogen radicals, hydroxide radicals, and carbon radicals. The reactions that lead to production of these materials are complex, but some examples are

$$C_3H_8 + 5/2O_2 \rightarrow CO + C_2 + 4H_2O$$

$$C_3H_8 + 7/2O_2 \rightarrow 3CO + 4H_2O$$

$$C_3H_8 + 5O_2 \rightarrow 3CO_2 + 3H_2O + H + OH$$

Thus, the hot gases in the inner cone are not fully oxidized; they can be viewed as a hot reducing atmosphere. The compounds and radicals, such as CO, C_2, and H, are called reducing species. In the outer cone, oxygen is entrained from the surrounding air, and many, but not all, of the reducing species are further oxidized to CO_2 and water.

Consider the profile of temperature in such a flame (Fig. 8–2, drawn to scale with Fig. 8–1). The exact temperature of the flame depends on the choice of fuel and oxidant and their relative proportion and flow rates. Some typical approximate flame temperatures are provided in Table 8–1.

Thus, a flame is a zone of high temperature that is produced by some complex chemistry. The secondary reaction products are frequently chemical reducing agents. The fact that a flame can be seen by eye is evidence that the reaction products are emitting some radiation in the visible region. The nature of this emission is illustrated in Figure 8–3.

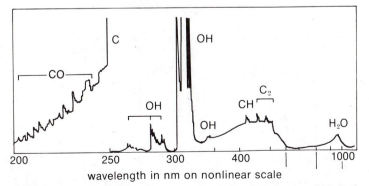

Figure 8–3. Emission spectrum of a flame burning a carbonaceous fuel. (From Instrumental Methods of Analysis, 4th Ed., by H. Willard, L. Merritt, and J. Dean. Copyright © 1965 by Litton Educational Publishing, Inc. Reprinted by permission of Wadsworth, Inc.)

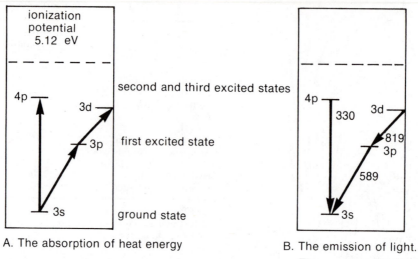

A. The absorption of heat energy

B. The emission of light.
The wavelengths are in nm.

Figure 8–4. From Instrumental Methods of Analysis, 4th Ed., by H. Willard, L. Merritt, and J. Dean. Copyright © 1965 by Litton Educational Publishing, Inc. Reprinted by permission of Wadsworth, Inc.

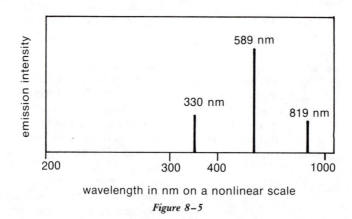

wavelength in nm on a nonlinear scale

Figure 8–5

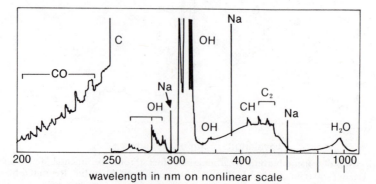

wavelength in nm on nonlinear scale

Figure 8–6. Adapted from Instrumental Methods of Analysis, 4th Ed., by H. Willard, L. Merritt, and J. Dean. Copyright © 1965 by Litton Educational Publishing, Inc. Adapted by permission of Wadsworth, Inc.

INTERACTION OF SALTS AND FLAMES

In studying FES, it is helpful to consider what happens to a salt solution when it is aspirated into a flame. The process is illustrated best diagrammatically (see Diagram 8–1 below). Sodium chloride (NaCl) has been chosen as a model compound. The NaCl solution is aspirated into the flame in the form of fine droplets of solution. The heat evaporates the solvent; this leaves the solid salt particles dispersed in the flame. The heat breaks down the solid particles into gaseous ions. The reducing species in the flame then chemically reduces the sodium ions to form gaseous sodium vapor. Some of the gaseous sodium atoms absorb enough heat to gain one of the excited states. After about 10^{-13} seconds, the excited sodium re-emits the energy as a photon of electromagnetic radiation. This process is illustrated diagrammatically in Figure 8–4.

The three most common wavelengths of radiation that sodium emits under these conditions are the 330 nm, the 819 nm, and the 589 nm lines (Fig. 8–5). The 589 nm line is the wavelength of maximum emission. Keep in mind that the radiation emitted from a sodium-containing flame is the combination of the wavelengths that are shown in Figure 8–5 and the flame spectrum that is illustrated in Figure 8–3. This combination is depicted in Figure 8–6. The greater the amount of sodium in the flame, the greater the intensity of the sodium emission. It is this concentration dependence that enables us to perform quantitative analysis by FES.

INSTRUMENTATION

Let us consider the instrumentation used in FES. There are two basic types of FES instruments, but the more widely used is the flame photometer (Fig. 8–7). Let's consider each part of the instrument. The burner has provisions for sample introduction. On most recently manufactured instru-

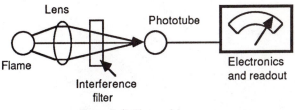

Figure 8–7. Flame photometer.

ments, a laminar flow burner is used (Fig. 8–8). The sample is drawn into the burner by the reduced pressure that is generated by the flow of oxidant through the nebulizer. This is shown in greater detail in Figure 8–9. Because of its momentum, the stream of sample solution that is drawn into the region of reduced pressure flows through the region and is promptly blown into a series of droplets by the oxidant flow. Once the droplets clear the nebulizer, they are hit by the fuel and oxidant mixture as it flows into the mixing chamber (see Fig. 8–8). This results in the production of a mist that in some burners is further chopped up by a series of propellers termed a flow spoiler. At this point, any of the solution that is not fine enough to be carried by the streaming gas falls to the bottom of the mixing chamber and then down the waste drain. The solution that is airborne enters the flame and starts the process outlined in Diagram 8–1. The net result is the emission of radiant energy.

A filter isolates the analyte's analytical line and rejects the flame spectrum. The isolated line is then caused to fall on a phototube or photomultiplier, where the photons emitted by the analyte are converted to electrical current for amplification and data processing.

A less common type of FES instrumentation uses a monochromator in place of the interference filter. You can synthesize the optical path in your mind by starting with Figure 8–7 and substituting a grating monochromator like the one in Figure 4–20 (p. 44). This type of flame emission spectrophotometer is generally used in research work.

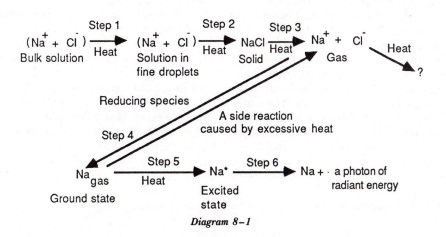

Diagram 8–1

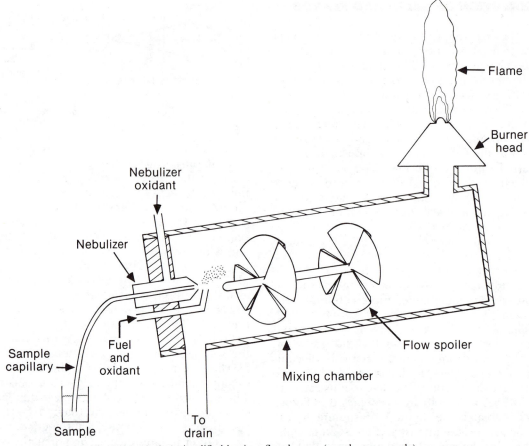

Figure 8–8. A simplified laminar flow burner (not drawn to scale).

QUANTITATIVE ANALYSIS

There are two common ways to do quantitative analysis by flame emission spectroscopy. First, it is possible to prepare standards that cover the expected range of unknown concentrations and to plot the emission intensities versus their concentrations. The calibration curve that results may or may not be linear; it depends on the concentration range that is covered. If the concentration range is fairly narrow, the curve will be linear. In either case, the curve that is obtained can be used to measure the analyte concentration in a sample. These calibration curves can be made by aspirating the most concentrated standard and adjusting the instrument to 100 units by means of the electronics. The other end of the curve is established by aspirating a blank and adjusting the instrument to 0 by means of the electronics. This step compensates for the flame emission that penetrates the filter or monochromator. With the instrument so adjusted, the other standards and samples will fall within the range of 0 to 100 arbitrary emission units. The emissions of

the standards can be plotted, and the plot can then be used to find the concentration of the analyte in the samples.

The second way to measure concentrations by flame emission spectroscopy is to build a multichannel flame instrument and use the internal standard method. Before considering the internal standard method in detail, first consider the multichannel instrument.

A multichannel flame photometer is made by building several instruments (such as the one shown in Figure 8–7) around the same burner. The resulting instrument is illustrated in Figure 8–10. This instrument has been designed to measure sodium while using cesium as the internal standard. This two-channel instrument has one flame but two photometers. One of the photometers has a sodium filter (589 nm), while the other has a cesium filter (852.1 nm). The sodium that comes into the flame acts more or less independently of the cesium, and vice versa. Thus, the sodium photometer responds to the sodium emission, and the cesium to the cesium emission.

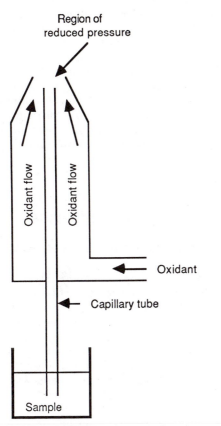

Figure 8-9. Pneumatic nebulizer.

With the two-channel flame photometer in mind, let's consider the internal standard method as it applies to a sodium analysis. Analyzing sodium samples that are in the 1 to 2 mmol/l concentration range can be done by mixing equal volumes of sample and 1.5 mmol/l cesium solution. The cesium internal standard dilutes the sodium sample to half its previous concentration, but the sodium sample also dilutes the cesium internal standard to half its previous value. If the resulting solution is aspirated into the flame, and the output of the two channels is measured, the following equation can be used to calculate the sodium concentration:

$$\text{Na conc.} = \left(\frac{\text{sodium channel output}}{\text{cesium channel output}} \right) (1.5 \text{ mmol/l}) \ (f)$$

Equation 8-1

where f is a response factor that is obtained by aspirating a solution that is prepared from equal volumes of 1.5 millimolar cesium and 1.5 millimolar sodium solutions.

$$f = \frac{\text{cesium channel output}}{\text{sodium channel output}}$$

The response factor takes into account that the two elements are not completely the same in the

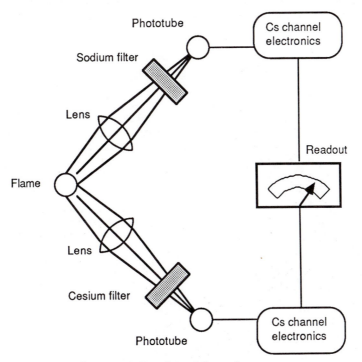

Figure 8-10. Two-channel flame photometer.

response. For example, if sodium were twice as responsive as cesium in its emission, the response factor would correct for the difference in response and would be 1/2. Equation 8–1 would become

$$\text{Na conc.} = \left(\frac{\begin{array}{c}\text{sodium channel}\\ \text{output}\end{array}}{\begin{array}{c}\text{cesium channel}\\ \text{output}\end{array}}\right)(1.5 \text{ mmol/l})\,(1/2)$$

This approach to concentration measurement works well when the analyte and internal standard are chemically similar and are present in the final solution in approximately equal concentrations. The approach has much to be said for it because anything that affects the flame does so to both the internal standard and the analyte. This leads to a lot of error cancellation. Thus, variations caused by changes in flame temperature, flow rates, aspiration rates, and reducing species concentration are all cancelled, as they affect the internal standard and the analyte equally.

On modern clinical flame photometers, Equation 8–1 is handled by the microprocessor. A sodium standard is diluted with the proper internal standard solution, and the resulting solution is aspirated. The microprocessor readout is then adjusted to read the correct sodium concentration. By this process, the product of the internal standard concentration and the response factor is programmed into the memory and is used in subsequent calculations.

The use of the internal standard method is very common in clinical analysis. Three- and four-channel instruments are very common. The three-channel instruments use lithium as the internal standard for measuring serum sodium and potassium. If lithium must be measured, potassium is made the internal standard. The four-channel instruments use cesium as the internal standard for measuring serum sodium, potassium, and lithium. Lithium is not a normal serum component but is present when it is being administered for the management of some types of mental illness. A four-channel instrument is illustrated in Figure 8–11.

SOURCES OF ERROR

The Solvent that Contains the Analyte

Although flame emission spectroscopy is capable of providing good analytical results, it does so only when great care is taken in the development of procedures. There are many sources of experimental error that can cause serious problems. A good place to start this discussion is by considering Diagram 8–1 (p. 99) and by realizing that anything that affects this progression of events will affect the output of photons from the analyte.

For example, consider step 1 in the diagram. The viscosity of the solution has an effect on the rate of aspiration and the extent of nebulization. The less viscous solutions will pass through step 1 with a greater percentage of solution getting to the flame, whereas the more viscous solutions will have a higher percentage going down the drain. The

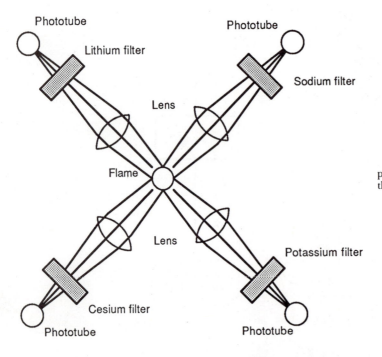

Figure 8–11. Four-channel flame photometer. The phototube outputs go to a microprocessor that does the math in Equation 8–1.

problem can be avoided if the standards and samples have the same solvent composition.

Nonaqueous solvent can markedly increase the emission of analyte photons from a flame. Standards and samples must have similar solvent composition. Solvents with a constant, high organic content are frequently used to obtain a lower limit of detection. The gain is not in orders of magnitude but is significant.

Chemical Interference

Next consider steps 2 and 3, in Diagram 8–1. When ions in solution find that the solvent is being rapidly evaporated, the cations grab the nearest anion and form very fine particles of solid material. If any of the analyte becomes attached to an anion that produces a thermally stable compound, step 3 in the diagram will become blocked, causing a decrease in analyte emission. This phenomenon is called chemical interference. An example of this type of problem is the reduction in calcium emission that always occurs when phosphate is present. It is thought that the phosphate reacts with calcium during step 2 to produce $Ca_3(PO_4)_2$. Calcium phosphate is thermally stable, so the normal amount of Ca^{+2} gaseous ions does not form in step 3. If an ion does not make it through step 3, it cannot emit photons in step 6. When the interfering anions are in the samples at varying concentrations, the stage is set for the production of some very bad results. The analyte concentrations calculated from the data will be low.

Chemical interference can be managed by one of four approaches. First, masking agents can be used. Using calcium analysis as an example, the inclusion of a lanthanum ion in the standards and samples will tie up phosphate and in so doing will allow calcium to pass through step 3 properly. The lanthanum ion is thought to react with the phosphate to form a more stable compound. As a result, the phosphate is made unavailable to the calcium. In any case, the inclusion of an excess of lanthanum in calcium standards and samples can significantly improve the accuracy of the results. One would assume that this improvement takes place in step 2 of the diagram.

A second way to manage chemical interference is to include an organic chelating agent in the standards and samples. The common textbook example of this approach is the protection of calcium from phosphate by the use of ethylenediaminetetraacetic acid (EDTA). The EDTA forms a strong chelate with the calcium ions. The chelate prevents the calcium from associating with phosphate during step 2. During step 3, the flame burns the organic EDTA off the calcium ions. This allows the calcium ions proper entry into step 4.

A third way to solve the problem of chemical interference is to separate the interfering ions from the analyte or to separate the analyte from the interfering ions. The separation is done before flame emission spectroscopy is attempted. Some separation methods will be considered in later chapters.

The fourth way to avoid some chemical interference problems is to use a hotter flame. The hotter flame can sometimes overcome a chemical interference problem by breaking up a compound. The disadvantage of using a hotter flame is that the higher flame temperature may reionize the analyte by reversing step 4 in Diagram 8–1. The hotter the flame, the higher the percentage of analyte reionization. For some elements, the higher temperature may overcome the chemical interference, but the loss of analyte gaseous atoms to reionization may partially destroy the gain.

Ionization Effects

While we are dealing with analyte reionization, keep in mind that some of it always happens and that good results are obtained only when the percentage of analyte reionization is the same for standards and samples. The problem is most acute when the analyte concentration is low. It can become such a problem that calibration curves may pass through zero emission before reaching zero concentration. Consider Figure 8–12. The deviation at low con-

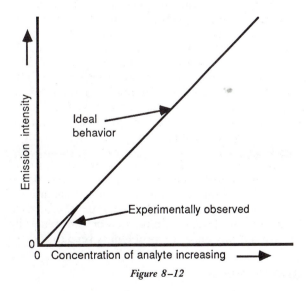

Figure 8–12

centration is clearly evident. It can be accounted for by assuming that in the dilute solution a greater percentage of the analyte is being reionized. The problem can be eliminated by adding about 1000 ppm of a noninterfering salt that is easily ionized to all of the standards and samples. This added salt is sometimes called an ionization suppressor. Ionization problems are a big source of error when the analyte is present in a sample that is high in salt but low in analyte concentration. This allows the analyte to emit radiant energy with ordinary efficiency. On the other hand, if the standard solutions do not contain any ionization suppressors, the analyte will be badly ionized. This will lead to low emission intensity for the standards. When samples of unknown concentration are analyzed under these conditions, the analysis will produce falsely elevated results. If this problem is to be avoided, the ionization suppressor must be present in the standards at about the same concentration as it is in the unknowns.

Spectral Interference

The next problem is that of spectral interference. This type of interference is independent of the diagram. There are two basic types of spectral interference. First, the flame itself emits radiation. The emission spectrum of a carbonaceous fuel oxygen flame is shown in Figure 8–3. If the element to be analyzed has its best emission line at a wavelength of strong flame emission, it may be necessary to use a different fuel or to use a secondary emission line from the element. Instrument manufacturers manage spectral interference for us in that they generally do not sell instruments that are configured in a way that will lead to problems with spectral interference.

The second type of spectral interference can be a more common problem. Some of the thermally stable compounds or ions can emit bands of radiation. The band spectrum of calcium hydroxide (CaOH$^+$), for example, covers an 80 nm range and does interfere with the 589 nm sodium line. The CaOH$^+$ ion emits 589 nm photons; no filter or monochromator can separate 589 nm CaOH$^+$ photons from 589 nm Na photons. If CaOH$^+$ exists in a flame during a sodium analysis, a high result is inevitable.

Several things can be done to manage this type of spectral interference. The use of a hotter flame might destroy the offending molecule. A separation procedure could be used to separate the analyte from the interference. Some answer can generally be worked out once the problem is recognized.

Self-Absorption

Still another problem in flame emission measurement is self-absorption. This process occurs at high analyte concentration and leads to deviation from the ideal on calibration curves (Fig. 8–13).

Self-absorption develops when the photon emitted by one atom of the analyte is absorbed by another atom of the analyte. The second atom re-emits the photon, but, in the meantime, it does not

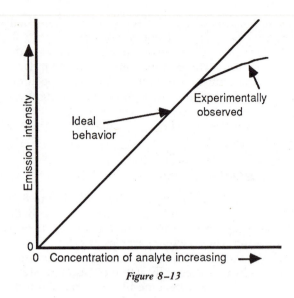

Figure 8–13

produce one of its own. In this way, two atoms of analyte that should produce two photons produce only one. The higher the concentration, the greater the self-absorption. This problem, once recognized, is easily solved by diluting the standards and samples to lower concentrations.

Matrix Effects

Most samples do not consist of analyte and distilled water. They are made up of analyte and many other compounds that are in a single solution. The part of a sample that is not the analyte is called the sample matrix. Frequently the sample matrix is the source of a chemical interference, a spectral interference, or an ionization problem. When this is the case, scientists are prone to speak of matrix effects.

Matrix effects can be quite a problem. Because of matrix effects, a good procedure for the analysis of an analyte in one matrix might be totally useless for the analysis of the same analyte in a second matrix. These effects are insidious. The best way to study their impact is to analyze samples by an independent method or by the use of the standard addition technique, which is discussed in detail in Chapter 9.

SUMMARY

In summary, FES is a good approach to the quantitative analysis of many metals. If procedures are well designed, quantitative analysis by FES is reliable. It is widely used in the clinical laboratory for sodium, lithium, and potassium analysis. The method is simple; the instruments are straightforward.

Problems   Use the following information as background for problems 1 through 4.

A sample of unknown lithium concentration is to be analyzed. Several standard solutions are prepared from lithium chloride and distilled water. The solution concentrations in milliequivalents per liter (mEq/l)are

Standard 1	0.25
Standard 2	0.50
Standard 3	1.00
Standard 4	1.50
Standard 5	2.00

When the five standards, along with a distilled water blank, are analyzed by flame photometry, the following standard curve results:

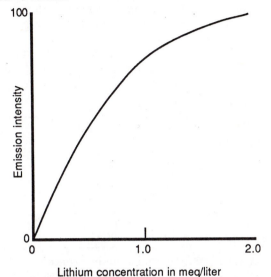

Lithium concentration in meq/liter

The sample of unknown lithium concentration is analyzed.

1. If the solvent for the sample contained 10 per cent ethanol, would you expect the analysis to yield a correct lithium concentration? If not, would the result be elevated or depressed relative to the correct value?

2. If the solvent for the sample contained 1000 ppm sodium chloride, would you expect the analysis to yield a correct lithium concentration? If not, would the result be elevated or depressed relative to the correct value?

3. If the solvent for the sample contained 10 per cent glycerol, would you expect the analysis to yield a correct lithium concentration? If not, would the result be elevated or depressed relative to the correct value?

4. If the solvent for the sample was distilled water, would you expect the analysis to yield a correct lithium concentration? If not, would the result be elevated or depressed relative to the correct value?

Use the following information as background for problems 5 to 8.

A sample of unknown lithium concentration is to be analyzed on a two-channel instrument that uses potassium as the internal standard. Five standards are prepared from lithium chloride and a solvent that contains 2 mEq/l of potassium chloride in distilled water. The standards solutions are of the following concentrations:

	Lithium Concentration (mEq/l)	Potassium Concentration (mEq/l)
Blank	0.00	2
Standard 1	0.25	2
Standard 2	0.50	2
Standard 3	1.00	2
Standard 4	1.50	2
Standard 5	2.00	2

The response factor was determined to be 1. Sufficient potassium chloride was added to the sample to give it a potassium concentration of 2 mEq/l, and it was then analyzed on the two-channel instrument.

5. If the solvent for the sample contained 10 per cent ethanol, would you expect the analysis to yield a correct lithium concentration? If not, would the result be elevated or depressed relative to the correct value?

6. If the solvent for the sample contained 1000 ppm sodium chloride, would you expect the analysis to yield a correct lithium concentration? If not, would the result be elevated or depressed relative to the correct value?

7. If the solvent for the sample contained 10 per cent glycerol, would you expect the analysis to yield a correct lithium concentration? If not, would the result be elevated or depressed relative to the correct value?

8. If the solvent for the sample was distilled water, would you expect the analysis to yield a correct lithium concentration? If not, would the result be elevated or depressed relative to the correct value?

9. Consider the following data:

Magnesium Concentration (ppm)	Observed Magnesium Emission
0	0.0
2	42.5
4	70.0
6	85.0
8	92.5
10	97.5
12	100.0

A sample of unknown magnesium concentration is analyzed. It produces an emission of 63.4 units. What is the magnesium concentration of the sample?

10. The magnesium sample in question 9 is treated with EDTA and is reanalyzed. The sample produces an emission of 80.2. Why would this higher value be obtained? Which analysis result would be the more accurate?

11. Consider the following data for a potassium analysis:

Potassium Concentration (ppm)	Emission Intensity
0	0.0
1	2.0
5	6.8
10	14.5
20	31.5
25	39.5
50	75.0
75	94.0
100	100.0

(a) Make a plot of potassium concentration versus emission intensity.
(b) Make a plot of the square root of the concentration versus the emission intensity.
(c) If a sample of unknown potassium concentration produces an emission intensity at 29.3, what is the potassium concentration?

Atomic Absorption Spectroscopy

Atomic absorption spectroscopy (AAS) is the form of absorption spectroscopy used to detect gaseous atomic metals. The method frequently relies on a flame to transform solutions of analyte into gaseous atoms. AAS is widely used for the quantitative analysis of metals in complex matrices; the limit of detection in favorable cases is on the order of a few tenths of a part per billion. The limit of detection does, however, vary greatly; it is dependent on the nature of the analyte and matrix as well as the instrument.

The technique has much in common with other forms of absorption spectroscopy in that the instrument consists of a source, a cell, a monochromator and a detector. The differences are first, that the sample cell in AAS is frequently a flame, and secondly, the source in AAS is frequently a line source instead of a continuous source. Beyond these differences, the measurements are typical of other forms of absorption spectroscopy.

SINGLE-BEAM AAS INSTRUMENTATION

The least complicated type of AAS instrument is the single-beam instrument. The optical path of a typical single-beam instrument is illustrated in Figure 9–1. Let's consider each component.

Hollow Cathode Lamp. Almost all atomic absorption instruments use a hollow cathode lamp

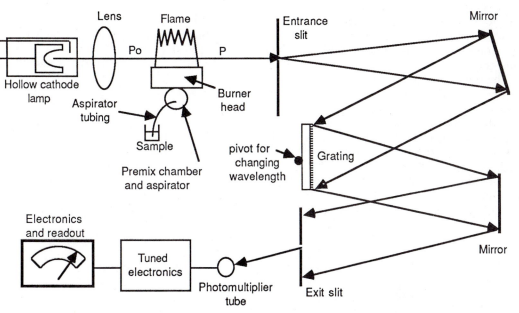

Figure 9–1. Single-beam atomic absorption spectrophotometer.

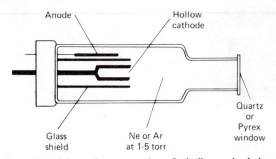

Figure 9–2. Schematic cross section of a hollow cathode lamp. (Reprinted from: D. A. Skoog, Principles of Instrumental Analysis, 3rd Ed., Philadelphia, Saunders College Publishing, 1985.)

(Fig. 9–2) as a source of radiant energy. The hollow cathode lamp is a good source of atomic line emission. The lamp is manufactured by machining or casting a hollow cathode from the metallic analyte of interest, for instance, calcium metal. The cathode is placed in a glass envelope that also contains a metal anode. The end of the tube is covered with a quartz window, and the tube is evacuated. A small amount of neon is bled into the tube, and the tube is sealed. When the hollow cathode lamp is connected to a suitable power supply, it emits radiant energy. The radiant energy is characteristic of the cathode element and neon. Using calcium as our example, let's consider how the lamp generates calcium line emission. A voltage that is of sufficient magnitude to cause the ionization of the neon gas is applied to the lamp. The process can be written

$$Ne \xrightarrow[\text{energy}]{\text{electrical}} Ne^+ + e$$

The voltage is applied so that the cathode is negatively charged. The electrons are accelerated toward the anode. They acquire considerable energy in the process. Because of their energy, the electrons generate more neon ions and electrons. The electrons generated are collected ultimately by the anode. The neon ions are attracted to the cathode, and as they move through the electrical field they gain energy. The accelerated neon ions impact on the cathode surface, causing metal atoms to be boiled from the surface of the cathode. Many of the atoms that are ejected into the gas phase are in the first excited state, and these excited atoms emit photons of radiation. The atoms then tend to redeposit onto the cathode surface. For a calcium cathode, the process could be written as follows:

The electrons in Equation 9–1 come from the external circuit. The calcium gaseous atoms in the excited state tend to be formed in the cavity of the hollow cathode. Most of the condensing calcium atoms in Equation 9–3 tend to redeposit in the cavity or, at least, on the hollow cathode surface. Some atoms, however, do deposit on the window or walls of the lamp envelope. The redeposited atoms on the window cause the lamp's output to decrease by absorbing the radiation that is generated in the cathode. When this deposit becomes excessive, the lamp must be replaced. Students frequently ask: "Why throw the lamp away? Why not increase the lamp current to produce the needed intensity?" The reason this will not work is less than obvious. When the lamp is run at a current that is too high, the cloud of atoms generated in Equations 9–1 and 9–2 gets so large that it cannot redeposit rapidly enough. When this happens, the lamp emission intensity drops instead of increasing. Using calcium as an example, the calcium atoms in the cloud absorb many of the calcium photons generated. This process is called self-reversal, and its onset limits the maximum current that can be used for the lamp. If the current that is applied becomes excessive, it is possible that the output of photons at the analytical wavelength will drop near zero.

In summary, the hollow cathode lamp is a line source that produces the major emission lines of the metal from which the cathode is made. In general, a different lamp is needed for each analyte that is to be measured, although a few alloy cathode lamps are multielement sources. The lamps have practical limitations on the emission intensity that can be generated. The limit of intensity is governed by design. The lamp also emits the line spectrum of the filling gas, which is frequently neon, although argon is sometimes used.

Burner and Flame. The next components in the optical path are the burner and flame. The flame, burner, and nebulizer in AAS measurements serve the same purposes that they do in flame emission spectroscopy (FES). (To review these topics, see Chapter 8, sections entitled *Interaction of Salts and Flames* and *Instrumentation*.) The flame acts on metal cations in solution to produce metal atoms. These atoms are in two populations: thermally excited atoms and ground-state atoms. FES measurements are based on the emission of the thermally excited atoms, whereas AAS makes use of the much larger ground-state population. The thermally ex-

$$Ne^+ + Ca_{(s)} + e \longrightarrow Ne_{(g)} + Ca^*_{(g)} \qquad 9\text{--}1$$

$$Ca^*_{(g)} \longrightarrow Ca_{(g)} + \text{a photon of 422.7 m radiation} \qquad 9\text{--}2$$

$$Ca_{(g)} \longrightarrow Ca_{(s)} \qquad 9\text{--}3$$

cited atoms are, in fact, a nuisance in AAS. AAS is frequently more sensitive than FES because the ratio of ground-state to thermally excited atoms in a flame is high. For example, at 2500°C, there is approximately 1 thermally excited sodium atom to 5000 in the ground state. For reasons beyond the scope of this book, the increased sensitivity is not always realized.

Monochromator. A monochromator of high quality is provided in an effort to reject the radiation that originates in the flame. This radiation includes thermally excited flame emission and thermal emissions of compounds and radicals as well as those of other elements that are part of the matrix. Most of these emissions do not occur at the analysis wavelength; therefore, the monochromator can reject most of the unwanted thermal emission intensity.

Unfortunately, the thermal emission of the analyte occurs at exactly the same wavelength as the photons that make up Po and P. Flame emission and matrix emissions also may have components at the same wavelength as Po and P. Because all of these emissions have the same wavelength as Po and P, the monochromator cannot reject them. For this reason, an additional system is required in order to make accurate measurements. The system works by pulsing the radiant energy that comes from the hollow cathode lamp. This is frequently done by connecting the lamp to an AC power supply. The lamp can emit radiation only when the cathode is electrostatically negatively charged. During the other half of the AC cycle, the lamp shuts down. In contrast to the pulsed emission from the lamp, the thermal emissions that originate in the flame are not pulsed. They are continuous; their intensities do not change rapidly with time. Thus, the monochromator passes pulsed radiation that is either P or Po and nonpulsed thermal emissions. The detector converts both of these components to electrical current, but the tuned electronics will accept only current that is pulsed at the pulse rate of the hollow cathode lamp; therefore, only P and Po are measured.

Detector. The last device in the optical path is a detector. The most common detector for AAS is the photomultiplier tube. Photomultiplier tubes are discussed in Chapter 4, **Visible Absorption Spectroscopy** (p. 49).

Calibration

Single-beam AAS instruments are calibrated by aspirating a blank and adjusting the electronics to read 0 absorbance. They are further calibrated by aspirating standard solutions. The resulting absorbance data are plotted against concentration in order to make a standard curve (Fig. 9–3). With

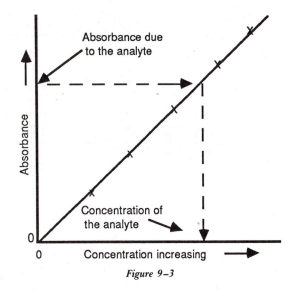

Figure 9–3

the instrument properly calibrated, samples are analyzed by determining their absorbances and using the curve to find the concentration. On instruments with microprocessors, the calibration information is received by the electronics. The electronics then calculates the results.

When working with a single-beam instrument, one must frequently adjust the calibration settings because of instrument drift. Although better electronics in recent years has reduced the problem somewhat, single-beam instruments are generally more inconvenient and time consuming to use because of the need to calibrate frequently. The advantage of single-beam instruments is their lower cost.

DOUBLE-BEAM AAS INSTRUMENTATION

Because of the instabilities caused by the drifting source intensity and the drifting electronics, double-beam AAS instruments were introduced. The double-beam instruments are much more stable. Consider the double-beam instrument in Figure 9–4.

The optical path in Figure 9–4 is much like the one in Figure 9–1. The new components are a sector mirror, like the one in Figure 4–48 (p. 54), and a half silvered mirror, like the one in Figure 4–49 (p. 55). These two mirrors, along with two fixed mirrors, make possible a reference optical path. The sector mirror mechanically pulses Po and P by alternately reflecting radiant energy through the reference path and the flame. The sector mirror rotates at about 600 rpm. Thus, both P and Po are known many times a second. The ratio of P to Po is determined by the electronics in order to find the

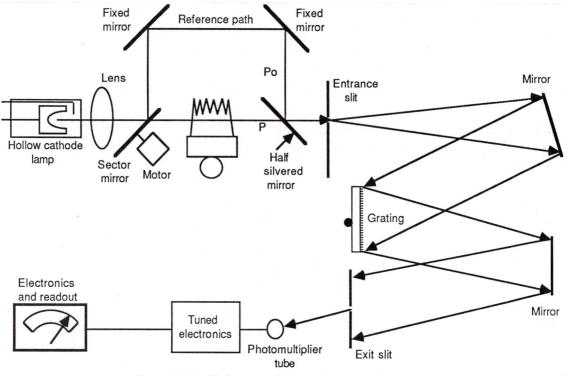

Figure 9–4. Double-beam atomic absorption spectrophotometer.

transmittance. The electronics then takes the negative log; the resulting value is the absorbance.

Calibration

The double-beam instrument must be calibrated before it can be used to measure analyte concentrations. The method of calibration is the same as that employed for single-beam instruments discussed previously in this chapter. Samples are analyzed by the same procedures as those used for the single-beam instrument. Double-beam insruments tend to be more stable; hence they tend to require calibration less frequently. Double-beam instruments are frequently more sensitive and are more convenient to use.

Even double-beam instruments are susceptible to some deviation from Beer's law. This deviation tends to occur at high absorbance values and results in calibration curves that have negative deviation. Consider Figure 9–5.

All atomic absorption instruments are susceptible to the error that is caused by radiation scattering from particulates in the flame. Radiation scattering is a problem when samples high in dissolved solid content are aspirated. Because the solids tend to escape destruction in the flame, they are present

to scatter radiation. The problem with radiation scattering is also present in flameless AAS, which is discussed in this chapter. In any case, the problem of radiation scattering can be handled by the use of a background correction system. Three such systems are discussed in this chapter.

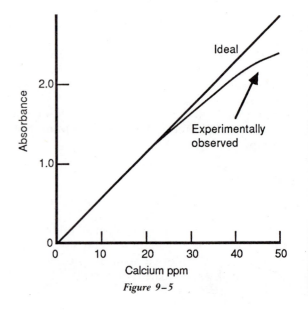

Figure 9–5

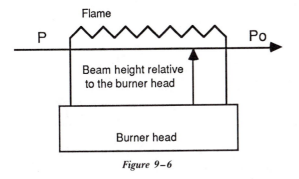

Figure 9–6

FLAME OPTIMIZATION

The beam of pulsed monochromatic radiant energy, Po, that is supplied to the flame is highly collimated and is much smaller than the flame (Fig. 9–6). Because Po passes through such a small part of the flame, the positioning of the burner relative to the beam can have a considerable effect on the sensitivity of the instrument (Fig. 9–7). Because of

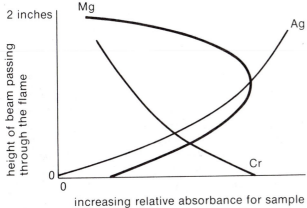

Figure 9–7. From: Atomic Absorption Spectroscopy by J. W. Robinson, Copyright 1966 by Marcel Dekker, Inc. Reprinted from Atomic Absorption Spectroscopy, p. 69, by Courtesy of Marcel Dekker, Inc.

this problem, the burner position on most instruments is adjustable. In order to obtain the highest absorbance possible for a given analyte concentration, it is necessary to find the proper burner height by systematically varying the burner height while measuring the absorbance at fixed analyte concentration. At each new burner height, the instrument must be adjusted to 0 absorbance with the blank.

The ratio of fuel to oxidant can also have a profound effect on the sensitivity for an element (Fig. 9–8). The best flow rate for the fuel can be established experimentally. Remember, however,

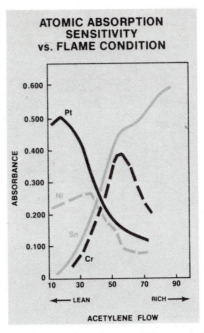

Figure 9–8. Atomic absorption sensitivity versus flame condition. (Courtesy of Perkin-Elmer Corporation, Norwalk, Conn. Reprinted with permission.)

that it may be necessary to readjust the burner height if the fuel flow rate is changed.

ODDS AND ENDS CONCERNING FLAME OPERATION

Most AAS instruments are run under fuel-rich conditions. This means that a vent system must be used to remove the toxic combustion products.

Make sure that the burner wire downs are properly secured. AAS burner explosions, although rare, do occur. The wires keep the burner parts that are blown free from hitting someone.

Follow the manufacturer's instructions concerning the cleaning of burners and aspirators. A dirty burner produces a noisy readout. A partly clogged aspirator can lead to erratic results.

FLAMELESS AAS

In the past few years there has been a lot of interest in flameless AAS. Flameless AAS is atomic absorption spectroscopy in which no flame is used. The flame is replaced by a new system for generating gaseous atoms. This increasing interest has been caused by the fact that much smaller samples can be analyzed by these techniques. Typical burn-

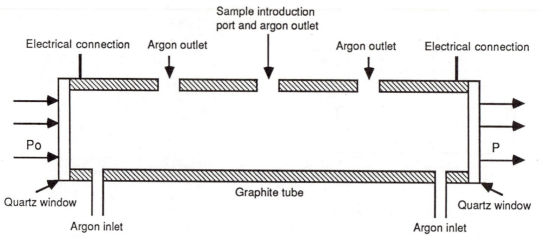

Figure 9–9. Simplified graphite furnace.

ers aspirate several milliliters of solution per minute; most of the solution that is aspirated goes down the drain. On the other hand, the flameless systems can function on as little as a few microliters of solution.

A second reason for the interest in flameless AAS is that the method provides between 10- and 1000-fold increases in absolute sensitivity. The most common device for doing flameless atomic absorption spectroscopy is the graphite furnace. A highly simplified graphite furnace is shown in Figure 9–9. The illustration shows only the central part of a graphite furnace. If the other components were illustrated, the drawing would be unduly complex. On the real furnace, there are provisions for protecting the graphite tube from atmospheric oxygen, and there are provisions for water-cooling the device in order to cool the tube rapidly. This is done so that the next sample can be introduced without a long delay for furnace cooling. The graphite tube is heated electrically with a high current–low voltage power supply.

In actual use, the burner is removed from the AAS optical bench, and the furnace is installed in its place. Samples are introduced with a microsyringe or in a small microboat. The sample is placed either on the tube wall or on a small platform that is suspended above the bottom of the tube. The power is then turned on, and, depending on the nature of the sample, the furnace is heated in stages. First the solvent is driven off. In the second stage, organic materials are charred. The furnace is then heated to a high temperature. At the elevated temperature, the graphite walls of the furnace interact with the sample. This interaction leads to the reduction of metal compounds and the generation of free metal. At the temperature of the furnace operation, the metals volatilize into the gas phase and absorb some of the Po. The argon flowing through

the tube keeps air out and carries the metal vapor out of the tube, but, while this is taking place, the absorbance versus time plot is being recorded by the instrument. The area under this curve is directly proportional to the mass of metal that caused the absorption peak. Once the metal vapor cloud is carried from the furnace, the furnace can be cooled for the introduction of the next sample.

Calibration

Calibration of the flameless AAS system is straightforward. Differing known amounts of analyte can be placed in the furnace, and the readout from these standards can be used to produce a standard curve. The absorption peak for a sample can then be compared with the standard curve in order to find the concentration. The actual math can be done by a microprocessor.

The use of the furnace generally requires a background correction system because smoke generally forms in the tube during the analysis. The smoke causes some of the Po to be scattered. This leads to falsely elevated absorbances. In some cases, thermally stable molecules can form that can also absorb some of the Po. Some background correction systems can correct for this problem as well.

Flameless AAS measurements are frequently less reproducible than those of normal AAS. Matrix effects can be very troublesome in flameless AAS. Still, the improved sensitivity is sufficient to interest many workers in using the flameless AAS approach.

BACKGROUND CORRECTION SYSTEMS

When solutions that have a high concentration of dissolved solids are aspirated into a flame, the flame cannot volatilize all of the solid material. This leads to the generation of particulate material in the flame. These particles cause some of Po to be scattered and lost. The result is an incorrectly elevated absorbance reading. Such a reading will cause an elevated result in an analysis. This problem is not alleviated by the use of a double-beam instrument. A background correction system is required.

When samples are analyzed with a graphite furnace, there is frequently a lot of smoke. The smoke causes the scattering of Po. As discussed in the previous paragraph, the overall effect of this scattering is a falsely elevated result. Thus, when a graphite furnace is used for samples that produce smoke, the instrument needs a background correction system.

The most common background correction system uses a continuous source for the correction. If the analyte absorbs between 185 and 420 nm, a deuterium arc lamp can be used for the correction. If the analyte absorbs in the visible region, a quartz-halide lamp can be used. The correction system provides for the measurement of the sum of the analyte absorbance and the apparent absorbance that is caused by the particulates. The sum is measured at the analytical wavelength, λ_A. In the next instant, the system measures the apparent absorbance that is caused by the particulates. This is done by measuring the apparent absorbance at an adjacent wavelength, λ_C. The instrument's microprocessor then, in effect, subtracts the apparent absorbance at λ_C from the absorbance at λ_A. The difference between these two values is the absorb-

ance caused by the analyte alone. The optical arrangement can be made in several ways.

For example, if the instrument in Figure 9–4 had its hollow cathode lamp replaced with the system illustrated in Figure 9–10, the instrument could do background correction. Sector mirror 1 is rotated two times slower than sector mirror 2. In this way, when the deuterium radiation is being reflected off of sector mirror 1, the instrument can measure Po and P for the deuterium radiation. If P does not equal Po, then radiation scattering is happening and the apparent absorbance for this scattering is stored in the computer. When sector mirror 1 turns another one half revolution, the hollow cathode radiation is routed to sector mirror 2. Sector mirror 2 rapidly sends radiation down the reference path and through the flame or furnace. The loss in radiation that takes place is caused by the sum of analyte absorption and scattering processes. The instrument calculates the absorbance for this measurement and subtracts the value for apparent absorbance previously stored in the computer. The resulting absorbance is the value for the analyte.

Upon first considering this system, one might wonder, "How can it function?" After all, the analyte is exposed to the deuterium correction system at the same time as the light scattering particles are. The reason the system works is that the monochromator bandpass is much greater than the bandwidth of the analyte's absorption. The small loss in deuterium intensity due to analyte absorption is so small relative to the total deuterium intensity examined by the detector that the loss is insignificant. When any particulates are present, the vast majority

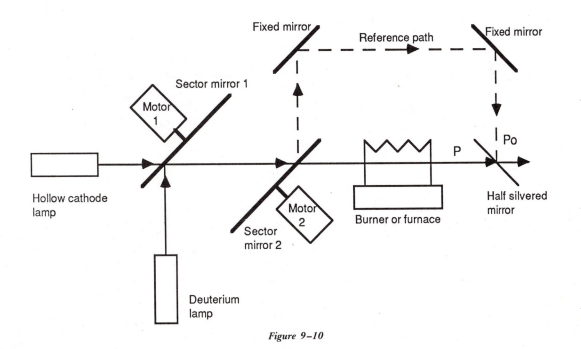

Figure 9–10

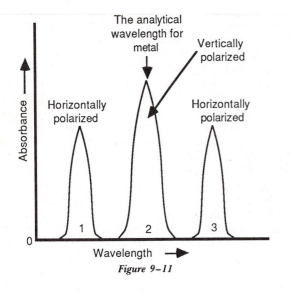

Figure 9–11

In addition, the population that absorbs vertically polarized radiation can do so only at the analytical wavelength λ_A, while the population that absorbs the horizontally polarized radiation can do so only at wavelengths slightly longer or shorter than λ_A.

If one were to measure the absorption spectrum of a gaseous atomic metal while the gas were in a strong magnetic field, one would obtain the spectrum shown in Figure 9–11. The atoms responsible for peaks 1 and 3 can absorb only horizontally polarized radiant energy. The atoms responsible for peak 2 can absorb only vertically polarized radiant energy.

Many optical configurations could be developed for Zeeman background corrected instruments. One of these configurations is illustrated in Figure 9–12. The instrument uses a dichroic filter* to remove the vertically polarized emission that is supplied by the hollow cathode lamp. The horizontally polarized emission from the lamp is allowed to fall on the flame or furnace. The horizontally polarized emission is at the analytical wavelength, λ_A. When the magnet is turned on, the atomic gas (analyte) population undergoes Zeeman splitting. The atoms that have the ability to absorb horizontally polarized radiation are shifted away from the analytical wavelength. Because of this fact, the horizontally polarized radiation at the analytical wavelength cannot be absorbed. This radiation does interact with the background however. The value of

of the loss in deuterium intensity is caused by the scattering that takes place in the flame or furnace. This means that the background correction is valid.

The second type of background correction system depends on the Zeeman effect. The Zeeman effect is the name applied to the complex effects that a magnetic field has on the absorption of radiant energy. When a population of gaseous atoms is placed in a magnetic field, the atoms interact with the field and a number of changes take place. First, part of the population is made incapable of absorbing vertically polarized radiant energy, while the other part of the population is made incapable of absorbing horizontally polarized radiant energy.

---

* A dichroic filter transmits horizontally polarized radiant energy while absorbing vertically polarized radiation.

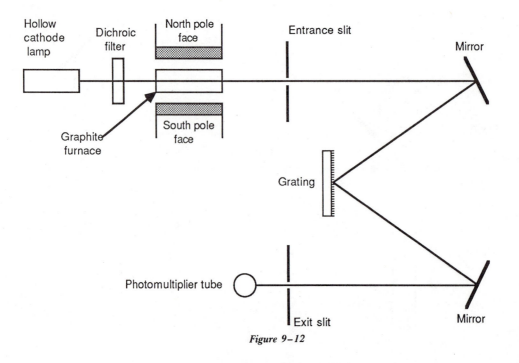

Figure 9–12

Po that is detected reflects the source intensity and the losses due to scattering from the particulates in the flame or furnace. In the next instant the magnet is turned off. The horizontally polarized radiation is now absorbed by the analyte and is scattered by the particulates. The value of P is detected by the detector and is ratioed with the previously determined Po. The resulting value of transmittance is used by the instrument's microprocessor to calculate the absorbance. The measurement of Po and P is done many times per second. The instrument is truly a double beam in time instrument. It compensates for source variations as well as for losses due to particulates in the flame or furnace.

The third background correction system uses the self-reversal property of the hollow cathode lamp in order to measure the needed correction. This is called the Smith-Hieftje background correction system. The instrument in Figure 9–4 would work well for this approach. This instrument measures P and Po while the lamp is operated at a normal working current. This provides an uncorrected absorbance. The lamp is then driven to a much higher current. The lamp self-reverses under these conditions and provides almost no intensity at the analytical wavelength. This is illustrated in Figures 9–13 and 9–14. The wavelength axis is the same in both figures. Much of the emission from the self-reversed lamp passes the monochromator, but not much of it is absorbable by the analyte. Thus, the values of P and Po for the self-reversed situation can be used to measure the scattering caused by smoke in the furnace. The apparent absorbance that is measured while the lamp is self-reversed can be subtracted from the absorbance that is measured when the lamp is not self-reversed. The value obtained is the corrected analyte absorbance.

It is hard to say which of the correction systems

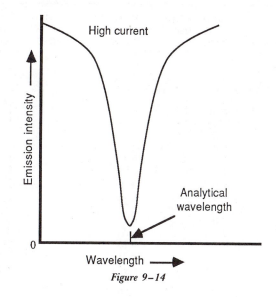

Figure 9–14

is best for your work. The information in the literature supplied by the various manufacturers is conflicting. Try each of these correction systems on your samples, and pick the approach that seems best to you. It is almost impossible to get by without a background correction system when you are using a furnace.

LIMITS OF DETECTION

Atomic absorption spectroscopy is used in the analysis of various metals in the clinical laboratory. Calcium is frequently analyzed on a flame-type instrument. Heavy metals, such as lead, copper, and cadmium, are frequently analyzed by graphite furnace. Tables 9–1 and 9–2 provide some limit of detection information. The data are supplied for comparison between flame and graphite furnace methods. These detection limits are impressive, but be mindful of matrix effects, and don't forget to take necessary precautions when working in the ppb or ppm concentration ranges.

PROBLEMS IN EXPERIMENTAL WORK

A number of problems that could lead to bad experimental results are discussed in Chapter 8. Unfortunately, these potential problems are still with us in AAS work, and they include chemical interferences, ionization effects, and the effect of viscosity on the rate of aspiration and nebulization. The sample matrix is frequently the source of chemical interferences. Spectral interference and self-absorption are seldom problems in AAS.

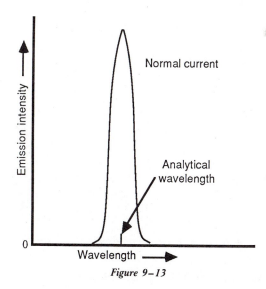

Figure 9–13

TABLE 9–1. Characteristic Concentrations and Detection Limits*

Element	Characteristic Concentration µg/ml	Detection Limit µg/ml	Notes	Element	Characteristic Concentration µg/ml	Detection Limit µg/ml	Notes
Bi	0.2	0.05		Os	1.0	0.1	N-A
Ca	0.01	0.001	N-A	P	120	40	N-A, EDL
Cd	0.01	0.0015		Pb	0.1	0.01	EDL
Co	0.05	0.005		Pd	0.05	0.01	
Cr	0.05	0.006		Pr	20	8.0	N-A
Cs	0.02	0.004		Pt	1.0	0.1	
Cu	0.03	0.003		Rb	0.05	0.009	
Dy	0.6	0.03	N-A	Re	8.0	0.8	N-A
Er	0.5	0.03	N-A	Rh	0.1	0.005	
Eu	0.3	0.02	N-A	Ru	0.4	0.08	
Fe	0.05	0.006		Sb	0.3	0.04	
Ga	0.8	0.08		Sc	0.3	0.05	N-A
Gd	20	2.0	N-A	Se	1.0	0.5	N-A, EDL, BC
Ge	1.0	0.2	N-A	Si	1.5	0.25	N-A
Hf	10	2.0	N-A	Sm	6.0	1.0	N-A
Hg	1.5	0.15		Sn	0.7	0.1	N-A, EDL
Ho	0.7	0.04	N-A	Sr	0.04	0.002	N-A
In	0.15	0.04		Ta	10	2.0	N-A
Ir	0.8	0.5		Tb	7.0	0.7	N-A
K	0.007	0.003		Te	0.2	0.03	
La	40	2.0	N-A	Ti	1.0	0.08	N-A
Li	0.02	0.002		Tl	0.2	0.02	
Lu	7.0	0.3	N-A	Tm	0.3	0.02	N-A
Mg	0.003	0.0003		U	100	40	N-A
Mn	0.02	0.002		V	0.7	0.07	N-A
Mo	0.3	0.02	N-A	W	5.0	1.0	N-A
Na	0.003	0.0002		Y	2.0	0.2	N-A
Nb	20	2.0	N-A	Yb	0.06	0.004	N-A
Nd	6.0	1.0	N-A	Zn	0.008	0.0008	
Ni	0.07	0.01		Zr	9.0	1.0	N-A

Notes: (1) N-A = nitrous oxide-acetylene flame; IS = Ionization suppressant 1000 ug k/mL; EDL = electrodeless discharge lamp; BC = background correction. (2) Characteristic concentration is that concentration which produces an absorbance of 0.0044. (3) Detection limit is that value which can be detected with a 95% confidence level.

*Data obtained on Varian AA-1475 (AAS [Flame] ppm).

(Courtesy of Varian Instrument Division of Varian Techtron Pty. Ltd, Australia. Reprinted with permission.)

TABLE 9–2. Detection Limits*

Element	Concentration ng/ml	Weight pg	Element	Concentration ng/ml	Weight pg
Ag	0.05	1.0	Li	0.13	2.5
Al	0.35	7.0	Mg	0.01	0.2
As[†]	0.4	8.0	Mn	0.035	0.7
Au	0.2	4.0	Mo	0.4	8.0
B	50	1000	Na	0.01	0.2
Ba	0.8	16	Ni	0.5	10.0
Be	0.04	0.8	P[†]	2000	40 000
Bi	0.35	7.0	Pb	0.15	3.0
Ca	0.03	0.6	Pd	0.6	12
Cd	0.15	0.3	Pt	4.5	90
Co	0.25	5.0	Rb	0.08	1.5
Cr	0.13	2.5	Sb	0.45	9.0
Cs	0.3	6.0	Se[†]	1.3	25
Cu	0.13	2.5	Si	1.3	25
Dy	1.8	35	Sn[†]	1.1	22
Er	3.8	75	Sr	0.15	3.0
Eu	0.8	15	Tb	0.18	3.5
Fe	0.1	2.0	Te	1.0	20
Ga	1.0	20	Ti	2.3	45
Hg	18	350	Tl	1.3	25
In	1.3	25	V	1.4	28
K	0.025	0.5	Zn	0.01	0.25

*Data obtained on Varian AA-975 (AAS [Furnace] ppb).

[†]Using chemical modifier.

(Courtesy of Varian Instrument Division of Varian Techtron Pty. Ltd., Australia. Reprinted with permission.)

Problems

1. Lithium can be determined by flame atomic absorption spectroscopy (FAAS) by using the 678 nm absorption line. This is generally done by diluting 1.00 ml of the patient's serum with 9.00 ml of deionized water. The relative absorbance of the patient's serum is then determined and compared with a standard curve. Consider the following data:

Li Concentration of Standards (mmol/l)	NaCl Concentration (mmol/l)	Relative Absorbance
0.25	14	100
0.20	14	81
0.15	14	61
0.10	14	41
0.05	14	20
0.00	14	0

The relative absorbance found for the patient's serum is 38. What is the patient's serum lithium concentration?

2. If the solvent for the sample contained 10 per cent ethanol, would you expect the analysis to yield a correct lithium concentration? If not, would the result be elevated or depressed relative to the correct value?

3. If the solvent for the sample contained 10 per cent glycerol, would you expect the analysis to yield a correct lithium concentration? If not, would the result be elevated or depressed relative to the correct value?

4. If the standard solutions in problem 1 were prepared with distilled water and lithium carbonate and if a serum sample were analyzed, would the experimentally determined value be high, low, or correct? Why?

5. Serum calcium may be determined by flame AAS by using its 422.7 nm absorption line. In some procedures 0.5 ml of serum is added to 9.5 ml of 4 per cent trichloroacetic acid. This precipitates the protein. The solution is centrifuged. The resulting clear solution is aspirated into the flame. Consider the following data:

Concentration Ca (mmol/l)	Relative Absorbance
0.0	0
1.5	43
2.0	58
2.5	72
3.0	85
3.5	100
deprotein serum	69

What is the serum calcium concentration?

6. Blood serum is known to contain a significant amount of phosphate ion. If the standards in problem 5 were prepared from distilled water and calcium carbonate and if a serum sample were analyzed, would the experimentally determined value be high, low, or correct? Why?

7. If the standards and serum sample in problem 5 were prepared for analysis so that they contained an excess of lanthanum^{+3}, would the resulting analysis provide a correct or an incorrect result? If the resulting value were incorrect, would it be high or low?

8. Outline an experimental method that should be capable of providing a good serum calcium determination.

9. Outline an experimental method that should be capable of providing an accurate serum lithium determination.

10. Urinary zinc can be determined by using the 214.8 nm zinc absorption line. In the analysis, the urine is diluted 1:1 with distilled water. Consider the following data:

Zinc Concentration (ppm)	Relative Absorbance
2.5	100
1.5	70
1.0	55
0.5	30

The relative absorbance of the diluted urine is 45. What is the urinary zinc concentration? Assume that the standards and sample have similar NaCl concentrations.

11. If the zinc standard in problem 10 were prepared from zinc nitrate and distilled water, would the resulting analysis of urinary zinc be high, low, or correct? Why?

12. A urine sample is diluted by half with distilled water. A flame AAS instrument is adjusted to 0 absorbance with a blank. The diluted urine sample is aspirated into the flame. It produces an absorbance of 0.32. Five ml of the diluted urine is mixed with 5.00 ml of 2.5 ppm zinc standard. The mixture is aspirated into the flame. It has an absorbance of 0.43. What is the urinary zinc concentration?

13. A serum sample is deproteinized by adding 0.5 ml to 4.5 ml of 4 per cent trichloroacetic acid. When analyzed by flame AAS, it produces an absorbance of 0.27. Three ml of the deproteinized solution is mixed with 2.0 ml of 3.5 mmol/l standard. The resulting mixture produces an absorbance of 0.32. What is the serum calcium concentration?

14. Serum lead is to be determined by graphite furnace atomic absorption spectroscopy. Standards are prepared with the following concentrations:

	Concentration (ppm)	Area Under Response Curve (mm^2)
Blank	0.00	0.00
Standard 1	0.25	5.4
Standard 2	0.50	10.9
Standard 3	1.00	25.2
Standard 4	2.00	48.2

The standards are loaded into the furnace and are analyzed. For each standard, the area under the response peak is given (see data). If the serum sample is analyzed in the same way and if the area under its response curve is 23.7 mm^2, what is the serum lead concentration?

15. Serum copper is to be determined by graphite furnace atomic absorption spectroscopy. Standards are prepared with the following concentrations:

	Concentration (ppm)	Area Under Response Curve (mm^2)
Blank	0.0	0.0
Standard 1	0.5	13.0
Standard 2	1.0	25.1
Standard 3	2.0	45.9
Standard 4	3.0	64.3

The standards are loaded into the furnace and are analyzed. For each standard, the area under the response peak is given (see data). If the serum sample is analyzed in the same way and if the area under its response peak is 49.3 mm^2, what is the serum copper concentration?

Radiation-Scattering Photometry: Turbidimetry and Nephelometry

Various spectrophotometric methods of analysis have been presented in previous chapters. These methods rely on either the absorption or the emission of electromagnetic radiation. Most of the methods are designed to measure the concentration of an analyte, but some are used for qualitative analysis.

In direct contrast to the previous chapters, this chapter deals with an important nonabsorptive method for the determination of concentration. The analysis method is called radiation-scattering photometry. Several important applications of radiation-scattering photometry can be found in the clinical laboratory. Let's begin this discussion by considering the scattering of radiant energy by particles in solution.

RADIATION SCATTERING BY PARTICLES IN SOLUTION

When a beam of monochromatic radiation of a nonabsorbable wavelength is allowed to fall on a suspension of particulate material, the direction of propagation of some of the photon in the beam is altered (Fig. 10–1). The change in the direction of propagation and the change in the intensity of the scattered radiation are dependent on many factors. Over the years, chemists have worked out the theoretical basis of the radiation-scattering process. Because it is complicated in its mathematics and offers little of practical significance, the theoretical basis will not be developed in this text. Consider, however, the general statements made in the next paragraph.

First, the wavelength of the transmitted, the scattered, and the incident radiation is the same. A second fact is that the intensity of scattered radiation frequently is not the same in all directions. In most clinical applications, the majority of the scattering is forward scattering. Forward scattering emanates from the half of the cell opposite the incident radiation. Third, the concentration of the particles in the suspension can be determined either by measuring the decrease in Po as it passes through a cell or by measuring the scattered radiation at a given angle. The choice of approach depends on the concentration of the suspension. If the concentration is high, P and Po can be measured effectively. This approach to the analysis is called turbidimetry. If the concentration is low, P does not differ enough from Po to provide an accurate measurement. In

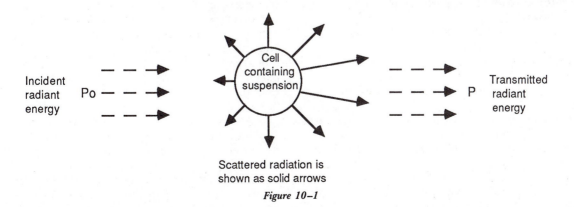

Incident radiant energy   Po

Cell containing suspension

P   Transmitted radiant energy

Scattered radiation is shown as solid arrows

Figure 10–1

that case, better results are obtained by measuring the scattered radiation. This technique is called nephelometry.

TURBIDIMETRY

When the concentration of a suspension of particles is to be measured by turbidimetry, the suspension is placed in an optical cell, and the cell is placed in an instrument similar to the one illustrated in Figure 10–2. The instrument allows for the

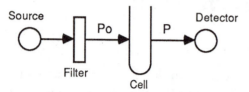

Figure 10–2. A simple turbidimeter.

measurement of Po and P. This makes it possible to calculate the concentration of the suspension. Consider the optical components shown in Figure 10–2.

The source is frequently a tungsten lamp, but it could be any other source of visible radiation. The filter may or may not be used. If the content of the cell is colored, the filter must be used to remove all absorbable photons from Po. Otherwise, the results would be a mixture of absorption spectroscopy and turbidimetry. This would lead to a falsely elevated measured concentration. The detector can be any detector that is sensitive to the wavelength of P. The most common detector in clinical turbidimeters is the semiconductor photodiode.

Keeping in mind the instrument shown in Figure 10–2, let us consider the mathematics of turbidimetry. It is very similar to the mathematics of absorption spectroscopy. The experimentally measured values of Po and P are related to the concentration of suspended particles by the equation

$$P/Po = e^{-fbc} \qquad \text{Equation 10–1}$$

where
  b is the cell path
  c is the concentration of radiation-scattering particles
  f is a constant for a particle of a given size at a given wavelength
Taking the log of both sides of Equation 10–1, one obtains

$$2.303 \log P/Po = -fbc \qquad \text{Equation 10–2}$$

Rearranging Equation 10–2, one finds that

$$-\log P/Po = \left[\frac{f}{2.303}\right] bc \qquad \text{Equation 10–3}$$

I call the term (f/2.303) the turbidivity and give it the symbol T. The turbidivity is to turbidimetric measurements what the absorptivity is to absorption measurements. The value of the turbidivity is a function of the particle size and the wavelength of the radiation used in the measurement. The value is not affected by the compound or compounds that compose the particles. Substituting T into Equation 10–3, one obtains

$$-\log P/Po = T bc \qquad \text{Equation 10–4}$$

Note that the term $-\log P/Po$ would be the absorbance if one were measuring absorbed radiation. One could call $-\log P/Po$ the turbidance, but, so far, no one does. It is sometimes called apparent absorbance. The use of the term turbidance should be adopted.

Turbidimetry can be performed on visible photometers or visible spectrophotometers. This is a common practice in academic laboratories.

Clinical Applications of Turbidimetry

One major clinical use of turbidimetry is in bacteriology. The typical bacterium is between 2 and 5 μm in diameter or length; therefore, bacteria are large enough to scatter visible radiation. Because of this fact, turbidimetry is widely used in automated instruments that measure antibiotic sensitivities. In these instruments, an optical path such as that shown in Figure 10–2 is used. They use turbidimetry to detect the condition of growth or no growth. The greater the amount of bacterial growth, the greater the amount of scattering that is detected.

In our laboratory, single-beam visible spectrometers are used at 500 nm to measure the number of bacteria per milliliter of broth. The instruments are standardized by analyzing samples of known bacterial content. A plot of the turbidance (apparent absorbance) versus the concentration of bacteria in the standards is made and then used to find the concentration of bacteria in samples of unknown concentration.

A second major use of turbidimetry is in coagulation analyzers. Blood cells are large enough to scatter visible radiant energy. Because of this fact, a number of instruments have been built that rely on the radiation-scattering process. These instruments are discussed in Chapter 24.

NEPHELOMETRY

When the concentration of a suspension of particles is low, it is better to measure it by nephelometry. In nephelometry, the suspension is placed in a cell, and the cell is placed in an instrument such as that illustrated in Figure 10–3. The instru-

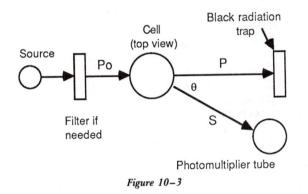

Figure 10–3

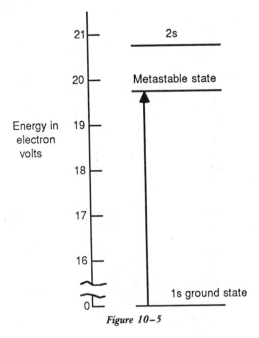

Figure 10–5

ment makes possible the measurement of the scattering intensity, S, which can then be related to the concentration. Sources used in such instruments include helium-neon lasers, quartz-halide lamps, and xenon lamps. Because the helium-neon laser is fairly common in nephelometry, it will be discussed at this time.

The helium-neon laser* is made by taking a fairly long, narrow bore tube and sealing between two mirrors a gas mixture of about 7 parts helium to 1 part neon (Fig. 10–4). The helium-neon gas pressure used in the tube is a few torr. A torr is equal to 1 mm mercury (Hg). The applied voltage is high enough to cause an electrical discharge through the gas. The electrical discharge is the source of energy that is converted to laser radiation.

*The term LASER is an acronym for light amplification by the stimulated emission of radiation.

In the process, helium atoms absorb electrical energy and are excited to an electronic state called a metastable state (Fig. 10–5). The metastable energy level is somewhat unusual in that its lifetime, by relative standards, is fairly long. The electrical energy can be stored for a time in these metastable helium atoms. The metastable helium atoms then collide with the neon atoms. When this happens, the energy of the metastable helium atom is given to the neon, and the helium returns to the ground state. The helium that has gone to the ground state then picks up more electrical energy and goes back to the metastable state, and the process starts again. The helium is said to be the pumping system of the laser.

The energy given to the neon excites the ground-state neon atoms to produce excited-state neon atoms. The favored state is the 5s orbital. This is shown in a highly simplified atomic orbital dia-

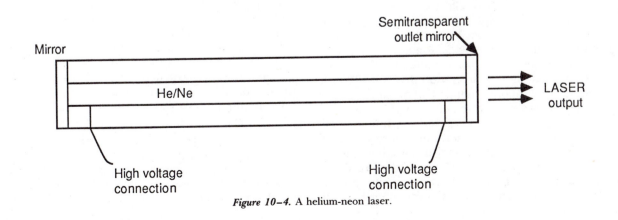

Figure 10–4. A helium-neon laser.

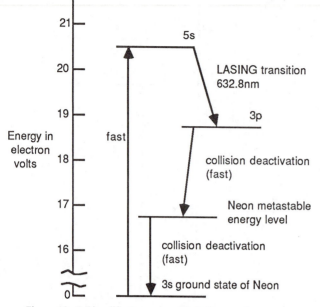

Figure 10–6. Simplified atomic orbital diagram for neon.

gram for neon in Figure 10–6. This process is very effective in producing large populations of excited-state neon atoms. This process is so efficient that there are more excited-state neon atoms than there are ground-state atoms. These excited neon atoms lose some of their energy by the emission of 632.8 nm photons. These photons are reflected back and forth by the mirrors. As they pass through the gas, they stimulate more emission. The reason for this stimulating effect is beyond the scope of this volume, but the effect is a generally observed phenomenon. This stimulated emission is called lasing. The 632.8 nm photons escape through the partially transmitting outlet mirror. The photons are then used in the instrument. The energy of the excited neon atoms that is not lost in the lasing process is lost by collision with the walls of the laser tube. The neon atoms can then reenter the process.

The filter may or may not be included in the instrument used to measure the scattering intensity (see Fig. 10–3), but frequently one is. The cell is of high quality and is similar to those used in fluorescent spectroscopy. The selected value of the angle of scattering, θ, may be between a few degrees and 90 degrees. A value that is somewhere between 30 and 70 degrees is most common. The value selected is a design consideration that is made with a specific application in mind. The radiation trap is provided in order to cut down stray radiation.

The formal mathematical relationship between the scattered intensity, S, and the concentration of suspended particles is seldom used. Let it suffice to say that at low concentrations a linear relationship exists between the scattering intensity and the concentration in most cases. On the other hand, if the concentration gets too large, a nonlinear calibration can develop.

Clinical Applications of Nephelometry

In the clinical laboratory, there are three main applications of nephelometry. First, the method is used in some blood cell counters to distinguish among different types of blood cells. It is also used to count the cells. The exact way in which this is done is discussed in Chapter 23.

The second application of nephelometry is the measurement of antigen-antibody complexes. The chemistry of these procedures is discussed in Appendix A. Several instruments use a radiation path like the one in Figure 10–3 in order to follow these immunochemical reactions. The angle θ on these instruments is in the 30 to 70 degree range. This range provides the optimum angle for the size of the particles generated in these types of reactions. The scattering developed in a sample is corrected for other sources of scattering by a blanking procedure. The scattering so corrected is compared with the scattering produced by standard solutions. These instruments are used to measure the serum level of many important proteins. One manufacturer lists kits for IgA, IgG, IgM, complement C_3, complement C_4, albumin, transferrin, α_2-macroglobulin, haptoglobin, α_1-antitrypsin, ceruloplasmin, c-reactive protein, orosomucoid, and rheumatoid factor. In addition, nephelometry is used for the determination of some drugs. These drugs are measured using nephelometry in conjunction with immunochemistry.

The third application is the analysis of enzymes that act on particulate substrates. The Coleman model 91 Amylase-Lipase Analyzer is an example of an instrument that does this type of analysis. Both amylase and lipase do their best work at water-substrate interfaces. For the analysis of amylase, the substrate is amylopectin. For the analysis of lipase, the substrate is olive oil. The reaction products that are produced when the enzymes act on these particles of substrate are too small to scatter radiation.

In both analyses, fairly stable suspensions of substrate are mixed with the sample containing the enzyme. The reduction in turbidity is then related to the enzyme concentration.

The Amylase-Lipase Analyzer is a typical nephelometer that uses white polychromatic radiation and photoelectric detection. It is calibrated by the use of stable gelatinous standards that are supplied by the company as part of the analysis kit.

SOURCES OF ERROR

In turbidimetry measurements, dirty cells cause losses in Po that show up as falsely depressed values for P. This leads to a falsely elevated value for the turbidance of the solution. The presence of an extraneous substance that is capable of scattering radiant energy leads to a falsely depressed value of P for the solution. This leads to a falsely elevated value for the turbidance. Stray radiation ultimately limits linearity of calibration at high values of turbidance. Fluorescent compounds that are excited to fluorescence by Po produce high apparent values for P. This causes a low value for the turbidance of a solution. A low turbidance value leads to a falsely depressed measured concentration of the particles in the suspension.

In nephelometry, dirty cells can cause problems. If the dirt decreases Po in a way that does not reflect radiant energy to the detector, a falsely low value for S will be detected. If radiant energy is reflected to the detector, a falsely high value for S will be measured. The presence of extraneous material in the cell that is capable of scattering radiant energy leads to an elevation in S. The presence of a fluorescent compound that is excited by Po leads to a falsely elevated value for S.

The chemical reactions that are used to prepare a sample for analysis are potential sources of error. These types of error are analysis-specific and hence cannot be discussed in a general way.

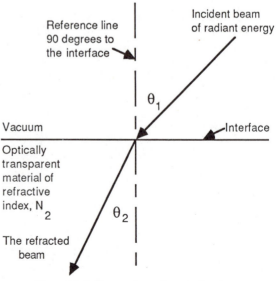

Refractometry of Liquids

The refraction of electromagnetic radiation, like the scattering of radiation, is a nonquantized process. The sample neither absorbs nor emits the radiation. Although the theoretical basis of refraction is known, it is very abstract, and its development is beyond the scope of this book. This chapter focuses on the practical concepts concerning how the instrumentation works.

The meaning of the verb to refract is best defined through an illustration. Consider the plate of glass shown in Figure 11–1. When the beam of

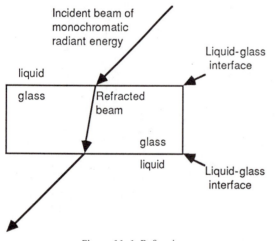

Figure 11–1. Refraction.

monochromatic radiation strikes the liquid-glass interface at an angle, the direction of propagation of the beam is altered. This change in direction of propagation is an example of refraction; the glass has refracted the beam of radiation. It is interesting to note that when the beam passes out of the glass, the process of refraction takes place again. The beam is refracted at the glass-liquid interface back to its original direction of propagation. There is,

however, a displacement of the beam relative to the original incident beam's position.

REFRACTIVE INDEX

When a beam of monochromatic radiation passes from a vacuum into any material, the speed of propagation decreases, and if the radiation strikes at an angle, refraction will occur. The reason for the decrease in speed of propagation is that the beam of radiation momentarily interacts with matter in a reversible way. The more dense the matter, the slower the momentary interaction. The interrelationship between the speed of propagation and the extent of refraction is not easily shown but is a demonstrated experimental fact. The mathematics of refraction is, however, easily seen and is illustrated along with the necessary reference lines in Figure 11–2.

Figure 11–2. The mathematics of refraction.

The refractive index of the optically transparent material is defined by the equation

$$\frac{N_2}{N_1} = \frac{\sin \theta_1}{\sin \theta_2} = \frac{V_1}{V_2} \qquad \text{Equation } 11-1$$

where N_2 is the refractive index of the optically transparent material, N_1 is the refractive index of a vacuum, V_1 is the velocity of electromagnetic radiation in a vacuum, and V_2 is the velocity in the transparent material. The refractive index of a vacuum has been arbitrarily assigned the value of 1, so Equation 11–1 becomes

$$N_2 = \frac{\sin \theta_1}{\sin \theta_2} \qquad \text{Equation } 11-2$$

In practice, the refractive index of materials is measured in air instead of in a vacuum. In that case, N_1 is 1.00027 at 25°C. This value is close enough to the vacuum value for the difference to be insignificant for most experimental purposes.

Clinical scientists are specifically interested in the refractive index of solutions. Refractive index values can be experimentally determined on the basis of Equation 11–1. This requires a knowledge of the angles θ_1, θ_2, and the refractive index of one of the materials.

The value of the refractive index of a solution is affected by three things. First, the wavelength of the radiation that is used to make the measurement affects the numerical value of the refractive index. For most work, the sodium emission line at 589.3 nm is used. This line is called the D line by emission spectroscopists and, consequently, refractive index values measured with it are called N_D values.

The second thing that affects the refractive index of a material is its temperature. The change in temperature of a solution produces a change in solution density, which in turn produces the change in refractive index. In practical terms, for water solutions, temperature variations of over 0.02°C will make *changes* in the refractive index of over 0.00001. For some work this may be an important matter.

The third thing that affects the refractive index of a solution is concentration. As the amount of solute changes, so does the solution density. The change in solution density, in turn, has an effect on the refractive index of the solution. This interrelationship allows for urine density determinations that are based on the measurement of refractive index.

CRITICAL-ANGLE REFRACTOMETERS

All the refractometers used in the measurement of urinary refractive index are critical-angle instruments. Let's look at the concept of the critical angle. Consider Figure 11–3. Note that the glass-liquid

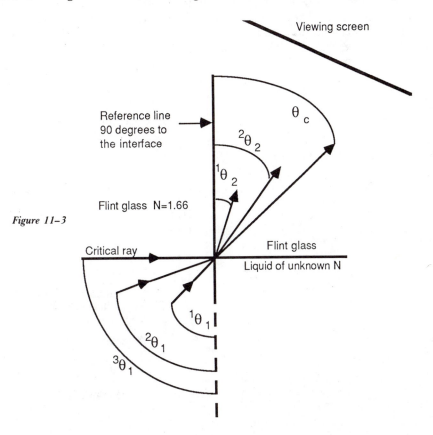

Figure 11–3

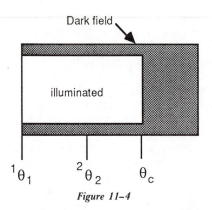

Figure 11–4

interface is arranged differently in Figure 11–3 than it was in Figure 11–2; the liquid is below the glass. Consider the monochromatic ray 1, which is incident on the interface at the angle of incidence $^1\theta_1$. The ray is refracted upon entry into the glass and is propagated in the direction shown with an angle of refraction of $^1\theta_2$. Now consider the behavior of ray 2. It has an angle of incidence $^2\theta_1$ and an angle of refraction $^2\theta_2$. Note that $^2\theta_1$ is greater than $^1\theta_1$, but $^2\theta_2$ is smaller than $^1\theta_2$. The trend is that the greater the angle of incidence, the smaller the angle of refraction. Now consider ray 3. Ray 3 is incident to the interface at an angle of 90 degrees. This is the highest angle of incidence possible and is called the critical ray. The refracted critical ray is propagated in the glass with the smallest angle of refraction possible for that system. The angle of refraction, θ_c, is called the critical angle.

Now, if the monochromatic radiation were made simultaneously available at all angles of incidence between $^1\theta_1$ and 90 degrees, and, further, if a screen were placed as shown in Figure 11–3, the radiation arriving at the screen would look like Figure 11–4. The screen would have a sharp, distinct demarcation between darkness and light. This demarcation would be at the critical angle, θ_c. If the screen had a scale associated with it that was calibrated in degrees, Equation 11–1 could be used to calculate the index of refraction of the liquid.

$$\frac{N_2}{N_1} = \frac{\sin \theta_1}{\sin \theta_2}$$

For flint glass, $N_2 = 1.66$, so

$$\frac{1.66}{N_1} = \frac{\sin \theta_1}{\sin \theta_2}$$

For the critical ray, $\theta_1 = 90°$ and $\theta_2 = \theta_c$, so

$$\frac{1.66}{N_1} = \frac{1}{\sin \theta_c}$$

or

$$N_1 = (1.66) \sin \theta_c \qquad \text{Equation 11–3}$$

Thus, Equation 11–3 could be used to calculate the refractive index of the liquid in Figure 11–3.

It would be difficult to build and use a device just like the one in Figure 11–3, but it is possible to build refractometers that make their measure-

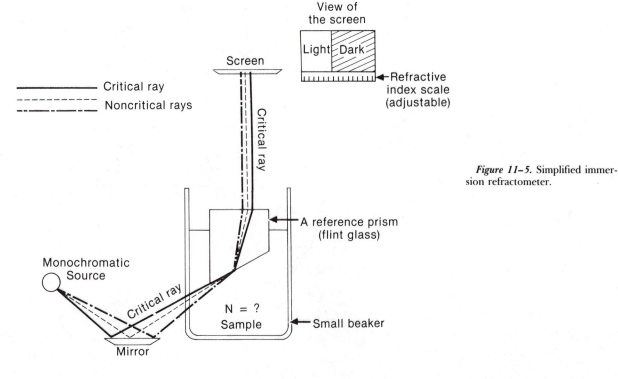

Figure 11–5. Simplified immersion refractometer.

ments on the same principle. Consider the simpli-fied immersion refractometer that is illustrated in Figure 11–5. Notice the similarities between the Figure 11–3 experiment and the setup in Figure 11–5. In each instance there is a monochromatic radiation source. The rays from the source are incident on the prism-sample interface. The refracted rays go to a screen with a refractive index scale. In the instrument, shown in Figure 11–5, however, the angle θ_c is not measured. A liquid of known refractive index is placed in the instrument, and the scale is adjusted to indicate the correct refractive index. The refractometer so calibrated is then used to measure the refractive index of the sample.

In common practice, many refractometers use a tungsten lamp as their source; this source is not monochromatic. When this is done, a "mono-chromator" is placed between the prism and the screen. This "monochromator," called an Amici compensator, allows only 589 nm radiation to reach the screen. Consider the illustration of the Amici compensator in Figure 11–6. The combination of the three prisms causes all wavelengths other than the 589 nm sodium wavelength to be deflected into the sides of the instrument case.

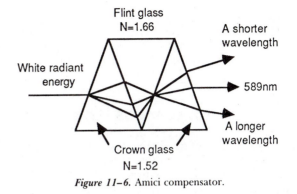

Figure 11–6. Amici compensator.

Contrary to the image projected by the drawing in Figure 11–5, radiation does not reach one spot on the reference prism and then refract. It refracts from many sites on the reference prism's face. These outputs are collected by a lens that focuses them into a sharp image.

Incorporating these new components into the refractometer in Figure 11–5, one obtains a refractometer that looks a lot like the Bausch & Lomb immersion refractometer (Fig. 11–7). The Bausch

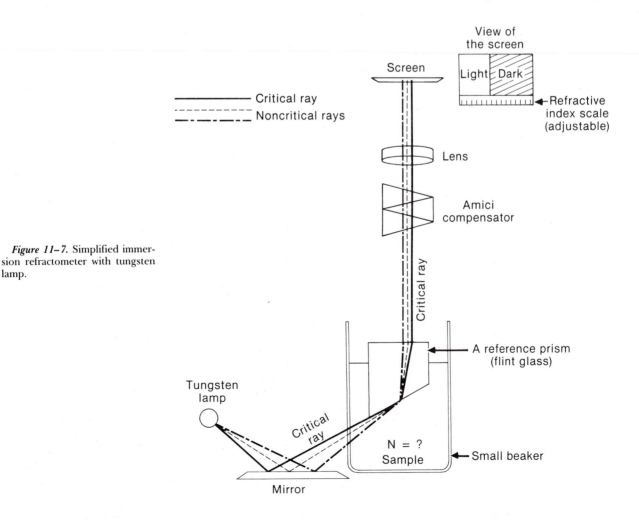

Figure 11–7. Simplified immersion refractometer with tungsten lamp.

& Lomb immersion refractometer is an effective, simple, dependable instrument that also serves as a good teaching example. Keeping it in mind will facilitate your understanding of the instrument that is used in clinical laboratories.

THE TOTAL SOLIDS METER

The total solids meter that is marketed by Reichert Scientific Instruments (formerly American Optical) has been widely used in the clinical laboratory for the measurement of the urinary specific gravity.

$$\text{urine specific gravity} = \frac{\text{density of the urine}}{\text{density of distilled water}}$$

The specific gravity of the urine is diagnostically valuable information.

In the early part of this chapter three things that affect refractive index values of solutions are outlined: (1) wavelength, (2) temperature, and (3) solution concentration. Previously discussed also is the idea that temperature and concentration produce their effect on refractive index because they change the density of the solution. Thus, it is not surprising that the refractive index of a solution can be correlated with the solution density and specific gravity. Note that refractive index measurements do not indicate what is in the solution; these measurements only respond to the density of the solution.

Consider the simplified optical path of a total solids meter (Fig. 11–8). The frosted coverplate entrains a thin film of urine between itself and the

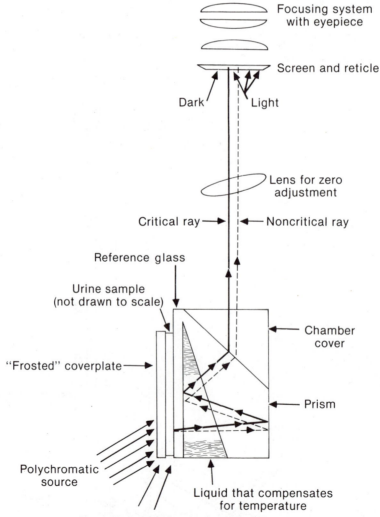

Figure 11–8. Optical path of a total solids meter.

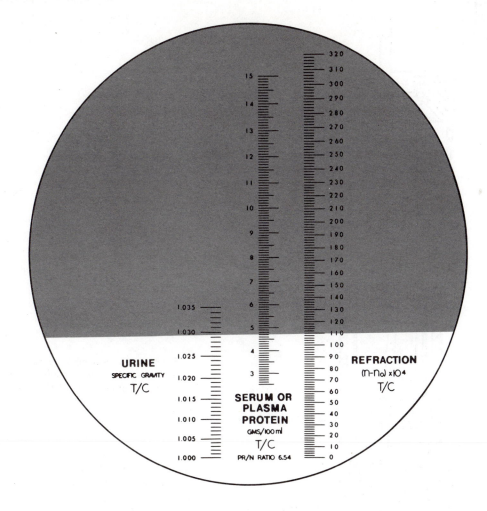

Figure 11–9. (Courtesy of Reichert Scientific Instruments, Buffalo, New York. Reprinted with permission.)

reference glass. The frosted surface receives the polychromatic source radiation and re-emits the radiation at all angles through the thin layer of urine. Two such rays are illustrated in Figure 11–8. The one at nearly glancing angle is drawn at an angle so it can be seen in the drawing, but it is really traveling down the urine layer at 90 degrees and is the critical ray. The other ray shown is correctly drawn and is a noncritical ray. After refraction in the reference glass, the two rays pass through a pair of glass-liquid interfaces. The liquid has been chosen because of its own refractive index and its variation with temperature. The refraction caused by this liquid corrects the path of the critical ray of our refracted beam to the path that it would travel if the temperature were at 20°C. This temperature

compensation feature allows for measurements between 15°C and 37°C. The one-drop urine sample quickly comes to temperature equilibrium with the instrument, and the system then corrects the refractive index to a value at 20°C. The system of prism, liquid, and chamber cover is roughly an Amici compensator. The screen and reticle, in the area of the critical ray, receive the 589 nm portion of the radiation. The screen has the appearance of a light field and dark field. The light-dark boundary is the critical angle information that is converted to refractive index values by the optical system. See Figure 11–9.

The urine specific gravity scale has been worked out by measuring the specific gravity of urine by other methods and then using the data to

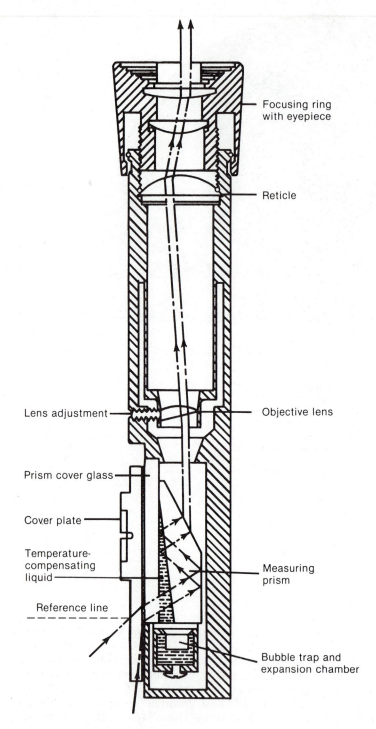

Focusing ring
with eyepiece

Reticle

Lens adjustment

Objective lens

Prism cover glass

Cover plate

Temperature-
compensating
liquid

Measuring
prism

Reference line

Bubble trap and
expansion chamber

Figure 11–10. Schematic of the total solids meter. (Courtesy of Reichert Scientific Instruments, Buffalo, New York. Reprinted with permission.)

set up the scale shown in Figure 11–9. The refraction scale is not in the common units but is

$$(N-No) \times 10^4$$

where No = 1.3330

No is the refractive index of distilled water at 20°C, and N is the refractive index referred to in the previous discussion.

The diagram of the instrument that is supplied by Reichert Scientific Instruments is shown in Figure 11–10.

Reflectance Photometry; Reflectance Densitometry

In recent years, interest in the analytical applications of reflected radiant energy has been increasing. Chemists who work with insoluble pigments have been interested in reflectance measurements because they allow the study of the color of pigments in a quantitative way. Other chemists have been interested in reflectance measurements because they offer a means of measuring concentration.

You have probably had an opportunity to see a rack of test tubes containing a set of standard solutions that would have been prepared in the conduct of a colorimetric determination. An example of such a set of test tubes is shown in Figure 12–1. Look at the tubes carefully; think about what you

know about these standards. First, the standards all contain the same analyte and chromogen. The only thing that differs from tube to tube is the analyte concentration. You also know that the tubes are being perceived by your eyes because of the radiant energy that is reflected off the tubes. The last thing that you know is that the tubes are noticeably different in appearance. This difference in appearance, when measured by an instrument in a quantitative way, is the basis of reflectance photometry.

TYPES OF REFLECTION

When a polished surface is illuminated with radiant energy, the radiant energy is reflected from the surface. The angle of incidence of the radiant energy is equal to the angle of reflection. This type of reflection is said to be specular reflectance. Specular reflectance is used to manage and direct radiant energy, but it is not used for the measurement of concentration, at least not in a direct way.

Diffuse reflectance is a second type of reflection. This type of reflection occurs at nonpolished surfaces. It is characterized by the fact that the reflected radiant energy tends to go in many directions (Fig. 12–2). This type of reflectance is what

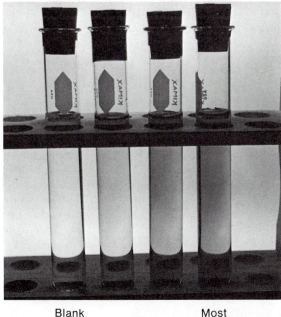

Blank   Most concentrated standard

Figure 12–1. Test tubes that contain a set of standard solutions. (Photo: Donald Suiter.)

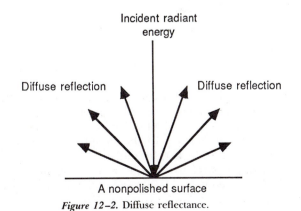

Figure 12–2. Diffuse reflectance.

131

makes barium sulfate or magnesium carbonate powder look white. If these powders are illuminated with white light, they diffusely reflect it in total, and our eyes perceive the compounds as white. On the other hand, potassium dichromate looks orange when illuminated with white light because the compound absorbs strongly in the area around 440 nm. The light not absorbed is diffusely reflected. Our eyes see the reflected light, and the powder is perceived as an orange solid.

THE MATHEMATICS OF REFLECTANCE MEASUREMENTS

Strictly speaking, the diffuse reflectance, R, of a solid is equal to the sum of all reflected photons divided by the intensity of the incident radiant energy. This can be written as an equation.

$$R = \frac{R_T}{Po} \qquad \text{Equation 12–1}$$

where R_T is the sum of all reflected radiant energy at the analytical wavelength, and Po is the total incident radiant energy at the analytical wavelength.

From the experimental point of view, it is difficult to measure the sum of all the reflected photons. Instead, most instruments detect a constant fraction of the reflected photons. This is done with an instrument called a reflectometer (reflectance densitometer). There are two typical layouts for reflectometers. The first is shown in Figure 12–3. The

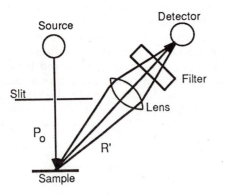

Figure 12–3. A reflectometer.

instrument has a source of radiant energy. The source is frequently a tungsten lamp or a quartz-halide lamp. The incident radiant energy is directed at 90 degrees relative to the sample surface. The fraction of the diffuse reflectance, R′, that is radiated at 45 degrees relative to the incident beam is collimated by a lens and is allowed to fall on an optical filter. This angle minimizes the detection of spec-

ular reflectance. The filter is selected so that the only radiant energy detected is at a wavelength where the analyte absorbs. The radiant energy that is transmitted by the filter falls on the detector. The detector measures the intensity of the "monochromatic" radiant energy that is reflected from the sample. The value measured is a fraction, 1/f, of the total reflectance, R_T, at the analytical wavelength.

A second optical arrangement is also used (Fig. 12–4). The instrument is much like the one in Fig-

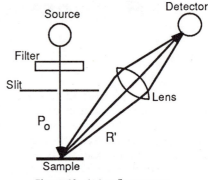

Figure 12–4. A reflectometer.

ure 12–3 except that the optical filter is before the sample instead of after. On at least one commercial instrument, a light-emitting diode is substituted for the source and filter. As before, the detector measures the intensity of the "monochromatic" radiant energy that is reflected from the sample. The value measured is a fraction, 1/f, of the total reflectance, R_T, at the selected wavelength. As before, this type of instrument provides a measured value defined as R′. The interrelationship between R and R′ is shown in Equation 12–2.

$$R = \frac{fR'}{Po} \qquad \text{Equation 12–2}$$

where $fR' = R_T$

In actual experimental work, it is common to use a highly purified powder of magnesium carbonate or barium sulfate or a white ceramic material as a reflectance standard. This is done in order to establish the instrument response when no absorbing substance is in the optical path. The reflectance of these white solids is frequently assumed to be 1. If this assumption is valid, Equation 12–1 becomes

$$1 = \frac{R_T \text{ std}}{Po} = \frac{fR' \text{ std}}{Po}$$

If we solve for f, we obtain

$$f = \frac{(1)Po}{R' \text{ std}} \qquad \text{Equation 12–3}$$

A second reference material is frequently used to calibrate a reflectometer. If a black sample is placed in the optical path, the instrument should produce a reading of 0 for the reflectance. If it does not do so, the intensity of the detected radiant energy, R_f, needs to be subtracted from measured values. The sources of R_f could be either stray radiation or flare. Flare is the specular reflectance that can sometimes be detected coming from black powder samples.

Once f for the instrument is known, and once R_f has been determined, it is possible to make reflectance measurements on samples. If a sample is placed in the optical path, R' can be measured. R for the sample can then be calculated. This is done by starting with Equation 12–2. First the flare correction is inserted into Equation 12–2.

$$R = \frac{f(R' - R_f)}{Po} \qquad \text{Equation } 12–4$$

If Equations 12–3 and 12–4 are combined

$$R = \left(\frac{(1)\, Po}{R'std}\right)\left(\frac{(R' - R_f)}{Po}\right)$$

$$R = \frac{(R' - R_f)}{R'std}\,(1) \qquad \text{Equation } 12–5$$

In many situations, R_f is 0. When that is the case,

$$R = \frac{R'}{R'std}\,1 \qquad \text{Equation } 12–6$$

Equation 12–5 converts the experimentally measured values, R', R_f, and R' standard to a reflectance value for the sample.

$$R = \frac{R'}{R'std}$$

Some authors speak of the per cent reflectance. Per cent reflectance is defined by Equation 12–7.

$$\%R = 100\,R \qquad \text{Equation } 12–7$$

The reflectance of a sample is not linearly related to the concentration of the absorbing analyte. It has been found, however, that a nonlinear but dynamic relationship exists between the negative log of R and the concentration of an analyte. The relationship is affected by many variables, but a typical curve looks like the one shown in Figure 12–5.

The negative log of R is called the reflection density and is given the symbol D_R.

$$D_R = -\log R \qquad \text{Equation } 12–8$$

If one assumes $R_f = 0$, then Equation 12–6 can be substituted into Equation 12–8. This results in

$$D_R = -\log\left[\frac{(R')}{R'\,std}\,(1)\right] \qquad \text{Equation } 12–9$$

When R_f does not equal 0, Equation 12–8 becomes

$$D_R = -\log\left[\frac{(R' - R_f)}{R'\,std}\,(1)\right] \qquad \text{Equation } 12–10$$

Some authors call reflection density D_{Rc}.

If the samples are highly uniform as to thickness, particle size, and refractive index, the standard curves prepared for a series of solutions of known analyte concentration are fairly stable and can be used for the determination of an analyte in a solution of unknown concentration.

In summary, in order to do an analysis by reflectance photometry, one must install the proper filter so that the detector monitors a wavelength that the analyte can absorb. Then one must

1. determine R' std for a white reflectance standard;
2. determine R' for several solutions of known analyte concentration;
3. determine R_f for a black standard of similar particle size;
4. determine R' for the solution of unknown concentration.

The data that are obtained can be worked up with Equation 12–9 or 12–10. The resulting values of D_R can be plotted as in Figure 12–5. The analyte

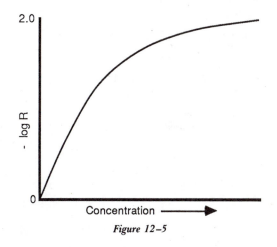

Figure 12–5

concentration of the sample can be determined graphically.

A computer can be used to do all of these calculations for you. With the application of suitable programs, the computer can do the calculation of the analyte concentration from D_R data.

URINE ANALYSIS BY DIPSTICK METHODS

Ever since the invention of litmus paper, analysts have wanted to develop methods that utilize strips of reagent-impregnated paper to do chemical analysis. The advantages of this approach are convenience and simplicity. The idea is to dip the impregnated paper into the sample. The dipping process supplies the solvent for the reaction; it also introduces the analyte. Once the strip is dipped, a color change tells the analyst the analyte concentration of the sample. Paper strips that have been impregnated with suitable chemical reagents are stored dry until needed.

Thinking of this type has led to considerable success in clinical analysis. The measurement of urinary glucose was the first method that used the "dip the strip" technology. Since that time, other methods for the analysis of other analytes have been developed. Let's use measurements of urinary glucose as our teaching example.

Consider a strip of filter paper that has been impregnated with glucose oxidase, horseradish peroxidase, potassium iodide (KI), and a buffer.* The paper would have the blue color of a robin's egg. If the strip were dipped into urine that contained no glucose, there would be no color change. If, on the other hand, the paper were dipped into urine that contained glucose, a series of reactions would start. The first reaction would be as follows:

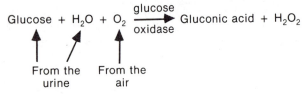

The hydrogen peroxide formed would then react with the KI as follows:

$$2\,H^+ + H_2O_2 + 3\,KI \rightarrow KI_3 + 2\,K^+ + 2\,H_2O$$
$$\text{Buffer}$$
$$\text{pH} = 5$$

---

*This is the method used by the Ames Division of Miles Laboratory on their Dipstix.

The KI_3 is a yellowish brown compound. The production of KI_3 is proportional to the original glucose concentration. The more glucose in the urine, the more the color of the paper changes from robin's egg blue to yellowish brown. If the filter paper were subjected to reflectance measurements at about 540 nm, the reflectance could be used to find the glucose concentration. This could be done with an instrument like the one shown in Figure 12–4. It would require that the instrument be calibrated with a white diffuse reflectance standard and that several glucose standards be run for establishing the calibration curve. A black standard could be applied to check for flare. The reflectance values obtained could be converted to D_R values by the use of Equation 12–9 or 12–10. The D_R values could be plotted against concentration, and the curve could be used to find the concentration of glucose in samples.

The Ames Division of Miles Laboratory and Boehringer Mannheim Diagnostics have developed this type of methodology to an impressive level. The Ames N-Multistix-SG reagent strip is a good example. Nine different swatches of impregnated paper are placed on the same dipstick. These swatches contain reagent to allow the measurement of urine specific gravity, pH, protein, glucose, ketone, bilirubin, blood, nitrite, and urobilinogen. The strips are made on Mylar plastic backing (Fig. 12–6). Each of these reaction zones absorbs urine when the strip is dipped. The strip is blotted off, but the water from the urine, in addition to the analyte, does allow the reaction to start. The blotted strip is then evaluated by reflectance photometry. The instruments that are used to evaluate these strips must have the capacity to make measurements at several wavelengths. The configuration of these instruments is similar to that of the reflectometer illustrated in Figure 12–4. The Ames instrument uses a filter wheel in conjunction with a tungsten source in order to manage the task. The Ames instrument mechanically moves the strip through the reflectometer. The instrument's microprocessor selects the proper filter to make the reflectance measurement at each reaction zone.

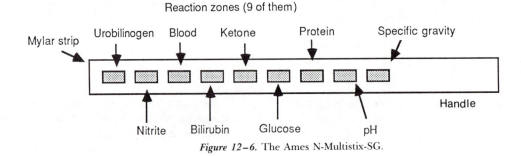

Figure 12–6. The Ames N-Multistix-SG.

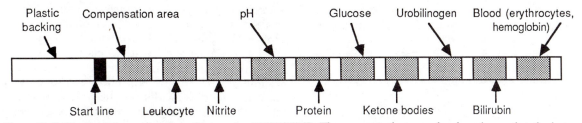

Figure 12–7. The Boehringer Mannheim Diagnostics CHEMSTRIP. The compensation area absorbs urine so that the instrument can correct for its color.

The Boehringer Mannheim Diagnostics CHEMSTRIP is illustrated in Figure 12–7. Boehringer Mannheim Diagnostics' instrument consists of eleven separate reflectometers that are arranged in a linear array. Each of the reflectometers consists of a light-emitting diode (LED) source and a phototransistor detector. There is one reflectometer for each of the nine test fields. The tenth reflectometer provides data that the instrument uses to correct for interference caused by the color of the urine. The eleventh reflectometer monitors and makes sure that the CHEMSTRIP is properly aligned in the instrument. In order to avoid stray light problems, only one of the eleven LED sources is turned on at a time. The reflectance of each test field is measured and stored in the instrument's computer. The computer uses the stored data to calculate the concentration of the nine analytes.

REFLECTANCE MEASUREMENTS USING "SLIDE" FORMAT

In addition to the dipstick approach, there is the slide configuration, which has drawn considerable interest. The slide configuration uses a matrix in the form of a thin layer that contains the reagents needed to do an analysis. The slides are stored dry. When the sample is applied to a slide, the water in the sample activates the reagents; the analyte is also introduced. The analyte then undergoes a series of reactions that lead to the production of a colored product. The amount of colored product is measured by reflectance photometry. Let's consider two examples of glucose "slides". All of the glucose "slides" use more or less the same chemistry. The chemistry was first developed by Trinder.

Kodak's version of this reaction is done on a slide that is illustrated in Figure 12–8. The porous spreading layer is made of cellulose acetate that contains TiO_2 as a reflective backing. The spreading layer has a void volume of about 80 per cent and is very uniform. When a drop of sample is placed on the spreading layer, it spreads the drop and meters it uniformly into the gelatin reagent layer.

The gelatin layer is about 100 μm thick; it is made up of individual layers of gelatin that contain the reagents. These layers are coated independently of one another and maintain their integrity. The color develops in the gelatin layer.

The plastic base is transparent and is provided to give the system some mechanical strength. The radiant energy from the reflectance photometer enters and leaves through the plastic base.

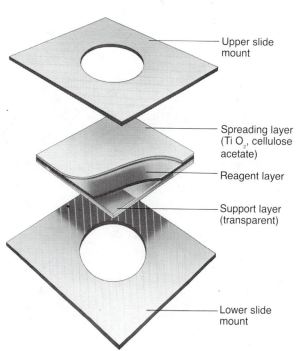

Figure 12–8. Kodak's glucose slide. (Reprinted courtesy of Eastman Kodak Company. Copyright © by Eastman Kodak Company.)

$$\text{Glucose} + 2H_2O + O_2 \xrightarrow[\substack{\text{oxidase} \\ pH=5}]{\text{glucose}} \text{Gluconic acid} + 2H_2O_2$$

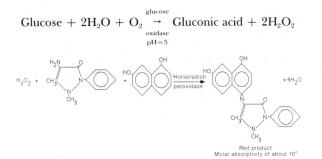

The Ames glucose reagent strip slide is illustrated in Figure 12–9. The sample is applied to the

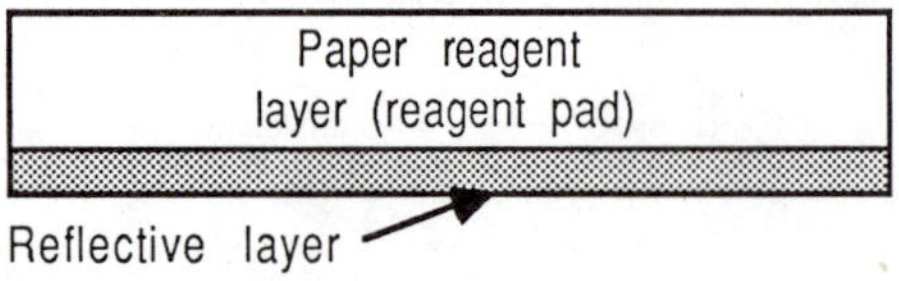

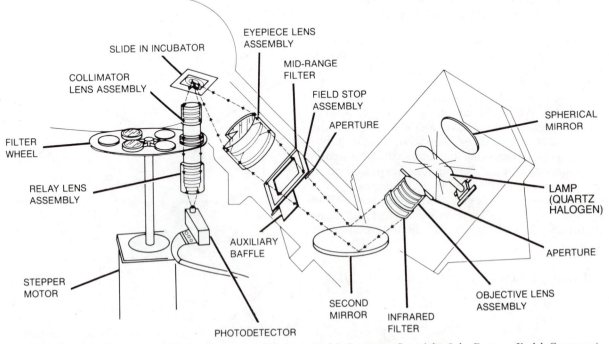

Paper reagent
layer (reagent pad)

Reflective layer

Figure 12–9. Ames version of the glucose slide.

reagent pad. The glucose diffuses with water into the paper reagent layer. The reactions that lead to color formation take place, and the pad is read by reflectance photometry. The radiant energy enters the slide through the film membrane.

REFLECTANCE PHOTOMETERS DESIGNED TO MEASURE THE REFLECTANCE OF "SLIDES"

The instrument that is used to measure the reflectance of Kodak's slides is illustrated in Figure 12–10. The drawing is fairly straightforward. The radiant energy from a quartz-halogen lamp is collimated and passed through an infrared filter. The

radiant energy is then directed to the slide. The reflected radiant energy is passed through a filter and is detected. The instrument knows what it is trying to determine, so it selects the proper filter from the selection on the filter wheel. Glucose, for example, is read with 540 nm radiant energy. The radiant energy is incident to the slide at 45 degrees; the diffuse reflectance is detected at 90 degrees relative to the face of the slide.

When no slide is in the read station, a white reflectance reference is exposed to the radiant energy beam. This reflectance standard is used by the instrument and its microprocessor for calibration purposes.

The way that the results are calculated is fairly complex. A typical calibration curve for glucose is illustrated in Figure 12–11. It is not necessarily a calibration from a current production instrument. Notice that the curve is not linear at any point. The nonlinearity of the curve is an experimental nuisance; it tends to lead to excessive calibration costs. Because of that fact, complex approximation algorithms are applied by the microprocessor. The so-called transform data that result are plotted in Figure 12–12. The plot of D_T versus concentration is linear.

The Ames reagent strips are read in an Ames Seralyzer. The optics of the Seralyzer is shown in

Figure 12–10. A reflectometer. (Reprinted courtesy of Eastman Kodak Company. Copyright © by Eastman Kodak Company.)

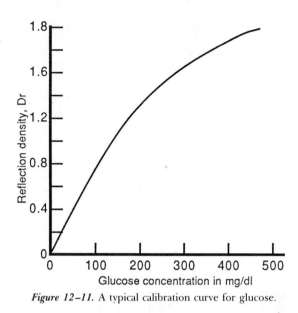

Figure 12–11. A typical calibration curve for glucose.

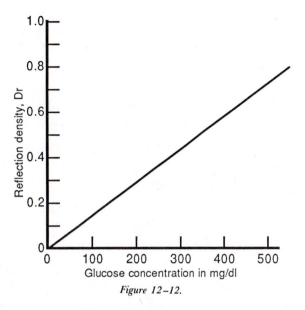

Figure 12–12.

Figure 12–13. This instrument uses a xenon lamp as its source. Xenon sources are discussed in Chapter 7. The instrument features an integrating sphere, which is a sphere that is coated internally with a white solid, such as barium sulfate ($BaSO_4$). The integrating sphere serves as a collector and flux averager of radiant energy. The radiant energy of the source arrives at the reagent pad after many reflections off the sphere. The diffuse reflectance arrives at the filter after many internal reflections. The selected wavelength passes through the filter onto the detector.

The Seralyzer uses its microprocessor to compensate for the nonlinearity of reflectance data.

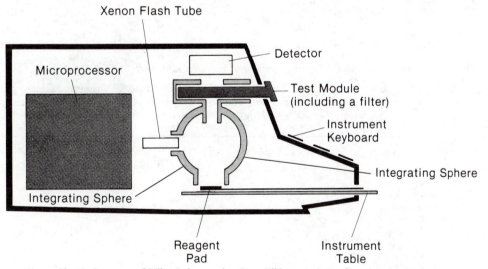

Figure 12–13. Courtesy of Miles Laboratories, Inc., Elkhart, Ind. Reprinted with permission.

Problems   1.   Glucose concentration is to be determined by reflectance photometry. Chromatography paper is impregnated with the proper chemicals, and a reflectance photometer is configured to operate at the proper wavelength. Flare is evaluated with a black standard and found to be 0. The instrument is adjusted to a readout of 100 units with a blank. With the instrument so adjusted, the following data are obtained:

	Glucose Concentration (mg/dl)	R'
Standard 1	500	3.16
Standard 2	400	4.47
Standard 3	300	7.94
Standard 4	200	14.10
Standard 5	100	28.20
Black flare standard	Infinite	0.00
Blank (reflectance standard)	0	100.00
Sample 1	?	5.30
Sample 2	?	29.70

What is the glucose concentration of the two samples?

2.   Total protein concentration is to be measured by reflectance photometry. Chromatography paper is impregnated with the proper chemicals, and a reflectance photometer is configured to operate at the proper wavelength. Flare is evaluated with a black flare standard and found to be 0. The instrument is adjusted to a readout of 100 units with a blank. With the instrument so adjusted, the following data are obtained:

	Percentage Protein (%)	R'
Standard 1	1	67.3
Standard 2	2	49.5
Standard 3	3	34.8
Standard 4	4	29.2
Standard 5	5	25.7
Standard 6	6	24.0
Standard 7	7	22.9
Blank (reflectance standard)	0	100.0
Sample 1	?	26.5
Sample 2	?	37.6

What is the protein concentration of the two samples?

3.   Urea concentration is to be determined by reflectance photometry. Gelatinous media are mixed with proper chemicals, and slides are prepared. A reflectance photometer is configured to operate at the proper wavelength. The instrument is used to obtain the following data:

	Urea Concentration (mg/dl)	R'
Standard 1	5	84.5
Standard 2	10	74.0
Standard 3	15	67.6
Standard 4	20	61.9
Standard 5	25	58.0
Standard 6	30	55.5
Standard 7	35	54.0
Standard 8	40	52.0
Black flare standard	Infinite	3.0
Blank (reflectance standard)	0	100.0
Sample 1	?	69.2
Sample 2	?	53.1

What is the urea concentration of the two samples?

13

Potentiometric Methods
of Analysis

Batteries that are used to measure solution concentrations* are called electrochemical cells or simply cells. These cells are always made from two half cells. A half cell always contains an electrode. The electrode at which reduction occurs is called a cathode. The electrode at which oxidation occurs is called an anode. The anodic half cell is the half cell that contains the anode and, as such, is the half cell in which oxidation is occurring. The cathodic half cell is the half cell in which reduction is occurring. The analysis methods that are based on the relationship between measured cell potentials and solution concentrations are called potentiometric methods. The most common example of such a method is the measurement of pH.

In clinical chemistry, potentiometric cells have been widely applied in the measurement of hydrogen ion, sodium ion, potassium ion, calcium ion, carbon dioxide, and urea. The electrodes that have been used for these measurements are somewhat complex. Before they are considered in detail, let's discuss some examples of less complicated electrodes.

TYPES OF CLASSIC ELECTRODES

The first type of classic electrode is the metal electrode. The zinc and copper electrodes discussed under *Batteries* in Chapter 1 are examples. If a metal electrode is placed into a solution that contains ions of the same metal, a reproducible, predictable potential forms at the interface. The electrode is said to be poised to the ions of the electrode metal because the activity of these ions does have a significant effect on the potential of the half cell. For example, a copper electrode could be used to measure the activity of copper ions. The copper electrode

---

*The electrodes actually measure activity. Activity can then be related to concentration. The concept of activity is developed in Appendix B.

would not be suitable for the measurement of other metal ions.

Another type of classic electrode is formed when a metal electrode is dipped into a solution that contains a slightly soluble salt of the electrode metal. This situation is illustrated by the silver–silver chloride system in Figure 13–1.

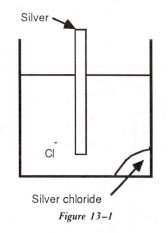

Figure 13–1

The potential of the silver electrode responds to the silver ion activity, but the silver ion is in equilibrium with the silver chloride (AgCl) solid. Because of the equilibrium, the chloride ion controls the activity of silver ion. An electrode of this type is poised to chloride ion by way of the common ion effect. This is mathematically shown for silver chloride as follows:

$$(Ag^+)(Cl^-) = K_{sp} = 1.80 \times 10^{-10}$$

$$(Ag^+) = \frac{1.80 \times 10^{-10}}{(Cl^-)}$$

Thus an increase in the activity of chloride ion decreases the activity of silver ion. This change in silver ion activity would then change the interfacial potential at the silver electrode.

From the practical point of view, it does not

matter where the silver chloride is located in the half cell. Once equilibrium is obtained, the silver ion activity will be the same in all cases. It is common to coat the silver electrode with a thin layer of silver chloride, and in that way the greatest convenience of use is obtained. The cathodic half cell reaction for the silver–silver chloride electrode is

$$AgCl + e \rightarrow Ag + Cl^-$$

A third type of classic electrode is formed when a noble metal such as platinum is placed in contact with a solution containing two different oxidation states of a different metal. For example, a platinum electrode in a solution that contains Fe^{+3} and Fe^{+2} is poised to the ratio of the concentrations of Fe^{+2} and Fe^{+3}. The cathodic half cell reaction is

$$Fe^{+3} + e \rightarrow Fe^{+2}$$

A fourth type of classic electrode is the gas electrode. In this electrode, a piece of porous platinum is set up so that a gas can be placed in contact with a solution. For example, the hydrogen gas electrode would be set up as in Figure 13–2. The po-

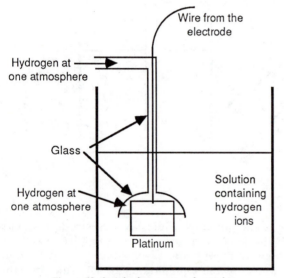

Figure 13–2. A hydrogen gas electrode.

rous platinum is saturated with hydrogen gas at 1 atmosphere pressure. The saturated platinum is also in contact with the solution containing hydrogen ions. The hydrogen can diffuse along the electrode surface from the hydrogen reservoir to that part of the electrode which is in contact with the solution. In this way, H_2 and H^+ ions are in contact with each other, and a potential develops just as if the H_2 were a metal. The cathodic half cell reaction is

$$2H^+ + 2e \rightarrow H_2$$

STANDARD ELECTRODE POTENTIAL

Unfortunately, it is physically impossible to measure the potential of a single electrode or half cell. In order to measure potential, it is necessary to use two half cells. This means that the sum of the potentials of two half cells is all that one can measure. If one can know only the sum of two half cells, it is impossible to know the true value of either. In order to measure at least the relative potential of half cells, electrochemists agreed to use the hydrogen gas electrode (Fig. 13–3) as a half

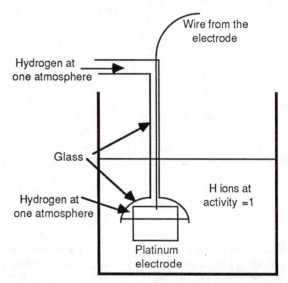

Figure 13–3. A normal hydrogen gas electrode (NHE). E° = 0.0000v.

cell against which to measure all others. They agreed to assume that the hydrogen electrode had a potential of 0.0000 volts. This value is based on a hydrogen ion activity of 1.000 and a temperature of 25°C. The electrode is called a *normal hydrogen electrode* and is abbreviated NHE.

The normal hydrogen electrode is used as a standard, and all other half cells are compared with it. In order to produce a table of the potentials of various other electrodes, scientists have made half cells from many other materials. These half cells are set up with the ions in the cell at 1 unit of activity, and the potential of each half cell is measured against the NHE. The resulting value is called the *standard electrode potential*, E°. An example of the electrochemical cell used for this measurement is shown in Figure 13–4. The salt bridge is prepared from a glass tube filled with saturated potassium chloride solution and a gelatinous material called agar. Its function is to make electrical contact between the half cells without affecting the measured potential. This salt bridge minimizes the junction potentials that tend to form at liquid-liquid or liquid-solid boundaries.

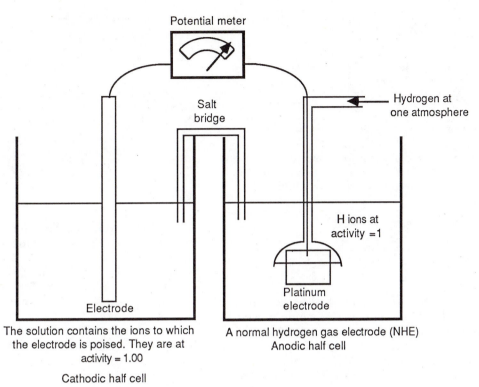

Figure 13-4. Electrochemical cell used to measure standard electrode potential.

Using such a cell (Fig. 13-4), the potential is measured with the potential meter. Remember that

$$E^0_{cell} = E^0_{anode} + E^0_{cathode}$$

but E^0_{anode} is defined as 0 V in this case. Thus, E^0_{cell} is said to be the standard potential of the cathodic half cell. The zero notation (E^0) denotes that the concentrations of all electroactive ions in the cell are at unit activity. If we measure the potential of many half cells in this way, we can tabulate the results to create a table of standard potentials (Table 13-1).

The negative potentials found in Table 13-1 indicate that the electrode would function as an anode in the cell shown in Figure 13-4. A positive potential indicates that the electrode would function as a cathode in the same cell.

These standard potential values can be used to find the standard potential of any electrochemical cell for which standard potential data exist. For example, let's calculate the standard potential of a cell that is made from a copper cathode and a zinc anode.

Anode	$Zn^0 \rightarrow Zn^{+2} + 2e$	$E^0 = 0.76$ volts
Cathode	$Cu^{+2} + 2e \rightarrow Cu^0$	$E^0 = 0.34$ volts
Cell reaction	$Zn^0 + Cu^{+2} \rightarrow Zn^{+2} + Cu^0$	$E^0_{cell} = 1.10$ volts

Notice that the zinc in the cell is being oxidized instead of being reduced. Because of this fact, the correct reaction for the zinc is the reverse of the reaction in Table 13-1, and the sign on the potential is also reversed.

As a second example, consider a cell with a

TABLE 13-1. Standard Electrode Potentials

Cathodic Half Cell Reaction	E^0 (volts)
$K^+ + e \rightarrow K$	-2.93
$Ca^{+2} + 2e \rightarrow Ca$	-2.87
$Na^+ + e \rightarrow Na$	-2.71
$Mg^{+2} + 2e \rightarrow Mg$	-2.37
$Al^{+3} + 3e \rightarrow Al$	-1.66
$Zn^{+2} + 2e \rightarrow Zn$	-0.76
$Pb^{+2} + 2e \rightarrow Pb$	-0.13
$2H^+ + 2e \rightarrow H_2$	0
$AgCl + e \rightarrow Ag + Cl^-$	0.22
$Hg_2Cl + 2e \rightarrow 2Hg + 2Cl^-$	0.268
$Cu^{+2} + 2e \rightarrow Cu$	0.34
$I_2 + 2e \rightarrow 2I^-$	0.54
$Hg^{+2} + 2e \rightarrow Hg$	0.79
$Ag^+ + e \rightarrow Ag$	0.80
$Cl_2 + 2e \rightarrow 2Cl^-$	1.36
$Au^{+3} + 3e \rightarrow Au$	1.50
$F_2 + 2e^- \rightarrow 2F^-$	2.65

E^0 data are from: W. M. Latimer, The Oxidation States of the Elements and Their Potentials in Aqueous Solutions, 2nd Ed., ©1952, renewed 1980. Adapted by permission of Prentice-Hall, Englewood Cliffs, New Jersey.

gold cathode and silver anode. The cathodic half cell reaction is

$$Au^{+3} + 3e \rightarrow Au^0 \quad E^0 = 1.50 \text{ volts}$$

The anodic half cell reaction is

$$3Ag^0 \rightarrow 3Ag^+ + 3e \quad E^0 = -0.80 \text{ volts}$$

Adding the anodic and cathodic half cell reactions so that the number of electrons produced equals the number of electrons used, we obtain the cell reaction

Cathode	$Au^{+3} + 3e \rightarrow Au^0$	$E^0 = 1.50$ volts
Anode	$3Ag^0 \rightarrow 3Ag^+ + 3e$	$E^0 = -0.80$ volts
Cell reaction	$Au^{+3} + 3Ag^0 \rightarrow Au^0 + 3Ag^+$	$E^0_{cell} = 0.70$ volts

Notice that although 3 moles of silver are required per mole of gold the potential is *not* multiplied by 3. The potential developed at an electrode is not affected by the number in front of the symbol in the cell reaction equation. Also note that the silver half cell reaction was multiplied by 3 in order to make the number of electrons needed by the gold equal to the number produced by the silver. This should always be done in order to produce a balanced cell reaction.

THE NERNST EQUATION

Keeping in mind the facts discussed in the previous section, consider cells in which the ions are not at unit activity. This situation is more true to reality. Consider the general cell reaction

$$aA + bB \rightleftarrows cC + dD$$

A German chemist named Walther Nernst found that the potential of such a cell could be predicted by the following equation:

$$E = E^0 - \frac{(2.303)\,RT}{NF} \log \frac{(C)^c\,(D)^d}{(A)^a\,(B)^b} \qquad \text{Equation 13–1}$$

where

E^0	would be the potential of the same cell if all the substances were at unit activity
R	is the molar gas constant (8.314 joules/(mole)(°K))
T	is the absolute temperature, 298°K (25°C)
N	is the number of moles of electrons transferred in the balanced cell reaction
F	is Faraday's constant (96,491 coulombs per equivalent weight)
( )	means activity; all solids and gases at 1 atmosphere have an activity of 1.

In order to gain a better understanding of the Nernst equation let's consider some cells and calculate the potential that those cells should produce. What will be the potential of the cell illustrated in Figure 13–5? Assume that the copper electrode is the cathode.

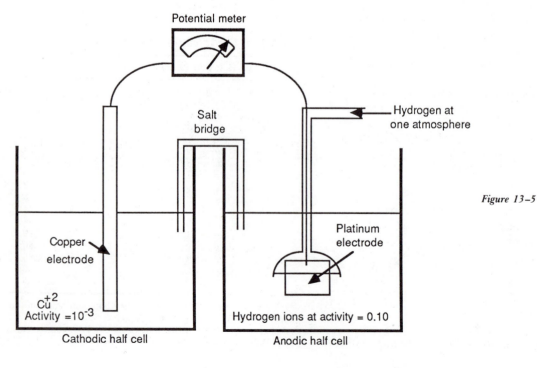

Figure 13–5

Potential meter

Salt bridge

Hydrogen at one atmosphere

Copper electrode

Platinum electrode

Cu^{+2} Activity $= 10^{-3}$

Hydrogen ions at activity = 0.10

Cathodic half cell

Anodic half cell

The Nernst equation is applied by first assuming that the ions are all present at unit activity and calculating the standard potential.

Cathode	$Cu^{+2} + 2e \rightarrow Cu$	$E^0 = 0.34$ volts
Anode	$H_2 \rightarrow 2H^+ + 2e$	$E^0 = 0$ volts
Cell reaction	$Cu^{+2} + H_2 \rightarrow Cu + 2H^+$	$E^0 = 0.34$ volts

The standard potential obtained is positive. This implies that the reaction as written is spontaneous. With the standard potential known, we can proceed with the Nernst equation.

$$E = E^0 - \frac{0.0591}{N} \log \frac{(C)^c (D)^d}{(A)^a (B)^b}$$

$$E = 0.34 \text{ volts} - \frac{0.0591}{2} \log \frac{(Cu)^1 (H^+)^2}{(Cu^{+2})^1 (H_2)^1}$$

Cu as a solid has an activity of 1.
H_2 as a gas at 1 atmosphere has an activity of 1.

Substituting these values into the preceding equation, we obtain

$$E = 0.34 \text{ volts} - \frac{0.0591}{2} \log \frac{(H^+)^2}{(Cu^{+2})}$$

Substituting the values from Figure 13–5,

$$E = 0.34 \text{ volts} - \frac{0.059}{2} \log \frac{(10^{-1})^2}{(10^{-3})}$$

and doing the math, we obtain

$$E = 0.31 \text{ volts}$$

Consider a second example (Fig. 13–6). What will be the reading on the potential meter? In order to solve this problem, assume that the zinc electrode is the anode and that the silver–silver chloride electrode is the cathode. (Generally, the anode is drawn on the left and the cathode on the right in these illustrations.) First, one calculates E^0,

Anodic reaction	$Zn^0 \rightarrow Zn^{+2} + 2e$	$E^0 = 0.76$ volts
Cathodic reaction	$2AgCl + 2e \rightarrow 2Ag^0 + 2Cl^-$	$E^0 = 0.22$ volts
Total cell reaction	$Zn^0 + 2AgCl \rightarrow 2Ag^0 + Zn^{+2} + 2Cl^-$	$E^0 = 0.98$ volts

The potential obtained is positive; this implies that the reaction as written is spontaneous. Now we use the Nernst equation.

$$E = E^0 - \frac{0.0591}{N} \log \frac{(Ag)^2 (Zn^{+2})(Cl^-)^2}{(AgCl)^2 (Zn)}$$

Remembering that solids have an activity of 1, we fill in the numerical value.

$$E = 0.98 - \frac{0.0591}{N} \log \frac{1^2 (Zn^{+2})(Cl^-)^2}{(1)^2 (1)}$$

The number of moles of electrons exchanged in the cell reaction is 2; the activity of Zn^{+2} is 0.1, while the activity of Cl^- is 0.2. The activity of the chloride is two times that of the zinc because $ZnCl_2 \rightarrow Zn^{+2} + 2Cl^-$. Performing the calculation

$$E = 0.98 - \frac{0.0591}{2} \log (0.1)(0.2)^2$$

$$= 0.98 - 0.0295 \log (4 \times 10^{-3})$$

$$= 0.98 - (0.0295)(-2.398)$$

$$= 0.98 + 7.06 \times 10^{-2}$$

$$= 1.05 \text{ volts}$$

Carefully note that the calculation of the activity of the ions in a solution has nothing to do with the exponent to which the concentration is raised. The exponent is given by definition (see Equation 13–1). The concentrations are figured from the experimental conditions.

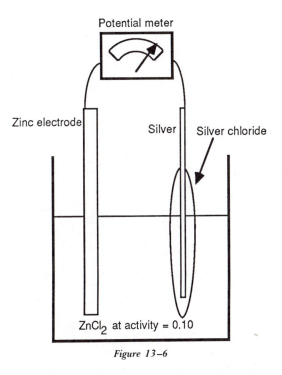

Figure 13–6

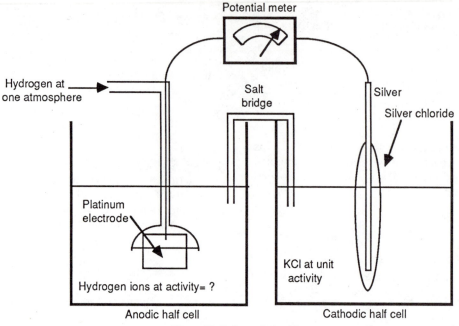

Figure 13–7. An archaic pH meter.

ANALYTICAL POTENTIOMETRY

Finally, we have the background to consider the analytical importance of potential measurements. Let us consider an early type of pH meter. The anode is a hydrogen electrode, while the cathode is a silver–silver chloride electrode (Fig. 13–7). The value of E^0 is calculated as usual.

Anodic reaction $\quad H_2 \rightarrow 2H^+ + 2e \qquad E^0 = 0$

Cathodic reaction $\quad 2AgCl + 2e \rightarrow 2Ag^0 + 2Cl^- \quad E^0 = 0.222$ volts

---

Total cell reaction $\quad H_2 + 2AgCl \rightarrow 2H^+ + 2Ag^0 + 2Cl^-$
$$E^0 = 0.222 \text{ volts}$$

The Nernst equation, applied to this cell, is

$$E = E^0 - \frac{0.0591}{N} \log \frac{(H^+)^2 (Ag)^2 (Cl^-)^2}{(H_2)(AgCl)^2}$$

Remembering that solids and gases at 1 atmosphere have an activity of 1, then

$$E = 0.222 - \frac{0.0591}{2} \log (H^-)^2 (Cl^-)^2$$

Further substituting the value for the activity of chloride ion, the Nernst equation reduces to

$$E = 0.222 \text{ volts} - 0.0591 \log (H^+)$$

Thus you can see that E is a function of H^+ ion activity. Any solution of H^+ ions can be investigated, and the ion activity can be found from the E values as measured in the cell shown in Figure 13–7. Also remember that

$$pH = -\log (H^+)$$

If this definition of pH is incorporated into the preceding equation, the potential measured is related to pH by the equation

$$E = 0.222 \text{ volts} + 0.0591 \text{ pH}$$

and

$$pH = \frac{E - 0.222 \text{ volts}}{0.0591} \qquad \text{Equation 13–2}$$

It is in this way that clinical chemists use electrochemical cells to measure the activity or concentration of various ions.

It is important to take note of a few things about the cell illustrated in Figure 13–7. First, the electrode that is poised to hydrogen ion is called the indicating electrode. This is because hydrogen ion is the analyte. The silver–silver chloride electrode is in contact with a solution of chloride ion; the activity of chloride ion is fixed and known. The silver–silver chloride electrode is said to be the reference electrode.

All electrochemical cells that are used for analytical potentiometry must have an indicating electrode and a reference electrode, so the equation for standard potential of a cell is frequently written

$$E_{cell} = E_{indicating\ electrode} + E_{reference\ electrode}$$

THE NERNST EQUATION AS APPLIED TO INDIVIDUAL HALF CELLS

Electroanalytical chemists frequently apply the Nernst equation to individual half cells. Let's explore this process by the use of the cell illustrated in Figure 13–7. The silver–silver chloride electrode in Figure 13–7 is known to function as a cathode in the cell. If the Nernst equation is applied to the silver–silver chloride reference electrode in Figure 13–7, we obtain

$$E_{reference\ electrode} = E^0 - \frac{0.0591}{1} \log \log (Cl^-)$$

$$E_{ref} = 0.222\ volts - \frac{0.0591}{1} \log (Cl^-)$$

As $(Cl^-) = 1$ in this example,

$$E_{ref} = 0.222\ volts - 0.059 \log (1) = 0.222\ volts$$

The same approach can be taken with the indicating electrode. The hydrogen electrode is the indicating electrode in Figure 13–7. The hydrogen electrode is functioning as an anode in this cell. Thus

$$E_{indicating\ electrode} = E^0 - \frac{0.059}{2} \log \frac{(H^+)^2}{H_2}$$

$$E_{ind} = 0\ volts - \frac{0.059}{2} \log \frac{(H^+)^2}{1}$$

$$E_{ind} = 0\ volts - 0.059 \log (H^+)$$

$$E_{ind} = 0 + 0.059\ pH$$

Recalling that

$$E_{cell} = E_{ind} + E_{ref}$$

then

$$E_{cell} = (0\ volts + 0.059\ pH) + 0.222\ volts$$

$$E_{cell} = 0.0591\ pH + 0.222\ volts$$

$$pH = \frac{E_{cell} - 0.222\ volts}{0.0591}$$

Notice that this equation is the same as Equation 13–2. This is just a second way of looking at the same cell potential. Consider another example (Fig. 13–8). The aluminum electrode is the anode in this cell. The anodic half cell reaction is

$$Al \rightarrow Al^{+3} + 3e \quad E^0 = 1.66\ volts$$

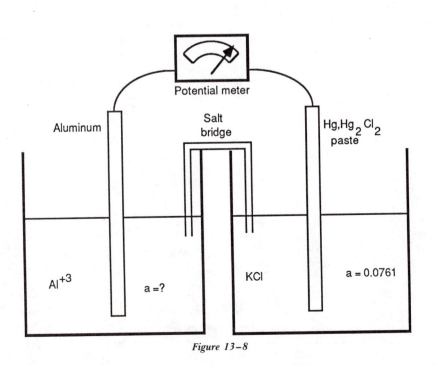

Figure 13–8

Thus the Nernst equation would be

$$E_{ind} = 1.66 \text{ volts} - \frac{0.059}{3} \log (Al^{+3})$$

$$E_{ind} = 1.66 \text{ volts} + \frac{0.059}{3} pAl$$

The mercury–mercury(I) chloride (Hg, Hg_2Cl_2) electrode is the cathode. The cathodic half reaction is

$$Hg_2Cl_2 + 2e \rightarrow 2Hg + 2Cl^-$$

$$E^0 = 0.268 \text{ volts}$$

$$E_{ref} = 0.268 \text{ volts} - \frac{0.059}{2} \log (Cl^-)^2$$

$$E_{ref} = 0.268 \text{ volts} - 0.59 \log (Cl^-)$$

$$E_{ref} = 0.268 \text{ volts} - 0.059 \log (0.0761)$$

$$= 0.334 \text{ volts}$$

Thus the overall cell potential would be

$$E_{cell} = E_{ind} + E_{ref}$$

$$E_{cell} = 1.66 \text{ volts} + \frac{0.059}{3} pAl + 0.334 \text{ volts}$$

$$E_{cell} = 1.994 \text{ volts} + \frac{0.059}{3} pAl$$

$$pAl = \frac{(E_{cell} - 1.99 \text{ volts})\, 3}{0.059}$$

In both examples, measured cell potentials could be related to the analyte activity or p value.

The reference half cell potential that was calculated for the Hg, Hg_2Cl_2 electrode is sometimes called a formal potential. These values are sometimes found in tables and are given the symbol $E^{0'}$. A mercury(I) chloride electrode, when dipped into saturated KCl at 25°C, has a formal potential of 0.242 volts for its cathodic reaction. The elec-

trode is called a saturated calomel electrode (SCE); calomel is the archaic name of Hg_2Cl_2.

THE POTENTIAL METER

Thus far we have not discussed the equipment that is used for measuring the potential values. *All equipment used for these measurements is set up such that the potential can be measured at essentially zero current flow.* The zero current flow stipulation is very important, because if potentials were measured while there is significant current flow, the cell reaction would change the composition of the solution next to the electrodes. For example, if in Figure 13–6 a significant current were flowing during the measurement of the potential, the concentration of the zinc ion next to the zinc electrode would be increased by the cell reaction, and the chloride ion next to the silver–silver chloride electrode would be depleted. This would cause the measured potential to be in error. For this reason, all devices measuring potential values must do so at zero or very nearly zero current flow. These instruments are complex electrical devices called vacuum tube voltmeters or field effect transistor (FET) voltmeters. The property that makes possible the measurements at nearly zero current flow is the extremely high internal resistance of these voltmeters.

THE COMPACT REFERENCE ELECTRODE

Modern reference electrodes are much more compact than the one illustrated in Figure 13–7. The entire reference half cell and salt bridge have been designed into a small self-contained half cell. The half cell and salt bridge unit in Figure 13–9b is equivalent to the larger assembly in Figure 13–9a. The potassium chloride solution in the commercial electrode of Figure 13–9b is generally saturated instead of being at unit activity. The asbestos

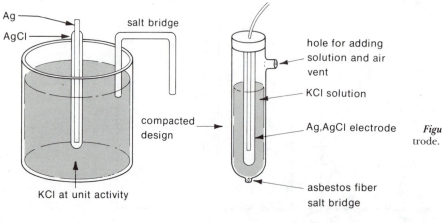

Figure 13–9. A, Classic reference electrode. *B*, Compact reference electrode.

Ag
AgCl
salt bridge
compacted design
KCl at unit activity

hole for adding solution and air vent
KCl solution
Ag,AgCl electrode
asbestos fiber salt bridge

a   b

fiber salt bridge serves the same function as the previous agar salt bridges. The value of the junction potential of the asbestos fiber salt bridge may be a little higher than that of the agar salt bridge. In that regard, the reference potential would be

$$E_{ref} = E^{0'} + E_j \qquad \text{Equation } 13\text{-}3$$

where $E^{0'}$ is the formal potential and E_j is the junction potential.

When the reference electrode is dipped into a solution, it is always adjusted so that the KCl solution has a pressure head of about 2 cm. This ensures that the KCl solution will be flowing out through the asbestos fiber at a rate of about 0.1 ml per day. This prevents the external solution from contaminating the internal solution.

The silver–silver chloride electrode and the calomel electrode are the two most common reference electrodes. The saturated calomel electrode is probably the most common; its formal potential is 0.242 volts based on a cathodic reaction.

MODERN pH ELECTRODES; THE GLASS ELECTRODE

Modern pH electrodes are somewhat different from the pH electrodes illustrated in Figure 13–7. Not only is the reference electrode more compact, but the hydrogen electrode has been replaced with a glass electrode (Fig. 13–10). The reference elec-

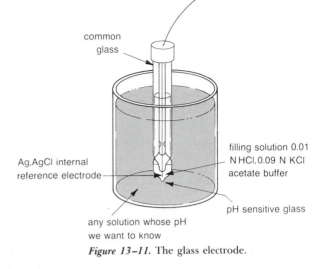

Figure 13–11. The glass electrode.

Figure 13–11 illustrates the glass electrode in greater detail. The glass electrode was developed on a commercial basis in the 1950s and has gained wide acceptance. The electrode is hydrogen ion–sensitive, and its response is predicted by the Nernst equation. The structure of the tip of the electrode is described in Figure 13–12.

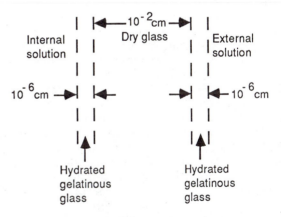

Thickness values are approximate

Figure 13–12. Structure of the tip of the glass electrode.

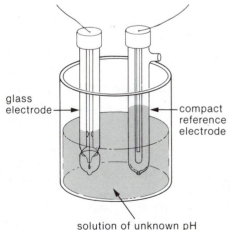

Figure 13–10. Modern pH electrodes.

trode serves as a reference potential source that is in contact with the unknown solution through the salt bridge. The potential of the glass electrode varies with pH; it is measured relative to the constant potential reference electrode. Any changes in potential are due to changes in the pH and as such, are measured by the pH meter.

The hydrated glass layer is completely necessary before the electrode can function. A new dry electrode must be soaked in a buffer solution for several hours before it will function properly. The hydrated layer is continuously dissolving from the surface, while the dry glass is continuously hydrating. Hence the hydrated layer is a dynamic layer that is continuously purified by dissolution. The dissolution process is slow; however, the 10^{-2}-cm dry glass layer can last for several years if treated well.

The hydrated gelatinous glass layers are known

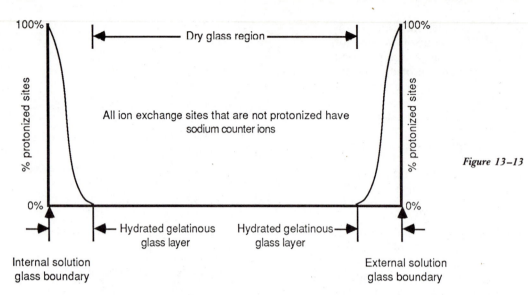

Figure 13–13

to undergo an ion exchange process of the following type:

$$Na^+(glass)^- + H^+(sol) \rightleftharpoons H^+(glass)^- + Na^+(sol)$$

where (glass)⁻ represents ion exchange sites in the hydrated glass.

If one plots the per cent of the ion exchange sites that are protonized versus position in the glass, one finds the situation illustrated in Figure 13–13. This plot is valid at pH values lower than approximately 9. It is thought that when the electrode is placed in an acid solution, protons from solution enter the external hydrated layer. In this way the electrode becomes positively charged with respect to the solution. This separation of charge produces the boundary potential that varies with pH. A similar boundary potential exists on the inside of the electrode. The potential that the internal glass layer develops is constant because the pH is constant. When one considers all the potentials associated with a glass electrode, one finds that

$$E_{ind} = E_{IR} + E_{IL} + E_A + E_{EL}$$

where

E_{IR} is potential of the internal reference electrode.

E_{IL} is potential of the internal hydrous gel layer.

E_A is the asymmetry potential. E_A is thought to be related to the difference in stress in the glass membrane.

E_{EL} is potential of the external hydrous gel layer.

In a given experimental measurement, the potential of the internal reference electrode, the potential of the internal hydrous layer, and the asymmetry potential are all constant. Thus

$$E_{ind} = E'_{constant} + E_{EL}$$
$$E_{ind} = E'_{constant} - 0.0591 \log H_3O^+ \qquad \text{Equation 13–4}$$
$$E_{ind} = E'_{constant} + 0.0591 \text{ pH}$$

Summarizing these ideas and referring to Figure 13–10, one can see that the voltage as read by the pH meter will be the sum of the voltages for the reference electrode and the indicating electrode.

$$E_{cell} = E_{ind} + E_{ref} \qquad \text{Equation 13–5}$$

Substituting Equations 13–3 and 13–4 into Equation 13–5, we obtain

$$E_{cell} = E'_{constant} - 0.0591 \log (H_3O^+) + E^{0'} + E_j$$

and combining the constants, we obtain

$$E_{cell} = E_{constant} - 0.0591 \log (H_3O^+)$$
$$E_{cell} - E_{constant} = 0.0591 \text{ pH}$$
$$\text{pH} = \frac{E_{cell} - E_{constant}}{.0591}$$

The values for the asymmetry potential and the reference electrode junction potential are not known but, within a reasonable time frame, they are constant. There is not a convenient way to measure them. It is common instead to calibrate the instrument with a buffer solution of known pH. If the electrodes of the pH meter are dipped into a solution of known pH, the meter can be adjusted so that it displays the proper value of pH. This works because E_{EL} is the only potential in the system that is not a constant but instead varies with pH. Thus, the glass electrode can be accurately calibrated to measure pH.

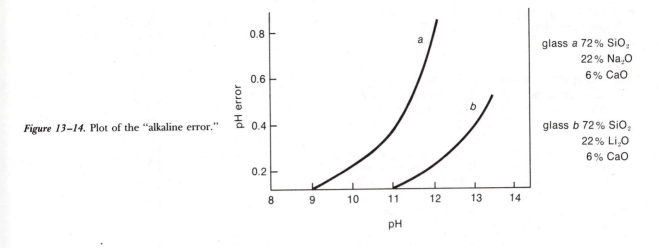

Figure 13-14. Plot of the "alkaline error."

glass *a* 72% SiO_2
22% Na_2O
6% CaO

glass *b* 72% SiO_2
22% Li_2O
6% CaO

When using a pH meter, remember that the results are temperature dependent and that the electrodes have temperature limitations. Check these limitations in the literature that is supplied with the electrodes.

The common glass electrode can provide accurate pH measurements from about pH 0.5 to about pH 10. That is, the increase in E is very nearly 0.0591 volts per pH unit; this change in potential for a tenfold change in hydrogen ion activity is said to be Nernstian response.

Above pH 9, a sodium glass membrane electrode starts producing erroneous readings. This is particularly true when the solution is being made basic with sodium hydroxide. A plot of this so-called alkaline error is illustrated in Figure 13-14.

The alkaline error in the context of the ion exchange equilibrium involved at the external hydrated layer is not unexpected. Remember the equilibrium

$$Na^+(glass)^- + H^+(sol) \rightleftharpoons H^+(glass)^- + Na^+(sol)$$

This equilibrium is the reaction that is responsible for making the external part of the external hydrated glass layer 100 per cent protonized. This seems to be necessary for the proper operation of the electrode. Sodium ion from the sodium hydroxide tends to repress this equilibrium and to leave a fair amount of Na^+ (glass)$^-$ on the external surface of the electrode. (Fig. 13-15). The electrode then becomes poised to sodium ion to some degree.

It is interesting to note that positive ions larger than sodium cause much less alkaline error. Thus it appears that the sizes of the spaces around the ion exchange sites in the hydrous layer cause a degree of selectivity. Cations smaller than those coming from the glass can cause an error of this type. Such information led to the development of lithium glasses. In these, lithium is the counter ion. The holes in the hydrated gel layer from which the lithium ions are exchanging are too small for sodium to enter effectively; thus the sodium error is much reduced. The lithium glass obtains a proton-

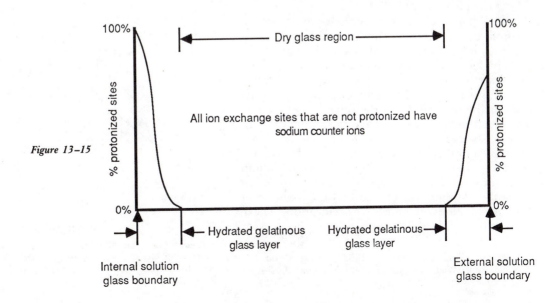

Figure 13-15

ized external hydrated layer in the following ion exchange reaction:

$$Li^+(glass^{-1}) + H^+(sol) \rightleftharpoons H^+(glass^{-1}) + Li^+(sol)$$

The fact that a glass electrode under high sodium activities can respond to sodium ion has led to research that brought about the development of many other "ion-specific" electrodes, which will be discussed in this chapter.

From the practical point of view, if one must measure pH at a high pH value, one must do something to manage the alkaline error. The use of lithium glass electrodes in conjunction with sodium or potassium hydroxide will minimize the error. Calibration of the pH meter with a buffer that is near the pH of the samples is very desirable. If these steps are taken, pH measurement up to about pH 13 can be made without great error.

ION-SELECTIVE ELECTRODES

Once the glass membrane electrode was discovered to be somewhat poised to sodium ion, scientists started on research programs to find ways of enhancing this property. It was found that by replacing some of the sodium oxide (Na_2O) and calcium oxide (CaO) with aluminum oxide (Al_2O_3), the glass membrane becomes fairly selective toward sodium ion. According to Willard, Merritt, Dean, and Settle,* the composition of sodium-selective electrode glass commonly is 11 per cent Na_2O, 18 per cent Al_2O_3, and 71 per cent silicon dioxide (SiO_2). The electrode looks just like a pH glass electrode, but it is highly poised to silver ion, hydrogen ion, and sodium ion. In its applications, the sodium ion activity is measured by keeping the pH and pAg constant. Once the pH and pAg are fixed, the potential of the electrode is controlled by the sodium activity. The electrode can also be used to measure pAg if the pNa and pH are held constant. Because it is sufficiently selective for sodium relative to potassium, the electrode is fairly useful.

A sodium electrode is illustrated in Figure 13–16. This electrode has a calomel reference electrode built into the top of the sodium electrode. A ceramic frit is fused into the side of the electrode and functions as the salt bridge for the reference electrode. The electrode response to varying sodium concentration is shown in Figure 13–17. The sodium type glass membrane electrode is used in the clinical laboratory for the measurement of serum sodium concentration. The electrode is calibrated by measuring the output of the electrodes for two different sodium solutions of known concentration. These values are used by the instrument's microprocessor

*H. Willard, et al., Instrumental Methods of Analysis, 6th Ed., Belmont, Calif., Wadsworth Inc., Litton Educational Publishing, Inc., 1981.

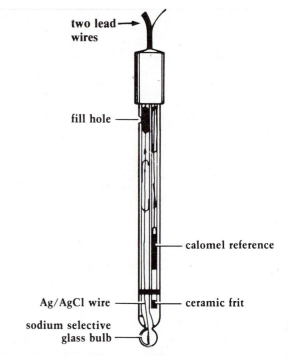

Figure 13–16. Construction of a sodium electrode. (Reprinted courtesy of Orion Research Incorporated, Cambridge, Massachusetts, U.S.A. "ORION" is a registered trademark of Orion Research Incorporated.)

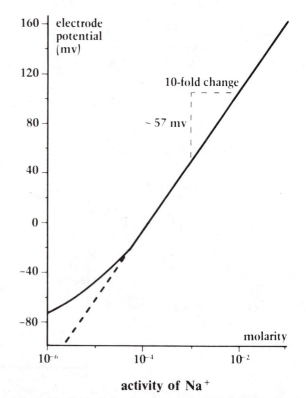

Figure 13–17. Typical sodium electrode calibration. (Reprinted courtesy of Orion Research Incorporated, Cambridge, Massachusetts, U.S.A. "ORION" is a registered trademark of Orion Research Incorporated.)

to establish the curve shown in Figure 13–17. The slope of the curve and the intercept are then used to calculate the sodium concentrations for the serum samples.

It is thought by electrochemists that the sodium ion–specific electrode develops its potential values in the same way that the pH glass electrode does. The same kind of explanation could be given here as was given for the pH glass electrode. The student should use the discussion for the pH glass electrode as a model and apply the same ideas to the sodium electrode situation.

It became obvious early in the research into ion-specific electrodes that glass would not be a satisfactory membrane for all electrodes. The research suggested that ion-exchange membranes should be capable of doing the same kinds of things that glass had done for pH and pNa measurements. This kind of thinking led to the development of many successful electrodes of the liquid ion exchange–membrane type. The theory of operation of pH glass electrodes also applies to these electrodes.

Two of these liquid–ion exchange electrodes are important in clinical chemistry. The calcium electrode is used to measure ionized calcium in blood serum. The indicating electrode for calcium is illustrated in Figure 13–18. The calcium ion exchanger is one of several organic phosphorus compounds. The electrode shows sufficient selectivity to make practical the measurement of calcium ion in the blood serum matrix. The response of the

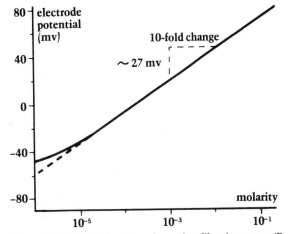

Figure 13–19. Typical calcium electrode calibration curve. (Reprinted courtesy of Orion Research Incorporated, Cambridge, Massachusetts, U.S.A. "ORION" is a registered trademark of Orion Research Incorporated.)

electrode is shown in Figure 13–19. The instrument is calibrated by standardizing the system with at least two standard solutions of known pCa. The microprocessor then uses the slope and intercept to calculate the calcium ion concentration in the blood serum.

A second ion exchange–membrane electrode that is important in clinical chemistry is the potassium electrode. The potassium electrode is specific for potassium ion and is capable of measuring serum potassium activities. The indicating electrode is shown in Figure 13–20. The valinomycin is an organic ion exchanger that is very selective for po-

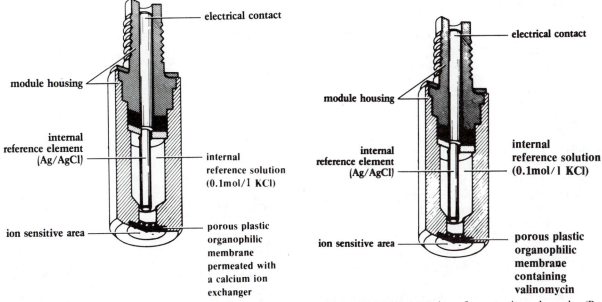

Figure 13–18. Construction of a calcium electrode. (Reprinted Courtesy of Orion Research Incorporated, Cambridge, Massachusetts, U.S.A. "ORION" is a registered trademark of Orion Research Incorporated.)

Figure 13–20. Construction of a potassium electrode. (Reprinted courtesy of Orion Research Incorporated, Cambridge, Massachusetts, U.S.A. "ORION" is a registered trademark of Orion Research Incorporated.)

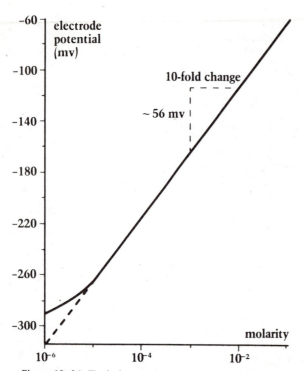

Figure 13–21. Typical potassium electrode calibration curve. (Reprinted courtesy of Orion Research Incorporated, Cambridge, Massachusetts, U.S.A. "Orion" is a registered trademark of Orion Research Incorporated.)

tassium ion. The potential formation model for this electrode is about the same as that for the pH–glass membrane electrode.

The electrode is frequently used in conjunction with the saturated calomel electrode. It is commonly standardized with two solutions of known pK. The microprocessor in the instrument uses these data to establish the slope and intercept of the curve in Figure 13–21. The serum potassium levels are then calculated from the potential that they produce at the calibrated electrodes.

In addition to glass electrodes and the liquid ion exchange–membrane electrodes, there are a number of successful solid-state electrodes. The chloride electrode is an example of such a device. The chloride electrode (Fig. 13–22) is used in conjunction with a reference electrode to measure sweat-chloride activities.

The electrode consists of a membrane of silver chloride (AgCl) that is mounted in the electrode body. The AgCl membrane interacts with the chloride content of the solution under investigation to produce a reproducible boundary potential. The electrical contact is made directly with the backside of the AgCl membrane. The electrode is calibrated as previously discussed. A typical response curve is shown in Figure 13–23.

In addition to the electrodes studied so far, there are several important gas-sensitive electrodes, such as the carbon dioxide electrode and the ammonia electrode. The carbon dioxide electrode (Fig.

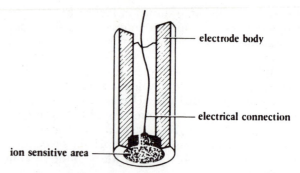

Figure 13–22. Construction of a chloride electrode. (Reprinted courtesy of Orion Research Incorporated, Cambridge, Massachusetts, U.S.A. "ORION" is a registered trademark of Orion Research Incorporated.)

13–24) is very important in clinical chemistry; it is used to measure the carbon dioxide level in blood. The indicating electrode consists of a glass electrode that is covered by a thin film of 5 mmol/l sodium bicarbonate ($NaHCO_3$) solution, which in turn is covered by a silicone rubber membrane. The reference electrode is a silver–silver chloride electrode. The measurements are such that temperature must be controlled. This is because the voltage change caused by the analyte is not great, and the effect of temperature is significant. This is in keeping with the fact that temperature is a variable in the Nernst equation.

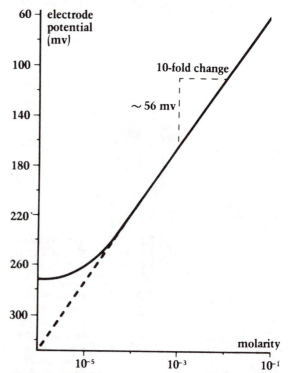

Figure 13–23. Typical chloride electrode calibration curve. (Reprinted courtesy of Orion Research Incorporated, Cambridge, Massachusetts, U.S.A. "ORION" is a registered trademark of Orion Research Incorporated.)

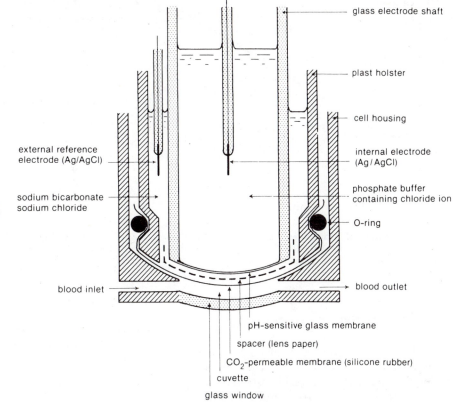

Figure 13–24. Schematic illustration of a $p\mathrm{Co_2}$ cell. The cell is temperature controlled. (Adapted from: N.W. Tietz, Textbook of Clinical Chemistry, Philadelphia, W.B. Saunders Company, 1986.)

The electrode functions because the CO_2 from the blood sample can permate the silicone rubber membrane and interact with the dilute $NaHCO_3$ solution.

$$CO_2 + 2H_2O \rightleftharpoons HCO_3^- + H_3O^+$$

The protons that are released by the CO_2 then interact with the glass electrode. In practice the system is calibrated with solutions of known CO_2 partial pressure. The calibration curve is shown in Figure 13–25.

The instrument's microprocessor is given the calibration data so that it knows the slope and intercept of the curve. In this way an apparent pH at the glass electrode can be used to calculate the partial pressure of CO_2 in the blood.

The ammonia electrode is used to measure blood urea concentrations. The electrode is a glass pH electrode that is covered with a suitable membrane permeable to NH_3. The enzyme urease is immobilized on the NH_3 permeable membrane. It catalyzes the reaction

$$\underset{\substack{\| \\ H_2N-C-NH_2}}{\overset{O}{}} + H_2O \xrightarrow{\text{Urease}} 2NH_3 + CO_2$$

The NH_3 passes through the membrane and then affects the moist pH electrode via the reaction

$$NH_3 + H_2O \rightleftharpoons NH_4^+ + OH^-$$

A typical response curve is shown in Figure 13–26.

All of these ion-specific electrodes have some interferences and various quirks. When preparing to use one of these electrodes, read the manufacturer's literature. It provides information concerning these problems.

There are many other ion-specific electrodes. I have discussed only the electrodes that are common in clinical chemistry.

The electrodes all measure ion activity. If concentrations are needed, suitable activity coefficients can be used to calculate molarities from activities. Information on activity coefficients can be found in Appendix B.

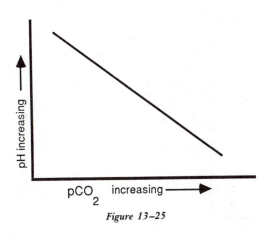

Figure 13–25

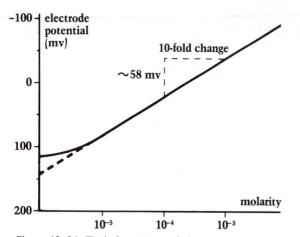

Figure 13–26. Typical response of the ammonia electrode. (Reprinted courtesy of Orion Research Incorporated, Cambridge, Massachusetts, U.S.A. "ORION" is a registered trademark of Orion Research Incorporated.)

CONCENTRATION CELLS

All of the potentiometric cells presented in this chapter have been built from two different electrodes. For example, the cell illustrated in Figure 13–5 has a copper electrode and a hydrogen gas electrode. In direct contrast to the cells previously discussed, there are potentiometric cells containing two electrodes that are exactly the same. Such a cell is illustrated in Figure 13–27; cells of this type are called concentration cells.

Even though the two electrodes are the same, the concentrations of chloride ion in the two half cells are different. This difference is sufficient to generate a cell potential that is greater than zero.

Consider the cell reactions that would take place if the cell were used as a battery. The anodic reaction would be

$$Ag + Cl^-_{anode} \rightarrow AgCl + e \quad E^0 = -0.222 \text{ volts}$$

The cathodic half cell reaction would be

$$AgCl + e \rightarrow Ag + Cl^-_{cathode} \quad E^0 = 0.222 \text{ volts}$$

The overall cell reaction would be

$$Cl^-_{anode} \rightarrow Cl^-_{cathode} \quad E^0_{cell} = 0 \text{ volts}$$

The Nernst equation could be used to calculate the potential of the cell in Figure 13–27.

$$E = 0 \text{ volts} - \frac{0.0591}{1} \log \frac{(Cl^-)_{cathode}}{(Cl^-)_{anode}}$$

If activity values from Figure 13–27 are substituted into the preceding equation, one finds

$$E = 0 \text{ volts} - 0.0591 \log \frac{(10^{-2})}{(10^{-1})} = 0.059 \text{ volts}$$

Concentration cells can be used for analytical potentiometry. If the cell in Figure 13–27 were to be used for chloride measurements, the half cell containing the highest concentration of chloride would be the anodic half cell. The half cell containing a fixed known chloride activity would be the reference electrode. If we were to assume that the solution of unknown activity was more concentrated

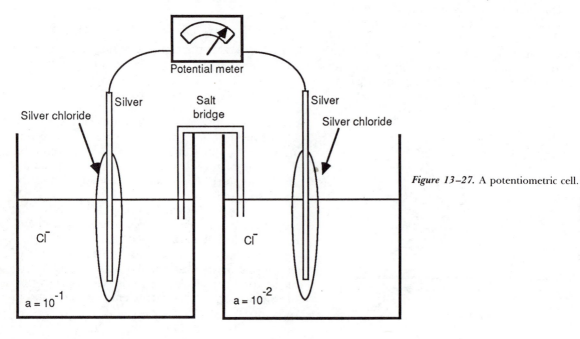

Figure 13–27. A potentiometric cell.

than the reference electrode–chloride solution, then the Nernst equation for the cell would be

$$E = -0.0591 \log \frac{(Cl^-)_{\text{reference}}}{(Cl^-)_{\text{indicating}}}$$

If the reference electrode chloride activity were substituted into the previous equation, experimentally measured E values could be used to calculate the activity of chloride in the indicating electrode.

Concentration cells do have some practical applications in some clinical analyzers.

PHYSICAL CONFIGURATIONS OF POTENTIOMETRIC ELECTRODES

In this chapter a number of electrodes have been illustrated. These electrodes have been primarily of the type that one would see in an academic teaching laboratory. This has been done in part to allow the student to work from the familiar toward the unfamiliar. The theory of operation of the potentiometric electrodes is the same no matter what their size and shape; let's consider some examples.

A good teaching example is the potassium ion–selective electrode. The common "academic laboratory type" potassium electrode shown in Figure 13–20 is nearly normal size and would be used with a reference electrode much like the one illustrated in Figure 13–9b. For the sake of comparison, con-

sider the potentiometric cell from a Beckman ASTRA (Automated Stat/Routine Analyzer Systems) clinical analyzer. Before considering the potassium electrode in the system, study the system in an overview (Fig. 13–28).

The ASTRA analyzer uses a sodium electrode as a reference electrode. The cell uses a liquid junction between the reference electrode and the indicating half cells. Between samples, new reference solution of known, fixed sodium concentration is pumped into the liquid junction. The indicating electrode for sodium measurements is a second electrode. Thus the sodium cell is a concentration cell. The cell is filled with 1.3 ml of buffer solution, and 50 μl of sample is added and mixed. The potential is then measured.

The ASTRA uses the sodium reference electrode for potassium measurements as well. The potassium electrode, which is a valinomycin-type electrode, is shown both assembled and disassembled in Figure 13–29. The electrode tip contains the porous plastic organophilic (hydrophobic) membrane that is impregnated with valinomycin. This tip can be removed and replaced. This is done every 5000 samples or every two months, whichever comes first. Notice that the electrode in Figure 13–29 is enlarged to about three times its real size.

For a further comparison, consider the valinomycin-potassium electrode from an EKTACHEM clinical analyzer. The EKTACHEM "slide" (Fig. 13–30) actually consists of two silver–silver chlo-

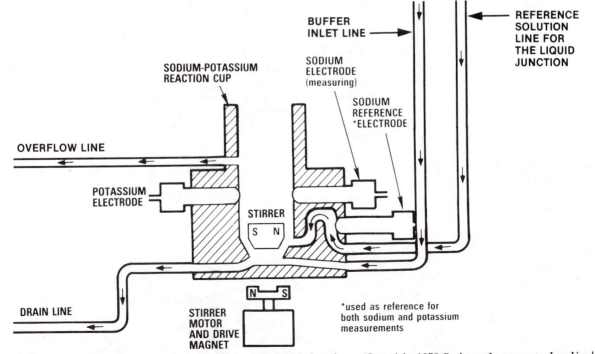

Figure 13–28. The sodium-potassium cell for an ASTRA clinical analyzer. (Copyright 1979 Beckman Instruments, Inc. Used by permission.)

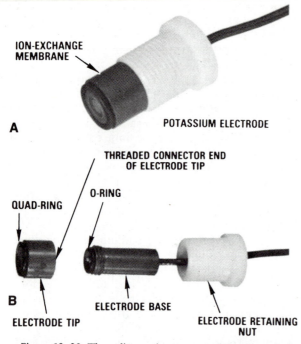

A

ION-EXCHANGE
MEMBRANE

POTASSIUM ELECTRODE

THREADED CONNECTOR END
OF ELECTRODE TIP

QUAD-RING   O-RING

B

ELECTRODE TIP

ELECTRODE BASE

ELECTRODE RETAINING
NUT

Figure 13–29. The valinomycin-type potassium electrode for the ASTRA analyzer. *A,* Assembled. *B,* Disassembled. (Copyright 1979 Beckman Instruments, Inc. Used by permission.)

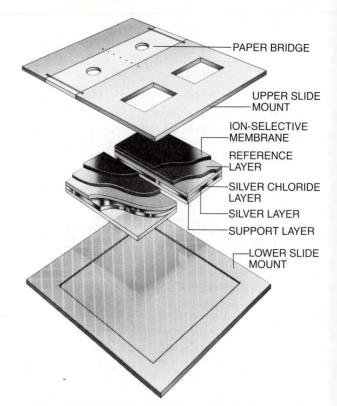

PAPER BRIDGE

UPPER SLIDE
MOUNT

ION-SELECTIVE
MEMBRANE

REFERENCE
LAYER

SILVER CHLORIDE
LAYER

SILVER LAYER

SUPPORT LAYER

LOWER SLIDE
MOUNT

Figure 13–30. Kodak's clinical chemistry slide for potassium. (Reprinted courtesy of Eastman Kodak Company. Copyright © by Eastman Kodak Company.)

ride reference electrodes—one valinomycin reference electrode and one valinomycin indicating electrode. The slide is about 2.4 cm by 2.8 cm and perhaps 2 mm thick. Before we consider the entire cell, let's look at a side view of the device (Fig. 13–31).

The plastic support is provided to give the slide some mechanical rigidity. The silver metal layer is extremely thin and is an electrical contact point to the potential meter. It is also an integral part of a silver–silver chloride reference electrode. The silver chloride layer is on top of the silver layer and is very thin. On top of the silver chloride layer is a gelatin layer that contains a carefully controlled concentration of potassium chloride. The three layers, silver, silver chloride, and gelatin, constitute a reference electrode. An organophilic film that is impregnated with valinomycin is on top of the gelatin layer. The potassium ion in the gelatin interacts with the valinomycin layer to form a boundary potential.

This is analogous to the potential formed at the internal hydrated glass layer in Figure 13–12. Resting on top of the valinomycin film is a filter paper with a "well" in its center. The sample is applied in this well. The potassium in the sample interacts with the valinomycin in the film and causes a boundary potential to form.

The situation is analogous to the potential that forms at the external hydrated glass layer in Figure 13–12. The filter paper serves as a salt bridge. This, along with the other half of the slide, can be seen in Figure 13–32.

In the actual use of this potentiometric cell, a reference solution of known potassium concentration is placed in the reference solution well. The

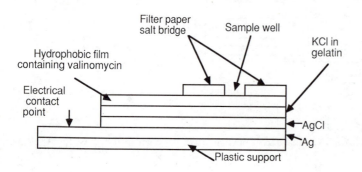

Filter paper
salt bridge

Sample well

KCl in
gelatin

Hydrophobic film
containing valinomycin

Electrical
contact
point

AgCl

Ag

Plastic support

Figure 13–31. Side view of Kodak's potentiometric slide.

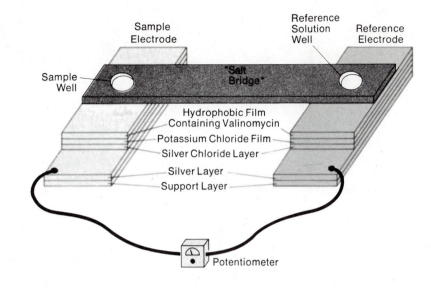

Figure 13–32. Diagram of Kodak's slide for potassium analysis. (Reprinted with permission from: B. Walter, Anal. Chem., *55*[4]:508A, 1983. Copyright 1983 by American Chemical Society.)

sample is applied in the sample well. The two solutions not only interact with the valinomycin films but also moisten the filter paper salt bridge and provide electrical contact between the two electrical arrays. Of the four electrodes in the system, the only one that is not experiencing a constant concentration is the top of the valinomycin film in the indicating valinomycin electrode. The electrochemical slides are calibrated by using a reference solution of known potassium concentration. This calibration is much like the adjustment of a pH meter. Once calibrated, the potentiometer provides potassium concentrations by direct readout or printout.

Thus you have seen three valinomycin electrodes of different sizes and shapes. The theory of operation and the potentials they produce are the same within the context of solution concentrations that they experience.

Problems   It is customary for electrochemists to write cells in a shorthand. In their system, salt bridges are given the symbol ‖. A phase boundary is given the symbol │ . The anode is always shown on the left side of the paper. thus the cell

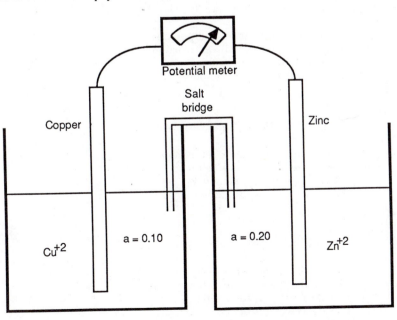

$$Zn \rightarrow Zn^{+2} + 2e \qquad E^0 = 0.763 \text{ volts}$$
$$Cu^{+2} + 2e \rightarrow Cu \qquad E^0 = 0.337 \text{ volts}$$

$$Zn + Cu^{+2} \rightarrow Cu + Zn^{+2} \quad E^0 = 1.1 \text{ volts}$$

would be written

$$Zn\,|\,Zn^{+2}(a = 0.20)\,\|\,Cu^{+2}(a = 0.10)|\,Cu$$

This way of writing cells serves their purposes well. In the following problems, however, the system has been modified slightly. The anode may or may not be on the left side of the salt bridge. Otherwise the symbolization is the same.

The data listed here are used in the problems that follow.

$$AgCl + e \rightarrow Ag + Cl^- \qquad E^0 = 0.222 \text{ volts}$$
$$Ag^+ + e \rightarrow Ag \qquad E^0 = 0.800 \text{ volts}$$
$$Hg_2Cl_2 + 2e \rightarrow 2Hg + 2Cl^- \quad E^0 = 0.268 \text{ volts}$$
$$Cu^{+2} + 2e \rightarrow Cu \qquad E^0 = 0.337 \text{ volts}$$
$$2H^+ + 2e \rightarrow H_2 \qquad E^0 = 0 \text{ volts}$$

Assumptions: There are no junction potentials.
All cell reactions are reversible.
There are no electrolytic cells in this problem set.

1. Consider the cell

$$Ag; AgCl\,|\,Cl^-(a{=}1)\,\|\,Cu^{+2}(a{=}1)\,|\,Cu$$

(a) What is the spontaneous cell potential?
(b) What is the spontaneous cell reaction?
(c) Which electrode is the anode?
(d) Which electrode is the cathode?

2. Consider the cell

$$Hg; Hg_2Cl_2\,|\,Cl^-(a{=}1)\,\|\,Ag^+(a{=}1)\,|\,Ag$$

(a) What is the spontaneous cell potential?
(b) What is the spontaneous cell reaction?
(c) Which electrode is the anode?
(d) Which electrode is the cathode?

3. Consider the cell

$$Ag; AgCl\,|\,Cl^-(a{=}1)\,\|\,Ag^+(a{=}1)\,|\,Ag$$

(a) What is the spontaneous cell potential?
(b) What is the spontaneous cell reaction?
(c) Which electrode is the anode?
(d) Which electrode is the cathode?

4. Consider the cell

$$Ag; AgCl\,|\,Cl^-(a{=}1)\,\|\,H_3O^+(a{=}1)\,|\,H_2(1 \text{ atm.})\,|\,Pt$$

(a) What is the spontaneous cell potential?
(b) What is the spontaneous cell reaction?
(c) Which electrode is the anode?
(d) Which electrode is the cathode?

5. Consider the cell

   $$Pt \mid H_2(1 \text{ atm.}) \mid H_3O^+ (a=1) \parallel Cl^- (a=1) \mid Hg_2Cl_2; Hg$$

   (a) What is the spontaneous cell potential?
   (b) What is the spontaneous cell reaction?
   (c) Which electrode is the anode?
   (d) Which electrode is the cathode?

6. Consider the cell

   $$Ag; AgCl \mid Cl^- (a=1) \parallel Cu^{+2} \; a=? \mid Cu$$

   (a) Derive an equation relating an experimentally measured potential value to the pCu of the copper solution.
   (b) Which electrode is the reference electrode?
   (c) Which electrode is the indicating electrode?
   (d) Which electrode is the anode?

7. Consider the cell

   $$Hg; Hg_2Cl_2 \mid Cl^- (a=2) \parallel Ag^+ (a=?) \mid Ag$$

   (a) Derive an equation relating an experimentally measured potential value to the pAg of the silver solution.
   (b) Which electrode is the reference electrode?
   (c) Which electrode is the indicating electrode?
   (d) Which electrode is the anode?

8. Consider the cell

   $$Ag; AgCl \mid Cl^- (a=0.1) \parallel Ag^+ (a=?) \mid Ag$$

   (a) Derive an equation relating an experimentally measured potential value to the pAg of the silver solution.
   (b) Which electrode is the reference electrode?
   (c) Which electrode is the indicating electrode?
   (d) Which electrode is the anode?

9. Consider the cell

   $$Ag; AgCl \mid Cl^- (a=?) \parallel Ag^+ (a=0.15) \mid Ag$$

   (a) Derive an equation relating an experimentally measured potential value to the pCl in the silver solution.
   (b) Which electrode is the reference electrode?
   (c) Which electrode is the indicating electrode?
   (d) Which electrode is the anode?

10. Consider the cell

    $$Pt \mid H_2(1 \text{ atm.}) \mid H^+ (a=?) \parallel Cl^- (a=2) \mid Hg_2Cl_2; Hg$$

    (a) Derive an equation relating an experimentally measured potential value to the pH in the solution.
    (b) Which electrode is the reference electrode?
    (c) Which electrode is the indicating electrode?
    (d) Which electrode is the anode?

11. Consider the cell

    $$Ag; AgCl \mid Cl^- (a=1 \text{ molar}) \parallel a=10^{-3} \text{molar } Cl^- \mid AgCl; Ag$$

    (a) What is the potential of this concentration cell?
    (b) Which half cell is the cathodic half cell?

12. Consider the cell

$$Pt \mid H_2(1 \text{ atm.}) \mid H^+(a=1) \parallel H^+(a=?)H^+ \mid H_2(1 \text{ atm.}) \mid Pt$$

Assume that $a=?$ is less than $a=1.00$.

(a) Derive an equation relating an experimentally measured potential value to the pH of the solution.
(b) Which electrode is the reference electrode?
(c) Which electrode is the indicating electrode?
(d) Which electrode is the anode?

13. (a) What is the potential of the electrode when immersed in 0.035 molar $AgNO_3$?
    (b) What is the potential of the silver electrode when immersed in 0.035 molar NaCl that is saturated with AgCl?

$$Ksp = 1.82 \times 10^{-10}$$

14. What is the potential of a platinum electrode that is immersed in

(a) A solution that is 10^{-2} molar in Ce^{+4} and 10^{-3} molar in Ce^{+3}?

$$Ce^{+4} + e \rightarrow Ce^{+3} \quad E^0 = -0.403 \text{ volts}$$

(b) A solution that is 1.32×10^{-2} molar in Fe^{+2} and 1.99×10^{-2} molar in Fe^{+3}?

$$Fe^{+3} + e \rightarrow Fe^{+2} \quad E^0 = 0.771 \text{ volts}$$

(c) A solution that is 0.20 molar in KI_3 and 0.01 molar in KI?

$$I_3^- + 2e \rightarrow 3I^- \quad E^0 = 0.536 \text{ volts}$$

15. A glass electrode for pH measurements and a saturated calomel electrode are immersed in a pH 5.25 buffer solution. The potential meter indicates that the cell potential is 0.333 volts. The buffer solution is replaced with a solution of unknown pH. The potential meter indicates that the cell potential is 0.426 volts. What is the pH of the solution?

16. A glass electrode for sodium measurements and a saturated calomel electrode are immersed in a 0.100 molar sodium nitrate solution. The potential meter indicates that the cell potential is 0.121 volts. The 0.100 molar sodium nitrate solution is replaced with an unknown. The potential meter indicates that the potential is 0.173 volts. What is the sodium ion concentration of the unknown?

17. A calcium ion–specific electrode and a saturated calomel electrode are immersed in a 0.100 molar calcium nitrate solution. The potential meter indicates that the potential is 65.3 mv. The calcium nitrate standard is replaced by a calcium unknown. The potential meter indicates that the cell potential is 43.2 mv. What is the pCa of the unknown?

$$pCa = \frac{E_{cell} - E_{constant}}{0.0591/2}$$

18. The following data were obtained using a sodium type (silver type) glass electrode and a saturated calomel reference electrode.

Concentration of the Standard (mmol/l)	Measured Potential (mv)
1	66.6
3	89.0
5	100.2
10	114.8
20	133.4

A sodium unknown produced a measured potential of 111.3 mv. What is the concentration of sodium in the unknown?

14

Voltammetry

In Chapter 13 the method of analysis called potentiometry is discussed. The idea that the voltage of an electrode can, in many cases, be used to measure the concentration of an analyte is presented. It is shown also that these voltage measurements must be made under conditions of essentially zero current flow.

In direct contrast to the material presented previously, this chapter concerns the analytical measurements that involve measuring currents that result when a voltage is applied to an electrochemical cell. This method of analysis is called voltammetry. There are many types of voltammetry; some types are important in clinical analysis. Solid state voltammetry is used to measure the partial pressure of oxygen in blood. Amperometry is used to detect the end point in coulometric chloride determinations. Some liquid chromatographs use voltammetry as a basis for detection. There are four or five other techniques that may find their way into the clinical laboratory in the future.

In order to gain an understanding of voltammetry methods it is convenient to consider a hypothetical experiment. The cell in the hypothetical experiment consists of two copper electrodes and a solution of copper sulfate. The cell is connected to a circuit that consists of a continuously variable voltage source and the associated meters that are necessary for the measurement of the applied voltage and the resulting current (Fig. 14–1).

The hypothetical experiment consists of measuring the current that flows while an applied voltage is slowly and continuously increased. The applied voltage and its relationship to time are shown in Figure 14–2.

Figure 14–3 shows the current versus voltage curve that would result if such a voltage ramp were applied to the cell. The curve has three segments. The first segment is from point A to point D. This segment is characterized by the fact that no electrode reactions are taking place in the cell. The gentle increase in current that is observed is the so-called charging current or ohmic current. The value of the charging current is predicted by Ohm's law.

$$V_{applied} = IR_{solution}$$

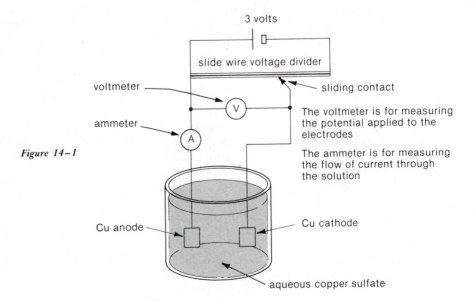

3 volts

slide wire voltage divider

sliding contact

voltmeter

The voltmeter is for measuring the potential applied to the electrodes

ammeter

The ammeter is for measuring the flow of current through the solution

Figure 14–1

Cu anode

Cu cathode

aqueous copper sulfate

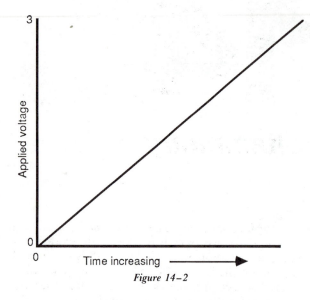

Figure 14–2

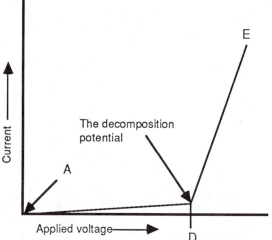

Figure 14–3. A hypothetical current voltage curve.

The processes that are occurring in the cell during this segment are discussed in detail in Chapter 16. The theoretical background, however, is not required for understanding the material in this chapter.

The second segment of the curve is the segment from point D to point E. This segment is characterized by a rapid increase in current. The rapidly increasing current results from the following cell reactions:

$$Cu^{+2} + 2e \rightarrow Cu \quad \text{Cathodic half cell reaction}$$

$$Cu \rightarrow Cu^{+2} + 2e \quad \text{Anodic half cell reaction}$$

The current value is controlled by the rate of the cathodic cell reaction. This rate is dependent on the concentration of copper ions in the solution, the rate of stirring, if any, and the applied voltage. The current continues to increase linearly with increasing voltage until point E on the curve. At point E, the solvent starts electrolyzing at the electrodes. This leads to a further increase in current.

Point D is said to be the decomposition potential of the copper (II) ion in the solution. The decomposition potential is the voltage that is required to initiate the cell reactions. Its value is a function of the concentration of the electroactive species. As predicted by the Nernst equation, its value shifts by 59/N mv for every tenfold change in concentration. The symbol N represents the number of electrons involved in the cell reaction. Thus in our example, the decomposition potential of the copper ion could shift by 29.5 mv for every tenfold concentration change.

Summarizing, electroactive species under the influence of an applied electrical potential may undergo an electrode reaction. The potential at

which the electrode reaction starts is called the decomposition potential. The value of the decomposition potential for a given electroactive species is concentration dependent. At a given applied potential, the value of the current is a function of concentration and the rate of stirring. If the rate of stirring is fixed, the value of the current is a function of concentration of the electroactive species. With this background in mind, consider some of the voltammetry applications that are important in the clinical laboratory.

THE OXYGEN (pO$_2$) ELECTRODE

One of the most important clinical applications of voltammetry is the oxygen electrode of Clark. The electrode system is depicted in Figure 14–4. A better name for the oxygen electrode is oxygen cell. It has a platinum microcathode as its working electrode and a silver–silver chloride macroanode as a reference electrode. Both are bathed in a supporting electrolyte of phosphate buffer that is saturated with silver chloride. The buffer also contains some potassium chloride. The pair of electrodes is charged with 0.65 v. The supporting electrolyte is separated from the blood sample by an oxygen-permeable polypropylene membrane. When the blood sample is placed in contact with the membrane, oxygen permeates the membrane and enters the supporting electrolyte. The amount of oxygen that enters the supporting electrolyte is directly proportional to the oxygen concentration in the blood that is in contact with the membrane. The oxygen that enters the supporting electrolyte is reduced as shown in the following reaction:

$$O_2 + 2H_2O + 4e \rightarrow 4OH^-$$

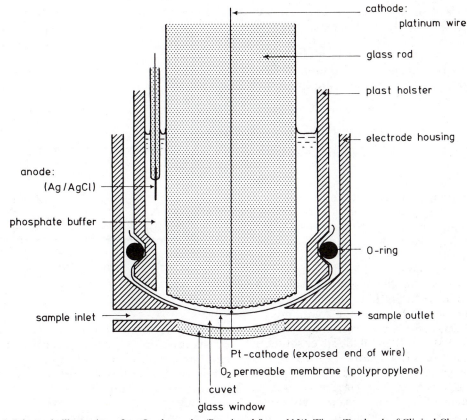

cathode:
    platinum wire

glass rod

plast holster

electrode housing

anode:
(Ag/AgCl)

phosphate buffer

O-ring

sample inlet   sample outlet

Pt-cathode (exposed end of wire)

O_2 permeable membrane (polypropylene)

cuvet

glass window

Figure 14–4. Schematic illustration of a pO_2 electrode. (Reprinted from: N.W. Tietz, Textbook of Clinical Chemistry, Philadelphia, W.B. Saunders Company, 1986.)

The current flow that results from this reaction is directly proportional to the partial pressure of oxygen in the blood. The actual oxygen partial pressure is determined by standardizing the cell with solutions of known oxygen concentration. In effect, the standardization process experimentally evaluates k in the equation

$$I = kC \qquad \text{Equation } 14\text{–}1$$

The value of k is made known to the microprocessor in the instrument. The pO_2 values for blood samples can then be found by measuring the current experimentally and calculating C. The polyethylene membrane is used to keep nonpermeating interfering substances away from the electrodes. The fact that oxygen can permeate the membrane is really a stroke of good fortune for clinical chemists.

AMPEROMETRY

An amperometric method of analysis can best be defined as any titration method that uses voltammetry in the detection of the end point. This definition includes several types of voltammetry. In order to gain an understanding of amperometric

methods, consider the titration of sodium chloride with silver nitrate. The titration reaction is

$$Ag^+ + Cl^- \rightarrow AgCl\downarrow$$

The amperometric end point–detection system is arranged as shown in Figure 14–5. A pair of silver

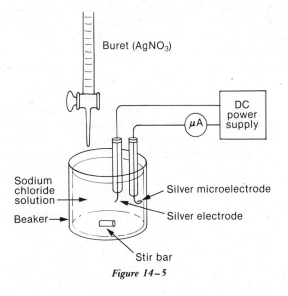

Buret (AgNO$_3$)

DC power supply

μA

Sodium chloride solution

Beaker

Silver microelectrode

Silver electrode

Stir bar

Figure 14–5

electrodes is placed in the solution that is to be titrated. The power supply in this experiment is adjusted to provide 0.25 v DC; this value is high enough to exceed the decomposition potential for silver ion at any reasonable concentration. When silver ion is present in the solution, the following reactions take place:

$$Ag^+ + e \rightarrow Ag \quad \text{Cathodic reaction}$$

$$Ag \rightarrow Ag^+ + e \quad \text{Anodic reaction}$$

The extent to which these reactions take place is mediated by concentration of silver ion in solution and the rate at which the solution is stirred. In a given experiment, the stirring rate is constant, so the only thing that affects the relative value of the current is the silver ion concentration.

Keeping this in mind, consider the titration of 40 ml of 0.10 molar chloride ion solution with 0.10 molar silver nitrate solution. If no titrant is added, the concentration of silver ion in the solution is 0. The current measured on the microammeter in Figure 14–5 is also 0. When 10 ml of the silver solution is added, 1 mmol of silver chloride forms and 3 mmol of chloride remains unreacted. The silver ion concentration is very low. The current measured on the microammeter in Figure 14–5 is also very low. If the volume of silver solution added is increased to 20 ml total volume, 2 mmol of silver chloride forms and 2 mmol of chloride remains unreacted. The concentration of silver ion is very low, and the resulting current is essentially 0. If the volume of silver solution delivered to the titration vessel is increased to 30 ml total volume, 3 mmol of silver chloride forms, and 1 mmol of chloride ion remains unreacted. The concentration of silver ion is very low, and the current is very nearly 0. At 40 ml of silver solution added, 4 mmol of silver chloride forms and the concentration of unreacted chloride and silver ion is very low. The current measured is very low. When 50 ml of silver solution is added, 4 mmol of silver chloride and 1 mmol of unreacted silver ion are in the cell. This unreacted silver ion causes the current to be fairly high. When 60 ml of silver solution is added, the solution contains 4 mmol of silver chloride and 2 mmol of unreacted silver ion. The current is twice as high as it is when 50 ml of titrant is added. If the measured current values are plotted against the volume of silver solution added, the titration curve shown in Figure 14–6 results. The dashed lines are extrapolations of measured data. The volume at which the values intersect is the end point. The reason for the shape of this curve is that there is essentially no

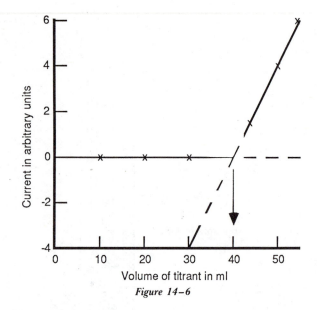

Figure 14–6

silver ion present in the solution until all the chloride is used up. As the titration is carried past the end point, silver ion builds up in the solution and is responsible for the increase in current at the last two experimental points.

This titration and the curve that is depicted in Figure 14–6 are the basis of the end point–detection method in the common chloride method of Cotlove. This method uses coulometry to generate silver ion for the titration of chloride, and it uses silver microelectrodes for the amperometric end point detection. The Cotlove method is discussed in detail in Chapter 15.

VOLTAMMETRY IN LIQUID CHROMATOGRAPHY

In some of the determinations that are done in liquid chromatography and in ion chromatography, it is possible to use voltammetry as a basis for detection. In these methods, a very stable power supply of appropriate voltage is connected to the very small electrodes in a flow cell. The flow cells that are commonly used have a volume of about 10 μl. Working electrodes of gold, silver, platinum, or glass carbon are available. The common reference electrode is silver–silver chloride.

The detector functions by being adjusted to a voltage sufficient to exceed the decomposition potential of the analyte of interest. Under these conditions, the current that results is directly proportional to the concentration of the analyte.

Constant-Current Coulometry; Coulometric Titrations

Coulometry is best defined as a titration that makes use of an electrochemically generated titrant. There are two types of coulometry—controlled-potential coulometry and constant-current coulometry. Controlled-potential coulometry requires rather elaborate equipment and has found only limited use in the clinical laboratory. On the other hand, constant-current coulometry requires the simplest of equipment and has found a great deal of use.

Constant-current coulometry is done by placing the sample that is to be titrated into a solution that contains a supporting electrolyte. Once this has been done, a suitable pair of electrodes is introduced into the solution, and a voltage is applied. The voltage causes an electrolytic reaction to occur at each of the electrodes. One of these electrode reactions produces the titrant.

Coulometric titrations have much in common with the classic titrations. The titration reaction must be quantitative and fast. The reaction must have reproducible stoichiometry. In addition, some means of detecting the end point must exist. End point detection can be done by visual indicators or by electrochemical means. Amperometry is a popular end point–detection method.

The underlying theory of classic volumetric titrations is the same as that of coulometric titrations. The experimental results can be calculated from the basic concept that the number of equivalents of titrant is equal to the number of equivalents of analyte. This assumption is, of course, valid only at a proper end point.

Keeping these facts in mind, let us consider the basic differences between coulometric and volumetric titrations. In volumetric titrimetry, the number of equivalents of titrant is determined from the volume of titrant used to reach the end point and from the known normality of the titrant. The product of these two values is equal to the number of equivalents of titrant used in the titration.

In coulometry the analyst is able to find the number of equivalents of titrant that has been produced by using Faraday's law. Faraday found many years ago that 96,491 coulombs of electricity always liberate one equivalent of compound or ion at each electrode. Mathematically this can be stated as

$$Y = \frac{Q}{F} \qquad \text{Equation } 15\text{--}1$$

where Y is the number of equivalents of titrant generated at the electrode;
Q is the number of coulombs of electricity that has been used to generate the titrant;
F is Faraday's constant, 96,491 coulombs per equivalent.
The value of Q is defined by a second equation

$$Q = IT \qquad \text{Equation } 15\text{--}2$$

where I is the electrical current (in amperes) that flows through the cell;
T is the time in seconds during which the current flows.

By combining Equations 15–1 and 15–2, we obtain Equation 15–3. It relates the equivalents of titrant to the experimental parameters, which are current and time.

$$Y = \frac{IT}{F} \qquad \text{Equation } 15\text{–}3$$

Thus, if we measure the current passing through the generating electrode and the length of time that the current flows, we can calculate the number of equivalents of titrant that is generated at the electrode. In this way the electrode replaces the buret, and the electron replaces the standard solution of volumetric titrimetry. Remember that equivalents react with equivalents, so the number of equivalents of titrant that is calculated by Equation 15–3 is also equal to the number of equivalents of analyte.

If we want to know the mass of the analyte, we need only multiply the number of equivalents by the equivalent weight of the analyte.

$$X = \frac{IT}{F} \times \frac{M}{n}$$

where X is the mass of the compound that was titrated;

M is the molecular weight of the compound titrated;

n is the number of equivalents per mole of the compound.

Thus, if a compound is titrated by an electrochemically generated titrant, all that must be measured is the electrical current that is generating the titrant and the length of time that the current flows. From these data the concentration of the analyte can be calculated.

APPLICATIONS OF CONSTANT-CURRENT COULOMETRY

Consider, for example, the titration of an acid with electrochemically generated hydroxide ions. The hydroxide ions can be generated at a platinum electrode via the following reaction:

$$2H_2O + 2e \rightarrow 2OH^- + H_2\uparrow$$

This reaction is a cathodic reaction. The anode in this titration is generally silver. The anodic reaction is

$$Ag \rightarrow Ag^+ + e$$

The cell in which the titration would be done is depicted in Figure 15–1. A suitable power supply is illustrated in Figure 15–2. The supporting electrolyte used is 0.1 molar sodium bromide. The major role of the supporting electrolyte is to get rid of the silver ions before they react with hydroxide ions. The reaction is

$$Ag^+ + Br^- \rightarrow AgBr\downarrow$$

The silver bromide precipitates on the surface of the silver electrode. The arrangement keeps the silver ions from interfering; it is periodically necessary to remove the silver bromide from the electrode surface. This can be done by soaking the

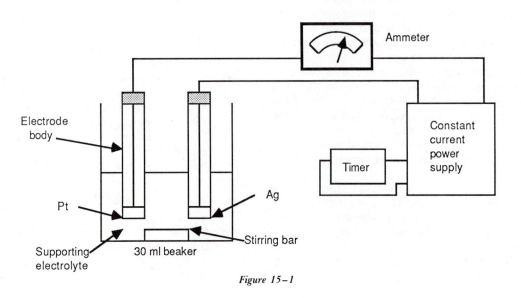

Figure 15–1

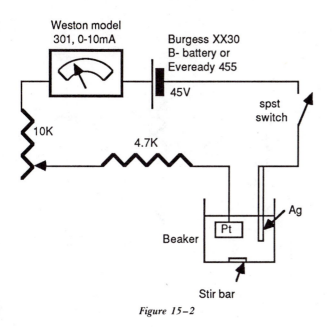

Figure 15–2

electrode in concentrated ammonia solution. Wear eye protection when working with ammonia. Do the work in a fume hood.

Using the preceding example as a basis, let's do an example calculation.

If 1 ml of sulfuric acid of unknown concentration is titrated to a phenolphthalein end point, and if this requires 3 minutes and 4 seconds of a current flow of 5 mA, what would be the sulfuric acid concentration?

$$X = \frac{ITM}{nF}$$

$$X = \frac{(5 \times 10^{-3} \text{ amp})(184 \text{ sec})(98.1 \text{ g/mol})}{(2 \text{ Eq/mol})(9.65 \times 10^{4} \text{ C/Eq})}$$

$$X = \frac{(9.20 \times 10^{-1} \text{ C})(9.81 \times 10^{1} \text{ g/mol})}{(2 \text{ Eq/mol})(9.65 \times 10^{4} \text{ C/Eq})}$$

Note the units.

$$X = \frac{(9.20 \times 10^{-1} \text{ C})(9.81 \times 10^{1} \text{ g/mol})}{(2 \text{ Eq/mol})(9.65 \times 10^{4} \text{ C/Eq})}$$

$X = 4.67 \times 10^{-4}$ g of sulfuric acid in the 1-ml volume of sample titrated.

A second example of a coulometric titration is the determination of serum chloride by the Cotlove method. In this procedure, a silver anode produces the titrant via the following reaction:

$$Ag \rightarrow Ag^{+} + e$$

The silver ion that is produced then titrates the chloride in the sample by the following reaction:

$$Ag^{+} + Cl^{-} \rightarrow AgCl \downarrow$$

The cathodic reaction takes place at a platinum cathode as follows:

$$2H_2O + 2e \rightarrow 2OH^{-} + H_2 \uparrow$$

The hydroxide ions that are produced would interfere with the chloride titration so they are destroyed by a highly acid supporting electrolyte. The supporting electrolyte is generally 10 per cent acetic acid that contains 0.1 mol of nitric acid per liter. The acetic acid is used to lower the solubility of silver chloride. A small amount of gelatin is included in the solution. It improves the reproducibility of the end point detection. The end point is detected by amperometry as was discussed in Chapter 14. Figure 15–3 is an illustration of a system that could be used for the chloride titration. Constant-current coulometry is a very common approach for the determination of serum chloride. The systems that are used in clinical laboratories are somewhat different from Figure 15–3 in appearance, but they are the same in principle.

The results of a serum chloride titration could be calculated via Equation 15–3, but they are more often calculated by ratio. In the ratio approach, a 0.10-ml portion of a 100 mEq/l chloride standard is titrated. Then 0.10 ml of the serum is titrated. And finally a blank of 0.10 ml distilled water is titrated. The three titration times are obtained at the same constant current, so the following equation is valid:

$$\text{Serum chloride concentration (mEq/l)} = \frac{(T_{Sample} - T_{Blank})}{(T_{Standard} - T_{Blank})}(100 \text{mEq/l})$$

This approach to serum chloride measurement is the basis of the chloride method on many clinical instruments and is reputed to be the most accurate method. Coulometry automates well; it is used, for example, as the chloride method on the newest of some of the automated systems.

Beyond this major clinical application coulometry is quite versatile; many titrants can be electrochemically generated. Chlorine, bromine, iodine, silver(I), iron(III), cerium(IV), hydrogen ion, hydroxide ion, titanium(III), copper(I), cadmium(II), and EDTA are commonly produced by coulometric methods.

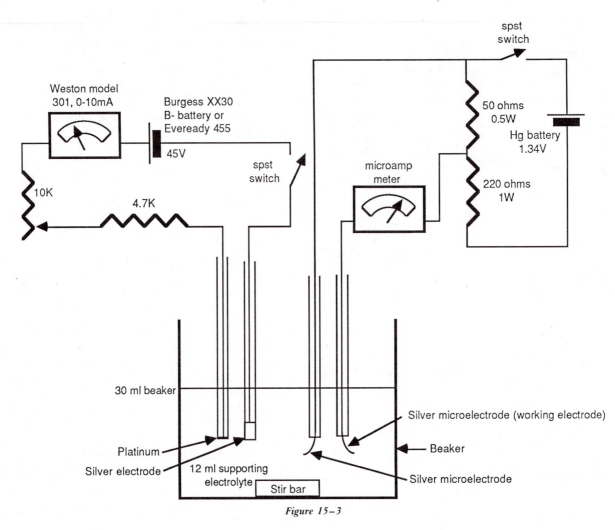

Figure 15–3

Problems

1. The following data are obtained using the circuit in Figure 15–3. The analyte titrated is chloride; the titration current is 6×10^{-3} amp.

   Blank titration time: 15.7 seconds
   Sample titration time: 119.8 seconds

   The sample volume is 0.10 ml of serum. What is the serum chloride concentration in milliequivalents per liter?

2. The following data are obtained using the circuit in Figure 15–3. The analyte titrated is chloride ion; the titration current is 6×10^{-3} amp.

   Blank titration time: 15.7 seconds
   Titration time for a 100 mEq/l standard solution: 123.7 seconds
   Titration time for a serum sample: 119.8 seconds

   The sample volume and volume of standard titrated are both 0.10 ml. What is the serum chloride concentration in milliequivalents per liter?

3. If the total volume of solution in the titration cell is doubled, what effect will it have on the blank titration time? (Refer to Chapter 14 if necessary.)

4. Acids can be titrated by coulometry. This is done by generating hydroxide ions at a platinum cathode. Phenolphthalein is used to detect the end point. If 0.50 ml of an acid sample is titrated to a phenolphthalein end point in 2 minutes, 37 seconds, what is the concentration of acid in the solution expressed in equivalents per liter? If the acid is a strong acid, what is the pH? Assume that the titration current is 5 mA.

16

Electrical Conductance Measurements

When a solution of an ionic solute is placed between two metal electrodes and a voltage is applied, there is a migration of positive ions toward the negative electrode and of negative ions toward the positive electrode. This migration of ions takes place even if the value of the applied voltage is below the value of the decomposition potential of the solute. This movement of ions constitutes an electrical current. Its value is proportional to the applied voltage, the type of ions present, and the number of ions available to carry the current through the solution. The current that flows under these conditions is a measure of the electrical conductivity, S, of the solution. The electrical conductivity of a binary solution is, over a limited range, a linear function of concentration.

The conductivity of a solution is equal to the reciprocal of the electrical resistance.

$$S = 1/R$$

The unit that is used in expressing conductance is the mho, or reciprocal ohm.

The inherent ability of an ion to conduct current is controlled to a large extent by the mobility of the ion. Large-diameter ions cannot move through a solution as rapidly as small-diameter ions can. Hence there is considerable variation in the ability of ions to conduct.

When there is more than one electrolyte in a solution, the conductivity of the solution is approximately equal to the sum of the values of conductance of the electrolytes present. This additivity of conductance of all ionic species in solution causes conductance measurements to be horribly nonspecific. This fact has limited the usefulness of conductance measurements.

On the other hand, electrical resistance and hence conductance can be measured very well. This fact makes conductance measurements very sensitive and is in large part responsible for the applications that do exist. Some conductivity instruments

are designed to provide the conductance, S, of a solution. The determination of S is straightforward and is all that is required in many applications. This is particularly true for differential conductance measurements. All that one must do to find S is to measure the electrical resistance of the solution of interest.

On the other hand, some instruments are designed to provide the specific conductance, K, of a solution. The specific conductance is defined as the conductance of a cubic centimeter of the solution when measured with one-square-centimeter electrodes. It is difficult to build a conductance cell that provides the specific conductance from a direct measurement. It is more common to use a conductance cell of indeterminate, but fixed geometry and to find a proportionality constant that relates S to the specific conductance. The proportionality constant that is used in the following equation relates S to K; it is called the cell constant, θ.

$$K = \theta S$$

The cell constant is found by placing a standard potassium chloride solution in the cell and measuring S. These solutions have known specific conductance values; some examples of solutions that could be used are shown in Table 16–1. From the

TABLE 16–1. Conductance of Solutions for Cell Calibration

Grams KCl per 1000 g of Solution in Vacuum		Specific Conductance at 25°C ohm^{-1} cm^{-1}
71.1352	1.00 Molar	0.111342
7.41913	0.10 Molar	0.0128560
0.745263	0.01 Molar	0.00140877

Adapted from: D.A. Skoog and D.M. West, *Principles of Instrumental Analysis*, 2nd Ed., Philadelphia, Saunders College Publishing, 1980.

known K and the measured S, θ can be calculated. Once θ is known for a cell, the cell can be used to find the specific conductance of other solutions.

THE MEASUREMENT OF CONDUCTANCE

The measurement of the conductance of a solution always involves the application of a voltage to a conductance cell. The conductance is then measured by a suitable electronic device. The power supply that is used for this purpose is generally an alternating voltage source that supplies from 2 to 10 volts at frequencies between 60 Hz and 1000 Hz. The reason for the use of alternating current (AC) will be discussed in this chapter.

The circuits that are used to measure conductivity are of several types. Since the conductance, S, is the reciprocal of the electrical resistance, a Wheatstone bridge can be used. Such a bridge is shown in Figure 16–1. The bridge is used just as is discussed in Appendix C. The variable resistor is adjusted to make the null detector indicate null. The variable resistor has a resistance scale associated with its sliding contact. At null the resistances of the conductance cell and the variable resistor must have the same value. The value of the resistance is read from the scale. The reciprocal of the resistance is S. The variable capacitor is provided to counterbalance the capacitance of the conductance cell.

A second circuit that can be used to measure conductivity is shown in Figure 16–2. By adjusting the variable resistor so that the meter gives the correct reading for the conductance standard, you

can calibrate this circuit to read the correct conductance. Once the circuit has been calibrated, the switch can be thrown to the conductance cell position, and conductance can be measured.

Keep in mind that conductivity values are quite temperature dependent. Conductance increases about 2 per cent per degree Celsius, so the temperature must be held constant in any measurement based on conductivity.

With this background in mind, let's consider the conductivity cell and the electrode processes associated with its operation.

THE CONDUCTANCE CELL AND ITS BEHAVIOR

Electrolytic conductance is measured in a conductance cell. The cells are manufactured from two electrodes that are held in a constant geometry relative to each other. The electrodes are generally made of platinum or gold. Various designs are used; three types, including a flow-through cell, are shown in Figure 16–3.

Let's consider the processes that take place at the electrodes when a voltage is applied. First, let's consider what happens when a direct current (DC) voltage that is below the decomposition potential of the electrolyte is applied. When such a voltage is applied to a conductance cell, the ions respond to the electrical field and start to migrate. The positive ions migrate toward the negative electrode, while negative ions are attracted to the positive electrode. Since the voltage is below the decomposition potential of the electrolyte, the ions migrate toward the opposite electrodes but cannot do anything on

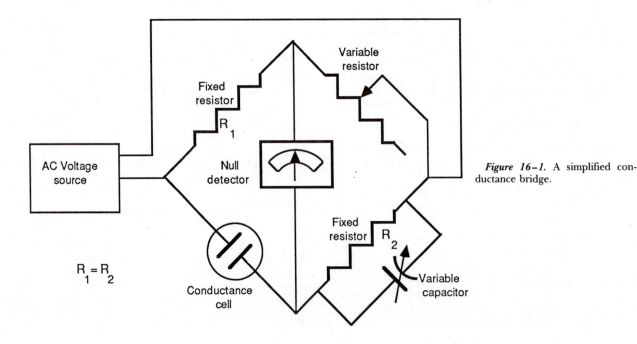

Figure 16–1. A simplified conductance bridge.

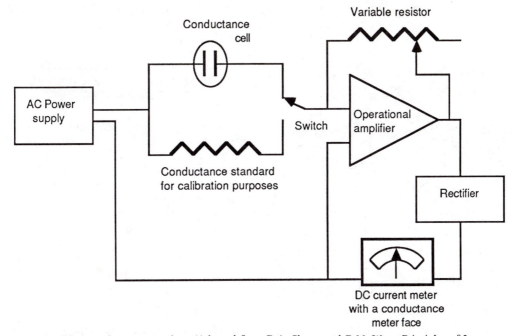

Figure 16-2. A simplified conductance monitor. (Adapted from D.A. Skoog and D.M. West, Principles of Instrumental Analysis, 2nd Ed., Philadelphia, Saunders College Publishing, 1980.)

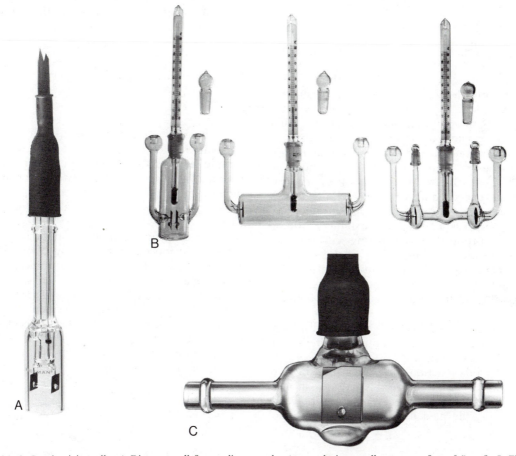

Figure 16-3. Conductivity cells. *A*, Dip-type cell for medium-conductance solutions; cell constants from 0.5 to 2. *B*, Fill type cell for laboratory work to contain the sample under testing and to be immersed in a temperature bath. *C*, Flow-through cell; cell constants from 0.01 to 0.2. (Courtesy of Beckman Industrial Corporation, Cedar Grove, New Jersey. Reprinted with permission.)

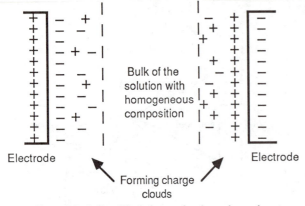

Electrode   Electrode

Forming charge
clouds

Figure 16–4. Simplified charge clouds at electrodes.

arrival; they just form charged clouds next to the electrodes (Fig. 16–4). This process continues until the charged clouds are thick enough to shield the bulk of the solution from the charge on the electrodes. When this occurs, the current in the cell and external circuit stops. This type of behavior means that electrical conductivity measurements are time dependent and that they have to be made very rapidly. Taking conductivity measurements rapidly could be done, but it would be expensive to do so.

The better approach is to use an alternating voltage that is below the decomposition potential of the electrolyte. As discussed in Chapter 1, alternating voltages are characterized by a rapid variation in instantaneous voltage and by the fact that they are positive for half of the time and negative for the other half. In conductance measurements, frequencies between 60 Hz and 1000 Hz are common. When an alternating voltage is used, the picture changes dramatically. During the first half-cycle, the charge clouds start forming as shown in Figure 16–4, but before the process gets very far, the polarity reverses and the charge cloud is "unformed" by means of an exactly opposite phenomenon. This process leads to the sustained production of a reproducible, alternating current that directly relates to the conductivity of the solution. Since the conductivity of the solution is a function of the total ionic concentration, it is possible in some cases to relate conductivity to ionic concentration.

It is not always possible to keep applied voltage below the decomposition potential of all ions in solution. When the decomposition potential is exceeded, electrode reactions occur, causing the current flows to be higher. The cell reaction that takes place during a half-cycle generally takes place in reverse during the next half-cycle. The additional current does not present a problem, as its value is also concentration dependent.

CLINICALLY IMPORTANT APPLICATIONS OF CONDUCTANCE MEASUREMENTS

One of the clinically important applications of direct conductance measurements is the assessment of the purity of distilled water. This type of measurement can be done with a relatively crude power supply and bridge. It is generally done by using a flow-through cell, such as the one in Figure 16–3. These instruments are frequently calibrated in units of specific conductance. The specific conductance of very pure water is 5×10^{-8} mhos. The presence of ionic impurities increases the specific conductance dramatically. Traces are said to cause up to a tenfold increase.*

A second major application of conductance measurements is the determination of blood urea nitrogen. The specificity of this measurement is obtained by the use of an enzyme, urease. The conductance of a solution of enzyme plus a few microliters of serum is the baseline conductance value. The following reaction increases the solution conductance:

$$H_2N-\underset{\underset{O}{\|}}{C}-NH_2 + 2H_2O \xrightarrow{\text{Urease}} (NH_4)_2CO_3$$

$$(NH_4)_2CO_3 \longrightarrow 2NH_4^+ + CO_3^{-2}$$

The rate of increase in conductance can be correlated with the amount of urea in the blood sample. The use of the very specific enzyme urease coupled with the extreme sensitivity of conductance measurements has made this method popular. This is an example of a differential conductivity measurement.

The third application of conductivity measurements is in ion chromatography. When ionic analytes are separated chromatographically, the extreme sensitivity of conductivity measurements can provide the basis for a very sensitive chromatographic detector. These measurements require constant cell temperature or at least electronic compensation for changes in cell temperature. The finer points of ion chromatography detection systems are discussed in Chapter 18.

---

*D. A. Skoog and D. M. West, Principles of Instrumental Analysis, 2nd Ed., Philadelphia, Saunders College Publishing, 1980.

17

Liquid Chromatography

ORIGIN OF LIQUID CHROMATOGRAPHY

Around the turn of the century, the Russian botanist Mikhail Tswett experimented on the separation of plant pigments. His experiments involved passing solutions of plant pigments through glass tubes that were filled with powdered chalk. He found that some of the pigments could be separated from each other; in fact, they appeared as colored bands in his packed tubes. Because of the colored bands, he called the separation method chromatography. The term is derived from the Greek words for color writing.

Let's consider his system in a little more detail, but let's do so with modern materials. Consider the empty column in Figure 17–1. If the column were filled with cellulose powder slurried in hexane, and if the powder were allowed to settle out into a bed, the resulting packed column could be used to sep-

arate plant pigments. The excess hexane would be allowed to drain through the cellulose powder and out through the stopcock. The hexane level, however, would not be allowed to drain below the column head. If this were allowed to happen, air would enter the packing and would disrupt the even flow of solvent through the column; this would lead to poor separations.

With the column packed and freed from excess hexane, the sample of plant pigments dissolved in hexane could be added. The amount of sample added would be small—only about 3 mm deep. The sample would then be allowed to drain into the column until the top of the liquid just disappeared into the top of the cellulose powder (Fig. 17–2).

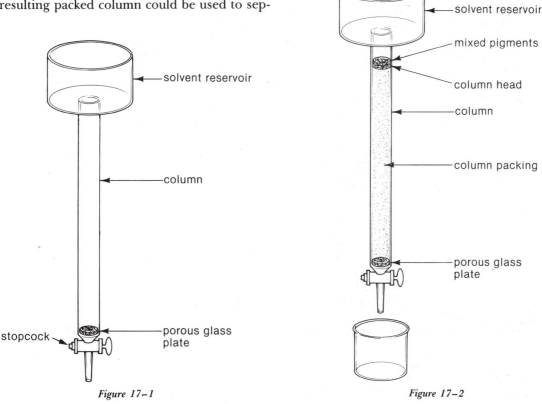

Figure 17–1

Figure 17–2

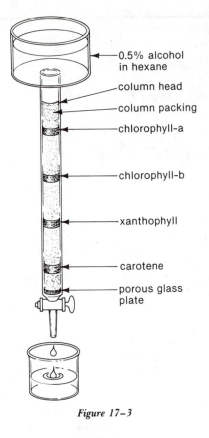

- 0.5% alcohol in hexane
- column head
- column packing
- chlorophyll-a
- chlorophyll-b
- xanthophyll
- carotene
- porous glass plate

Figure 17–3

column and could be collected in a beaker or a graduated cylinder. If the liquid that had been eluted from the column were collected in fractions of 1 ml, and if the fractions were analyzed by visible spectroscopy, a plot of absorbance versus elution volume (or fraction number) could be made (Fig. 17–4). The resulting plot is called a chromatogram.

The chromatogram is the standard readout for all chromatographic instruments. On modern instruments, the chromatogram is produced by automatically plotting the output from a concentration detector versus time. Chromatograms are important to scientists because they can be used for qualitation and quantitation of the various constituents in a sample. Qualitation is based on the retention time of the sample constituents, while quantitation is based on the area under the peaks. These topics will be covered in much greater detail in subsequent sections of this chapter.

A number of points from this example should be discussed and retained for further use. First, the alcohol-hexane solution serves as a mobile phase in this separation. All chromatographic systems have a mobile phase. Second, the cellulose powder serves as a stationary column packing. All chromatographic systems have a stationary packing. Third, the plant pigments separate from one another because they interact differently at the packing–mobile phase interface. The pigment that was eluted first interacted the least amount of time with the stationary packing, whereas the one that came out last interacted for a longer time with the stationary packing. These statements can be generalized to all chromatographic systems. Fourth, the sample is introduced in the form of a thin layer of solution so that it enters the column in a thin, uniform band. All chromatographic systems require that the sample be introduced in a thin, uniform band.

Beyond these generalities, one must recognize

With the sample loaded onto the column, a solution of 0.5 per cent alcohol in hexane would be added to the column reservoir, and the stopcock would be opened. The alcohol-hexane solution would flow through the column and out through the stopcock. After a time, the chlorophyll-a, chlorophyll-b, xanthophyll, and carotene would be separated into bands of color (Fig. 17–3).

If the alcohol-hexane solution were passed through the column for a sufficient amount of time, each of the pigments would be eluted from the

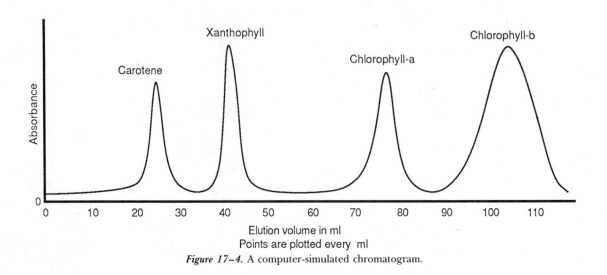

Figure 17–4. A computer-simulated chromatogram.

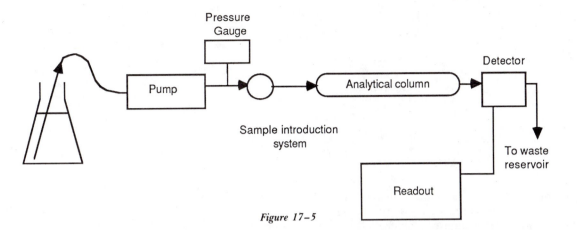

Figure 17–5

that this is a lengthy experiment. The column requires several hours to separate the pigments. The absorbance measurements on the collected fractions require several more hours because different wavelengths have to be used for the carotene, xanthophyll, and chlorophylls.

MODERN LIQUID CHROMATOGRAPHY

In the mid-1960s, some fine scientists who had considerable knowledge of gas chromatographic theory turned their attention to the development of modern liquid chromatography. The theoretical overlap between gas chromatography and liquid chromatography is considerable, so the scientists made very rapid progress. The fruits of their labor were very rewarding. Liquid chromatography has gained wide acceptance in the industries that deal with nonvolatile but soluble compounds. In clinical laboratories, the applications have been mainly in drug analysis, particularly in therapeutic drug monitoring and in the determination of drugs of abuse.

LIQUID CHROMATOGRAPHIC INSTRUMENTATION

Mobile Phase Delivery Systems

Mobile Phase Reservoir. Consider the flow diagram of a simple modern liquid chromatograph (Fig.17–5). Let's consider each part. The mobile phase reservoir can be anything from a liter E-flask to a heated vessel with a reflux condenser. If the mobile phase is one that tends to dissolve a lot of air, the heated system may be needed to degas the liquid. Sometimes helium bubbling through a mobile phase serves the purpose of a refluxing reservoir. Helium is effective in removing dissolved gas from a mobile phase, but it is a little expensive. Water and alcohols are prone to problems with dissolved gas. These dissolved gases tend to form bubbles in the detector and to cause spikes on the chromatograms. Modern pumps have greatly decreased the problem of gas bubbles.

Reciprocating Pump. Most current liquid chromatographs use a reciprocating pump or pumps (Fig. 17–6) to provide the constant flow of

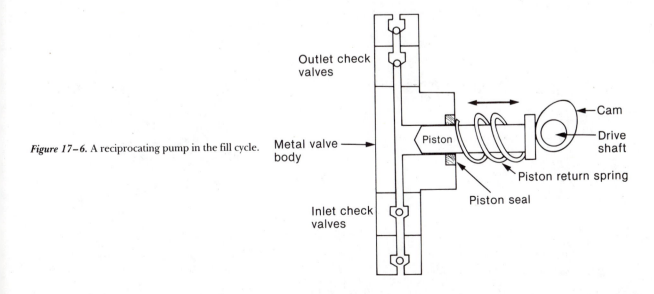

Figure 17–6. A reciprocating pump in the fill cycle.

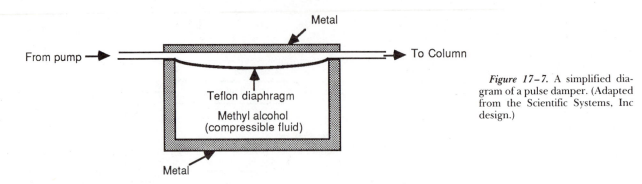

Figure 17–7. A simplified diagram of a pulse damper. (Adapted from the Scientific Systems, Inc design.)

mobile phase to the chromatographic column. The drive shaft is rotated either directly or indirectly by a motor. This rotates the cam, which in turn drives the piston into the valve body or alternately lets the return spring return the piston to the filled position. When the piston is returning to the filled position, the inlet check valves open to allow mobile phase to be drawn in. During this part of the stroke, the outlet check valves are closed. This situation is the one shown in Figure 17–6. When the cam rotates into the pump cycle, the inlet check valves close and the outlet check valves are opened by the surging mobile phase. Depending on how fast the cam is rotated, different flow rates are obtained.

This type of pump has a problem with pressure and flow pulses during its operation. Much work

has gone into the engineering of these pumps to cut down on the flow rate pulses. Some instruments use two or three of these pumps in tandem, each of them at a different point in its cycle. Although this really cuts down on the pulses, it also increases cost. On most instruments, a pulse damper is placed between the pump and the sample injection system. These pulse dampers have different designs, but they all provide a flexible metal device

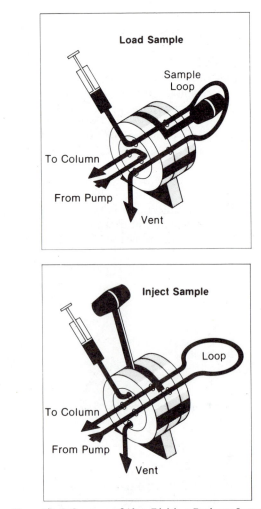

Figure 17–8. Sample introduction system.

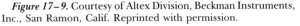

Figure 17–9. Courtesy of Altex Division, Beckman Instruments, Inc., San Ramon, Calif. Reprinted with permission.

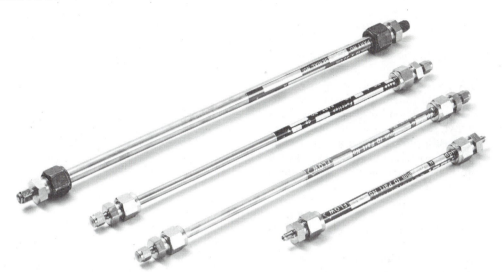

Figure 17–10. Courtesy of Altex Division, Beckman Instruments, Inc., San Ramon, Calif. Reprinted with permission.

to absorb some of the energy. An example is shown in Figure 17–7. As a pressure surge arrives, the pressure compresses the diaphragm, which in turn compresses the methyl alcohol. When the surge passes, the alcohol returns the energy, through the diaphragm, to the mobile phase. In this way the pressure surges and flow-rate surges are reduced.

Sample Introduction System

The sample introduction system may be any of several devices. When the pressure of the system is not excessive, say 2000 psi or less, samples can be injected into the instrument through a silicone rubber septum. Generally a 10-μl syringe is used. The septa are good for about 50 injections. Such a system is illustrated in Figure 17–8. The length of the syringe needle should be selected with care. It needs to be positioned as shown. The needle should not touch the packing retainer but should be near it.

For sample introduction at higher pressure or for automated operation, a sampling valve is required. The sampling valve shown in Figure 17–9 is a syringe-loaded model. It also comes in a model that is pneumatically activated. Many different sizes of sample loop are available. The sample loop is filled as shown in the top half of the figure. The content of the loop is carried to the column as shown in the bottom half of the figure. Notice that the handle position in the two parts of the figure is different. A number of different types of valves are available. Special syringes are needed for the syringe-loaded models. Do not use syringes with sharp pointed needles.

Columns for Liquid Chromatography

The next device in the flow diagram in Figure 17–5 is the column. A number of different types of columns have been used in liquid chromatog-

raphy. The most common column is a 4.6-mm ID stainless steel tube with heavy side walls. In my experience, the most common length is 25 cm. Many 15-cm columns are in use, however. A photograph of four columns is shown in Figure 17–10.

The packing that is used in most of these columns consists of surface-modified body-porous silica microspheres. There are other materials as well, but they are not as common. An electron photomicrograph of a body porous silica microsphere is shown in Figure 17–11; it is 5 to 6 μ in diameter and has a surface area of 350 m^2/g.

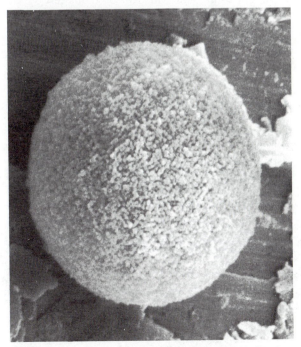

Figure 17–11. Porous silica microsphere. (Courtesy of Du Pont Company, Wilmington, Del. Reprinted with permission.)

For separation that can be done by adsorption chromatography, these porous silica products can be used without modification. It is more common, however, to modify the surface of the silica by chemical reactions. Frequently a monolayer of octadecyl groups ($C_{18}H_{27}$) is chemically bonded to the surface of the porous silica microspheres. This is done through a dimethyl silyl linkage. This produces a product with much different chromatographic characteristics. Sometimes octal, amino, nitrile, phenyl, or methyl groups are used instead of octadecyl groups. Each of these groups, when chemically bonded to the silica, produces a new packing material with different properties. These modified silica packings are said to have bonded stationary phases.

These silica-based packing materials are susceptible to mechanical shock. Because of this fact, it is common to allow several minutes for changes in the mobile phase flow rate. If mobile phase composition is to be changed, it should be changed slowly.

These column packings are easily clogged by particulate material. If your samples are not clean, it is necessary to filter them through a millipore filter before injection. Sometimes a short precolumn is used in lieu of this filtration step. It is replaced when necessary.

It is helpful to pass mobile phase through a small, silica-filled column before it reaches the analytical column. This saturates the mobile phase with silica and prevents the loss of silica packing from the analytical column. The solubility of silica is low but finite. Columns can be damaged by loss of packing that is caused by dissolution.

Liquid Chromatographic Detection

The next device in the flow path in Figure 17–5 is the detector. The most popular type of detection in liquid chromatography makes use of absorption spectroscopy. Many different instrument types have been used to make the absorption measurements—single-beam instruments, double-beam instruments, photometers, and spectrophotometers. There are many different design approaches. Thorough knowledge of the material presented in Chapters 2, 3, and 4 is prerequisite to understanding the following discussion.

Photometers. First, let's consider the photometer. There are two common approaches; the first is shown in Figure 17–12. A second popular approach is shown in Figure 17-13. The instruments depicted in these figures have much in common. They differ in that the one shown in Figure 17–12 does not have a reference flow cell. In both instruments a reference radiation path is provided. The reference detector provides the value for Po; the working flow cell and detector provide the value

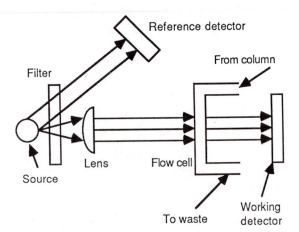

Figure 17–12. A fixed wavelength double-beam photometer for liquid chromatographic detection.

for P. The electronics forms the ratio of P to Po and takes the negative log of the result. The absorbance value is then presented on the readout.

A number of different sources have been used in these instruments. Deuterium lamps and low-pressure mercury lamps are common. Some instruments use a phosphor-coated mercury lamp to provide wavelengths that the mercury lamp cannot supply.

Interference filters are common in these instruments; quartz optics are always provided. Silicone photodiodes are the common detectors.

The internal volume of the cells is typically about 10 μl with a 1-cm cell path. Other sizes are manufactured as well. The design of these flow cells varies. Some are temperature controlled; in some the cell width is tapered. These kinds of features are included to control the impact of changes in refractive index. If some of the radiation that should reach the detector is lost because it is refracted away from the detector, one can obtain a poor baseline.

Spectrophotometers. In addition to the photometers that are so common, there are some liquid

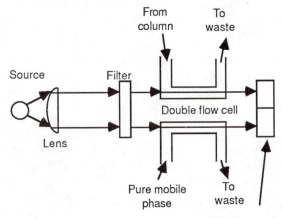

Figure 17–13. A fixed wavelength double-beam photometer for liquid chromatographic detection.

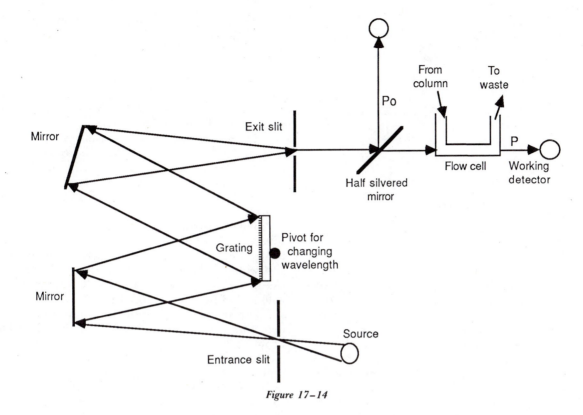

Figure 17–14

chromatographic (LC) detectors that are high-quality spectrophotometers with flow cells. A typical optical path of such an instrument is shown in Figure 17–14. Some instruments (Fig.17–15) use a double flow cell, and some are single beam. The deletion of the reference paths in Figures 17–12 and 17–14 would provide diagrams of single-beam instruments.

Long discussions take place concerning which of these detector types is best. The manufacturers of single-beam instruments say that the excellent stability of their instruments makes the reference

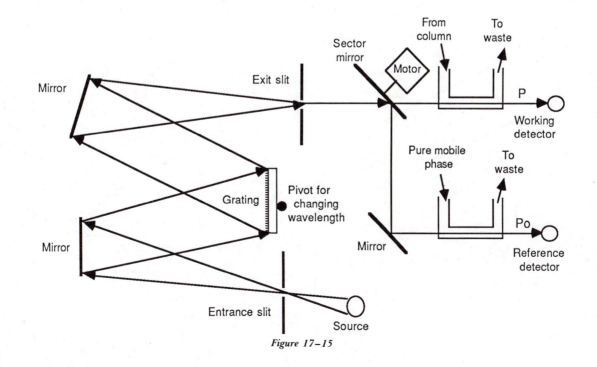

Figure 17–15

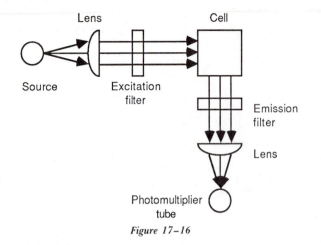

Figure 17–16

beam unnecessary. The manufacturers of double-beam instruments disagree with that statement. We can be sure, however, of two things: The photometers are much more common than the spectrophotometers, and all of the instruments are designed for good stability.

Many of these instruments produce chromatograms that have baseline noise of about 0.00003 absorbance units. The spectrophotometers generally have slightly more noise. Many of the instruments have as their most sensitive scale 0.002 absorbance units, full scale.

The spectral bandwidth of these instruments is generally in the 2 nm to 10 nm range. For the many compounds that can be detected in the ultraviolet range, these instruments do provide the basis for good detection. The chromatograms that result are plots of absorbance versus retention time.

Molecular Fluorescence. A number of compounds of interest fluoresce. This is especially true of drugs and biochemicals. The extreme sensitivity of fluorescence, coupled with the number of important fluorescent compounds, has made fluorescent detection fairly common. Thorough knowledge of the material presented in Chapter 7 is prerequisite to the following discussion.

The instruments that are available have a variety of designs. There are filter fluorometers (Fig. 17–16). There are instruments with monochromator excitation systems and filter emission systems (Fig. 17–17). There are instruments with two monochromators (Fig. 17–18).

These instruments use a number of different sources for ultraviolet operation. One sees xenon lamps, deuterium lamps, and mercury lamps with and without phosphorus. All of these instruments use flow cells of small volume. I have read of cell volumes of between 5 and 30 μl.

For samples that fluoresce, these fluorescent detectors are very convenient and sensitive. The problems seen in fluorescent spectroscopy are potential sources of problems in liquid chromatography. Even dissolved oxygen in the mobile phase may quench fluorescence in some cases.

The chromatograms that result from the use of these detectors are plots of the fluorescent emission intensity versus the retention time.

Voltammetry. The third common approach to detection in LC work makes use of voltammetry. In these methods, a very stable power supply of appropriate voltage is connected to the microelectrodes in a flow cell. Such a flow cell is illustrated in Figure 17–19.

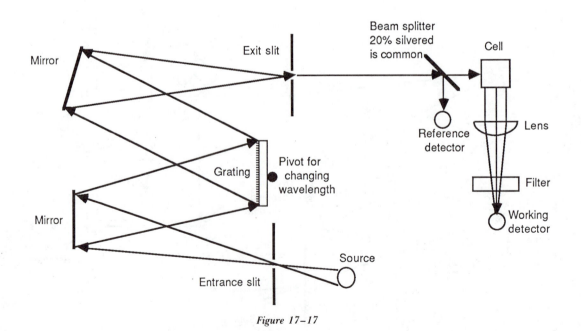

Figure 17–17

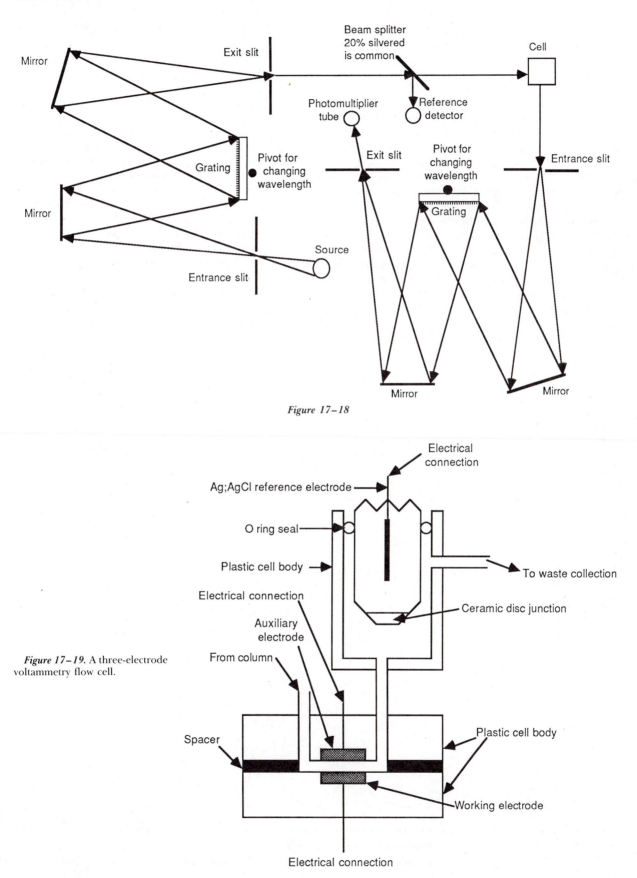

Figure 17–18

Figure 17–19. A three-electrode voltammetry flow cell.

The flow cells that are commonly used have a volume of about 10 μl. Working electrodes of gold, silver, platinum, glassy carbon, or carbon paste are available. The common reference electrode is a silver–silver chloride electrode.

The detector functions by being adjusted to a voltage sufficient to exceed the decomposition potential of the analyte of interest. Under these conditions, the current that flows is directly proportional to the concentration of the analyte.

The three-electrode cell configuration is necessary for chromatographic detection in most cases. This is true because the three-electrode system is capable of providing a more stable baseline and a greater linear response range.

The reason for the improved performance of these three-electrode cells is best understood by first considering the two-electrode cells that are discussed in Chapter 14. If the electrodes in a two-electrode cell experience constant voltage, then the equation

$$I = kC \qquad \text{Equation } 14–1$$

is valid. This equation is used to relate the experimentally measured current to the analyte concentration. In some work the requirement for constant potential is easily met. Unfortunately, this is not the case in liquid chromatography because the resistance of the solution that is passing through the detector can change during the experiment. This changing resistance has an effect on the potential

of the working electrode. The relationship is shown in the following equation:

$$E_{\text{working electrode}} = E_{\text{applied}} - E_{\text{reference electrode}} - IR_{\text{solution}}$$

The changing solution resistance causes E_{working} to vary and, hence, prevents Equation 14–1 from being completely valid. The way to remedy this problem is to use the three-electrode cell.

In the three-electrode system, the potential of the reference electrode is measured relative to that of the auxiliary electrode. Any variation in the measured value relative to the known value for the reference electrode is due to the IR drop across the cell. If this variation is, in effect, applied to the value of E_{applied}, then the proper value of E_{working} is obtained. Thus the auxiliary electrode is used to compensate for any variations in solution resistance. This keeps the working electrode at constant potential no matter what the resistance or current does. This greatly improves the stability and linearity between current and concentration of the electroactive species.

These detectors are sometimes called amperometric detectors. They are really voltammetry detectors. Amperometry is a titration–end point method.

THE CHROMATOGRAM

Now that considerable background concerning liquid chromatographic instruments has been pre-

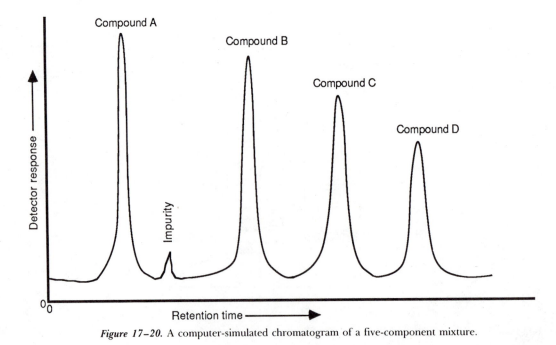

Figure 17–20. A computer-simulated chromatogram of a five-component mixture.

sented, it is time to expand our discussion of quantitation and qualitation. A brief overview of these topics is presented at the beginning of this chapter. Consider the chromatogram in Figure 17–20. It will be used as a teaching example in the discussion of quantitation.

Quantitation

For all detectors that are used in chromatography, at least over a reasonable range, there exists a linear relationship between peak area and the amount of material that is responsible for causing the peak. Thus, by injecting suitable amounts of pure analyte, the analyst can work up a standard curve for the analyte. The standard curve can then be used to find the concentration of the analyte in samples of unknown concentration. For example, if 1, 2, 3, 4, and 5 units of compound B from Figure 17–20 were introduced into the chromatograph, one after another, the readout would be five chromatograms with single peaks. The retention time of each of the peaks would be the same, but the area under the peaks would be different. If one were to plot the area under the peaks for the five standards, one would obtain a standard curve (Fig. 17–21).

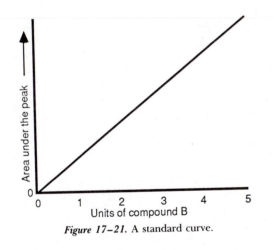

Figure 17–21. A standard curve.

If a sample of unknown B concentration were chromatographed, and if the area under the B peak were determined, and, further, if the area were located on the standard curve in Figure 17–21, the amount of compound B in the sample could be determined graphically (Fig. 17–22). The mass of compound B in our example (Fig. 17–20) would be about 2.4 units. The same approach to quantitation could be used for other sample constituents.

Frequently the amount of compound in a sample is calculated on the basis of the response of one standard. This can be done, once linearity is known to exist, by employing the following equation:

$$C = \left(\frac{\text{Response of sample}}{\text{Response of standard}}\right)(\text{Concentration of standard})$$

Equation 17–1

A third way that quantitation is done is by including an internal standard in the samples. For example, compound C in Figure 17–20 could be an internal standard. Internal standards are compounds that are chemically similar to the analyte but are not present in the sample until added by the analyst. The analyst adds a known amount of internal standard and then chromatographs the mixture. If Figure 17–20 were the resulting chromatogram, and if compound B were the analyte, the area under peaks B and C would be determined.

The amount of compound B in the sample could be calculated by the following equation:

$$\text{Amount of compound B} = \left(\frac{\text{Area under the B peak}}{\text{Area under the internal standard peak (C peak)}}\right)\left(\begin{array}{c}\text{Amount of internal standard added}\end{array}\right)(f)$$

Equation 17–2

where f is a response factor that is determined experimentally by injecting equal masses of compounds B and C. Under these specific conditions

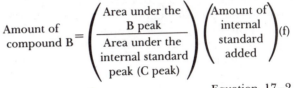

$$f = \frac{\text{Area under the C peak}}{\text{Area under the B peak}}$$ Equation 17–3

For example: If a mixture of 1 mg of compound C and 1 mg of compound B produces a chromatogram showing 14 cm² of area for the C peak and 7 cm² of area for the B peak, the indication is that the detector is twice as responsive for C as it is for B, and the factor f is equal to 14/7 or 2.

A microprocessor or laboratory computer is frequently built into these instruments. The com-

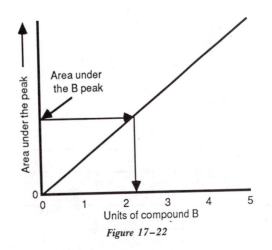

Figure 17–22

puter or microprocessor, given the slope from the plot in Figure 17–21, can calculate the amount of compound B in the sample from the area under the peak. It would use Equation 17–4 in this calculation.

$$\text{Amount of compound B} = \frac{\text{Area under B peak of sample}}{\text{Slope from Figure 17–21}}$$

Equation 17–4

The computer or microprocessor, given the data, could also use Equation 17–1 to calculate the results. It can, of course, do internal standard calculations as well.

The question might be asked, "How do we find the areas under peaks?" There are two common ways. First, for symmetric peaks, peak heights are proportional to peak areas. Hence peak heights are frequently used in lieu of properly determined peak areas. Peak heights, more so than peak areas, are affected by changes in operating conditions, so the use of peak heights is not the best, but is still common.

A second common way to obtain the values for peak areas is by the use of a microprocessor or computer that integrates the area under the peaks. These systems, when used with the proper software, are very good. The value of the peak area is seldom read out; more often the amount of analyte in the sample is calculated from the data and read out.

A third device that is common on some older recorders is a mechanical integrator. These are being replaced rapidly and will not be discussed in this book.

Qualitation

Qualitative analysis is frequently done by chromatographic techniques. The procedure for doing qualitative analysis can be shown by considering a sample chromatogram (Fig. 17–23). The chromatogram consists of three peaks; the retention time of each is measured from the point of injection to the appearance of the peak maximum. The retention times are labeled for future use.

t_M is the retention time of a nonretained compound

t_{RN} is the retention time of compound N

t_{RP} is the retention time of compound P

From the practical point of view, qualitation is done by determining the retention time of known pure compounds that are suspected of being constituents of the sample. These values are then compared with the retention times of the peaks produced by the sample. This approach is valid only if the chromatographic *conditions are identical* when all the measurements are taken, and if the same chromatographic column is used in taking all the measurements. Prior investigation into the nature of the sample must also guarantee that no two compounds found in the sample have the same retention time. This can be done by separating representative samples on several different types of columns. If all the samples produce no ambiguous results when analyzed on the different columns, you can be fairly certain that the samples do not contain compounds with identical retention times. If this has been proved, then retention times can be used for qualitation.

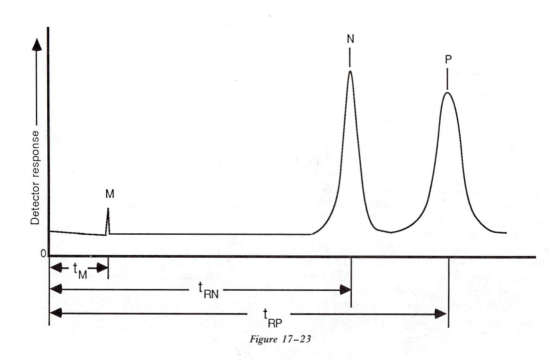

Figure 17–23

When data that have been obtained on different instruments must be compared, it is better to use the relative retention instead of the retention time. The relative retention is less affected by small differences in operating conditions. When the relative retention is to be calculated, a specific material is designated the standard against which the others are measured (see Fig. 17–23). The relative retention is given the symbol α and is equal to

$$\alpha = \frac{t'_{RN}}{t'_{RP}} \qquad \text{Equation 17-5}$$

where $t'_{RN} = t_{RN} - t_M$ (from Fig. 17–23)

and $t'_{RP} = t_{RP} - t_M$ (from Fig. 17–23).
In this example, compound N would be used as the standard of reference. The value of α for compound P could be used for qualitation just as retention times were previously used.

It is interesting to note that some chromatographic data systems do not subtract t_M from the retention times. The relative retentions are hence uncorrected and make comparison of data from one instrument to the next more difficult.

One other technique that is used for qualitation is the so-called spiking method. This method is a little on the quick and dirty side, but it is used. If a compound is suspected of being a constituent of a sample, the sample can be chromatographed, and the remaining part of the sample can be "spiked" by adding some of the pure suspected compound. If the peak of interest is bigger after spiking, one can say that the pure compound and the one in the sample could be the same compound. If a new peak shows up after spiking, the two are definitely not the same. The spiking technique does not allow one to verify whether the added compound and the one in the sample have identical retention times. This is the weakness in the method.

FACTORS THAT AFFECT SEPARATIONS

So, putting it all together, the mobile phase supply system drives the mobile phase through the sample introduction system and on through the rest of the instrument. The column separates the sample into its components, and the detector detects the instantaneous concentration of the various sample constituents as they are eluted from the column. The chromatograms that result are, in effect, plots of the instantaneous concentrations of various sample constituents versus time. Keeping this in mind, let's consider the operation of the column in more detail.

Columns. In the early days of modern liquid chromatography, scientists spent a lot of time talking about how columns separate things. At one extreme were the separations done on columns made of silica gel, alumina, or cellulose powder. These separations seemed to be based on adsorption of sample molecules to the polar surfaces of the packing. Samples separated because the various constituents had different polarities and, hence, interacted differently with the very polar surface of the packing. The sample constituents that were adsorbed least were eluted first.

At the other extreme were column packings made by coating a polar organic compound, such as 2-cyanoethyl ether, on small glass beads. These columns were used in conjunction with a nonpolar mobile phase, such as hexane, to separate various compounds. It was thought that sample constituents separated on these columns because the compounds had different partition coefficients relative to the hexane-ether interface. Thus a compound whose partition coefficient favored solubility in the stationary phase was slower to elute from the column than a compound whose partition coefficient favored solubility in the mobile phase.

More recently, the distinction between these two extremes has become clouded. A silica packing will do certain separations. The same silica, after having a monolayer of $(CH_3)_3Si-$ bonded to its surface, has different properties and does different separations. It is hard to talk about this packing as one in which there is a liquid stationary phase. I am hard pressed to provide for you a good physical model for the mechanism of the separations that occur on bonded "stationary phase" columns. The good news is that knowledge of the mechanism is not required in order to do separations; an analyst can do good chromatographic work while standing back and wondering what the mechanism of separation might be.

In recent years some new ways of discussing columns have come into use. Talk about liquid-solid or liquid-liquid chromatography is being replaced by talk about normal phase and reverse phase liquid chromatography. The use of a mobile phase that is less polar than the stationary phase is called normal phase liquid chromatography. The use of hexane as a mobile phase and 2-cyanoethyl ether as a stationary phase is an example of this. The use of a mobile phase that is more polar than the stationary phase is called reverse phase liquid chromatography. Reverse phase separations that are made on bonded stationary phase columns now account for about 75 percent of all liquid chromatography.

Mobile Phase Selection. Let's consider some examples of normal and reverse phase separations. First, consider a normal phase separation that can be done on cellulose powder. This separation is the same one that was discussed at the beginning of this chapter, that is, the separation of chlorophyll, xanthophyll, and carotene. It would be done at

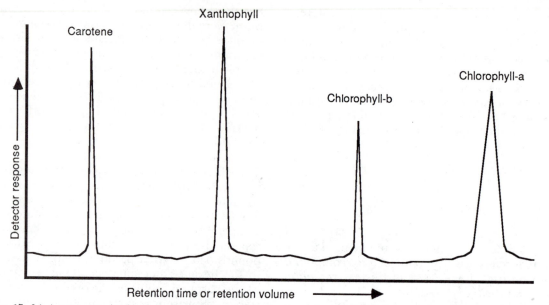

Figure 17–24. A computer-simulated chromatogram of plant pigments. Conditions: flow rate 1.0 ml/minute; mobile phase 0.5% ethanol in hexane; packing cellulose powder.

room temperature with a flow rate of 1 ml/min. If the separation were attempted with pure hexane, the retention times of the compounds would be very long. On the other hand, if 0.5 per cent ethanol in hexane were used as the mobile phase, the chromatogram in Figure 17–24 would be obtained.

Next, consider the same separation, done under identical chromatographic conditions, except for a slight change in the mobile phase. The mobile phase in this instance is hexane containing 2 percent ethanol (Fig. 17–25). Notice that the retention time of each pigment has decreased. This has taken place because the mobile phase has become more polar. In normal phase systems this is a general trend. The concentration of the polar compound in the mobile phase must be selected carefully so that the retention times are reasonable and the separations are good. In normal phase work, a more polar mobile phase decreases the retention time.

Next let's consider an example of a reverse

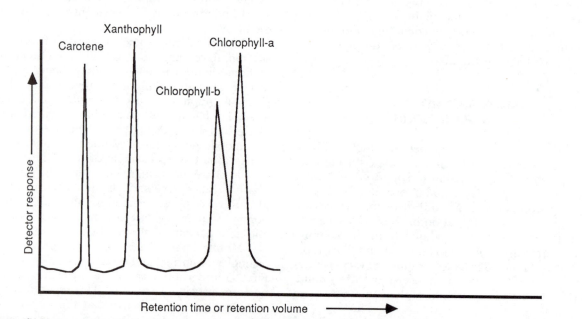

Figure 17–25. A computer-simulated chromatogram of plant pigments. Conditions: flow rate 1.0 ml/minute; mobile phase 2.0% ethanol in hexane; packing cellulose powder.

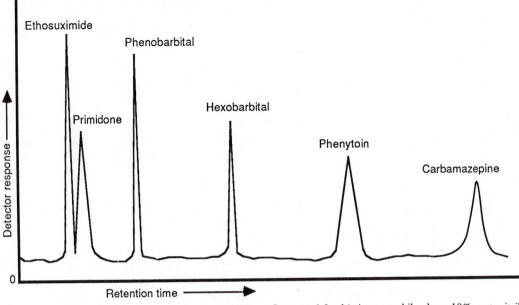

Figure 17–26. A computer-simulated chromatogram. Conditions: flow rate 1.0 ml/minute; mobile phase 10% acetonitrile in water; column octyldecylsilyl groups chemically bonded to silica.

phase separation. The separation of some important anticonvulsants has been done on $C_{18}H_{37}Si$–modified body-porous silica microspheres. The flow rate of the mobile phase in our example is 1.7 ml/min; the column temperature is 50°C. The mobile phase consists of various concentrations of acetonitrile in water. Remember that reverse phase methods use a mobile phase that is more polar than the stationary phase. Consider the separation illustrated in Figure 17–26. The mobile phase in this separation is 10 percent acetonitrile in water.

Next consider Figure 17–27. This separation is identical to the one in Figure 17–26, except for the mobile phase. Now consider Figure 17–28. As the concentration of acetonitrile increases, the retention times decrease, and some resolution of

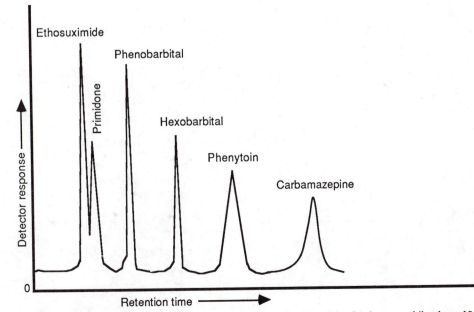

Figure 17–27. A computer-simulated chromatogram. Conditions: flow rate 1.0 ml/minute; mobile phase 15% acetonitrile in water; column octyldecylsilyl groups chemically bonded to silica.

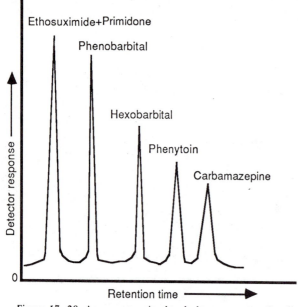

Figure 17–28. A computer-simulated chromatogram. Conditions: flow rate 1.0 ml/minute; mobile phase 20% acetonitrile in water; column octyldecylsilyl groups chemically bonded to silica.

TABLE 17–1. Mobile Phase Solvents

Solvent	Relative Polarity (ϵ°)	Approximate UV Cutoff (nm)
Hexane	.01	200
Chloroform	.40	245
Acetone	.56	330
Acetonitrile	.65	190
Ethanol	.88	200
Methanol	.95	200
Water	Large	160

Data are from: R. W. Yost, L. S. Ettre, and R. D. Conlon, Practical Liquid Chromatography, Norwalk, Conn. Perkin-Elmer Corporation, 1980.

however, the mobile phase components should generally have low viscosity. High-viscosity components just make higher operating pressure necessary. Higher operating pressure leads to a reduction in column life and an increase in the frequency of leaks. Table 17–1 contains some examples of commonly used mobile phase solvents.

The compounds listed in the table are highly soluble in each other and provide a wide range of polarities by making binary mixtures. The information on UV cutoff is important from the point of view of detector compatibility.

All of the separations considered thus far in our discussion have been done with a binary mobile phase of fixed composition. These separations are called isocratic methods. Isocratic methods are convenient and are done on less expensive chromatographs, but they do have some problems. Consider Figure 17–26; look at the nice separation

peaks is lost. This is a common trend in reverse phase work; the most polar mobile phase is the least powerful in its ability to elute compounds.

The exact choice of mobile phase composition depends on several considerations. First, the mobile phase must have the proper polarity. This has been shown in the previous examples. Beyond polarity,

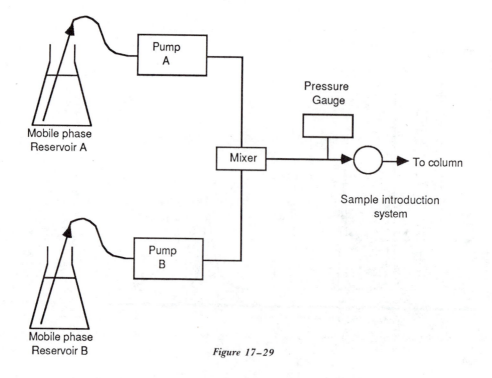

Figure 17–29

between ethosuximide and primidone. Notice, however, the long retention time of carbamazepine. Now consider the short retention time of carbamazepine in Figure 17–28, but also notice the fusion of the ethosuximide and primidone peaks. These facts suggest that starting the separation with a mobile phase containing 10 per cent acetonitrile allows for the separation of the ethosuximide and primidone, but that by increasing the concentration of acetonitrile to 20 per cent during the separation, we can cause the carbamazepine to be eluted more quickly. Thinking of this kind has led to the development of so-called gradient elution liquid chromatography.

In gradient elution work, the polarity of the mobile phase is systematically changed during the course of the separation. This can be done in several ways. One of the common ways is illustrated in Figure 17–29. The system with a pair of independent pumps can be used to produce the needed concentration of acetonitrile in water. In order to provide a lot of versatility, the pumps are microprocessor controlled. A common gradient profile is shown in Figure 17–30. Many other gradients are possible. Gradient elution procedures do provide the best separations of complex mixtures. Disadvantages of these procedures are that the equipment is expensive and the time required to return to the original conditions for the next injection is considerable.

Stationary Phases. In addition to the effect of the mobile phase composition on the separations, the choice of stationary phase can have a profound effect as well. An example of this kind of effect is shown in Figure 17–31. The exact reason for the change in chromatograms is somewhat obscure. The difference in polarity of the two stationary

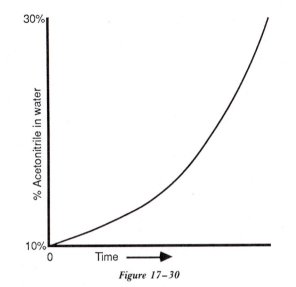

Figure 17–30

phases is part of the reason but is not the whole story. The changes are not entirely predictable. The vendors of columns have suggestions for the type of stationary phase that can be used.

Operating Conditions. In addition to the effects of stationary phase and mobile phase selection, there are other experimental variables that affect separations. In making the following remarks, I am assuming that the mobile phase composition and the stationary phase are fixed.

From the gross point of view, the flow rate does have a considerable effect on the retention time of sample constituents. The faster the flow rate, the shorter the retention time. The faster flow rate may or may not improve peak shape, but the flow rate–retention time correlation is invariant. The effect of flow rate on peak shape is discussed in the next

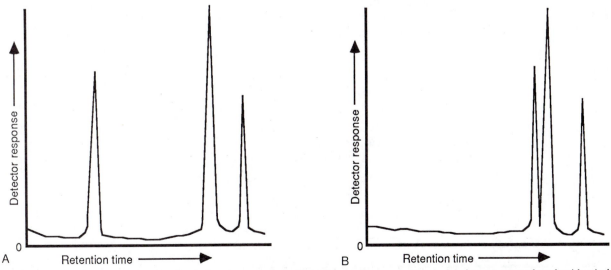

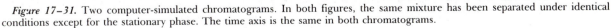

Figure 17–31. Two computer-simulated chromatograms. In both figures, the same mixture has been separated under identical conditions except for the stationary phase. The time axis is the same in both chromatograms.

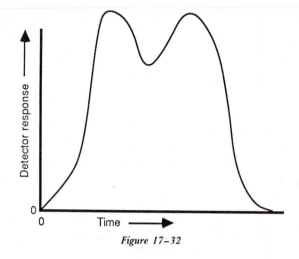

Figure 17–32

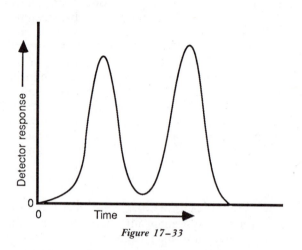

Figure 17–33

section of this chapter. Constant flow rate is a must if reproducible results are desired.

A second gross observation is that increasing column temperature will decrease retention times for all constituents of a sample. There is also a trend toward slightly better separations with increasing column temperature. There are practical limits, however. Column temperatures above 80°C are relatively uncommon. Column temperature must be held constant if separations are to remain reproducible from day to day.

Peak Shape and Resolution. From time to time in the course of chromatographic work, a chromatogram like the one in Figure 17–32 is obtained. This chromatogram is characterized by excessive peak width that has led to poor resolution. The same two compounds, when separated on a higher

quality column, might produce a chromatogram such as the one in Figure 17–33. The two compounds have the same retention time in both chromatograms, but the peak widths are smaller in Figure 17–33; hence, resolution is improved. When one encounters a chromatogram that is more similar to the one in Figure 17–32 than to the one in Figure 17–33, it is important to know that the operating conditions do affect the shapes of peaks. This is particularly true if the column is of high quality, yet the separations are still poor.

It may be possible to improve separation quality to some extent by finding better operating conditions for the instrument. This can be done by systematically studying the chromatographic peak shape while varying the operating conditions. This procedure calls for the selection of one of the an-

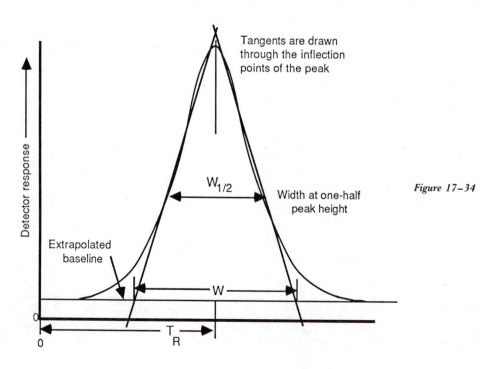

Tangents are drawn through the inflection points of the peak

$W_{1/2}$

Width at one-half peak height

Extrapolated baseline

W

T_R

Figure 17–34

alytes to be used as a model compound. The model compound is then chromatographed under different experimental conditions. The peaks obtained are used to measure the retention times and either the peak widths or the peak widths at one-half peak height. This is shown in Figure 17–34. These values can be used to measure the number of theoretical plates via the equations

$$N = 16\left(\frac{T_R}{W}\right)^2 \qquad \text{Equation 17–6}$$

$$N = 5.55\left(\frac{T_R}{W_{½}}\right)^2 \qquad \text{Equation 17–7}$$

where N is the number of theoretical plates
T_R is retention time
W is peak width
$W_{½}$ is peak width at one-half peak height
Both equations give about the same result.

The number of theoretical plates, N, is a measure of relative peak sharpness. The higher the value of N, the better the peak quality. After all, at fixed retention time, large values for N are produced by small values for peak width. The value of N is frequently large. The large numbers are inconvenient to graph, so the value of N is often divided into the length of the column. This gives a number called the height equivalent of a theoretical plate, HETP.

$$\text{HETP} = \frac{\text{column length}}{N} \qquad \text{Equation 17–8}$$

HETP values are small and easily graphed. Small values of HETP come with large values of N. Thus, separations of good quality are associated with small HETP values.

With this background in mind, let's talk about improved separations. If the column flow rate is varied, and HETP values are determined for our model compound, the data shown in Figure 17–35 will be found. The curve in the figure tells us that an optimum flow rate does exist for the operation of the column. If you are not at the optimum and your separations are poor, you could go to the optimum flow rate and improve your results.

The column temperature also has an effect on HETP. Generally, higher column temperatures produce smaller HETP values. This approach to improved peak shape is limited by other experimental considerations; you can go only so high before other problems develop.

Smaller samples generally lead to improved separations; HETP values get larger with increasing sample size.

The choice of stationary phase will affect the separation quality. Not only will HETP values vary, so might the elution order. Generally, a stationary phase that is similar in polarity to the samples is used. This topic is illustrated to a greater extent in Chapter 19.

SUMMARY

An introduction to liquid chromatographic methods has been provided in this chapter. The general instrument and its various components have been described. Methods for obtaining the best separations that a column can produce have been discussed. It has also been shown that the data produced can be used for qualitation and quantitation.

I would like to point out that this information is a start. The practice of liquid chromatography is an art as well as a science. In order to become a practicing liquid chromatographer, you need to supplement the information provided in this chapter with a great deal of experience in the laboratory.

Several works are available that handle this topic more extensively than I have. I highly recommend that you read the following books:

1. *Practical Liquid Chromatography* by Yost, Ettre & Conlon. Published by Perkin-Elmer Corporation.
2. *The LDC Basic Book on Liquid Chromatography* by Schram. Published by Milton Roy Company.
These books will be helpful as you try to make the transition from student of analytical chemistry to practitioner of liquid chromatography.

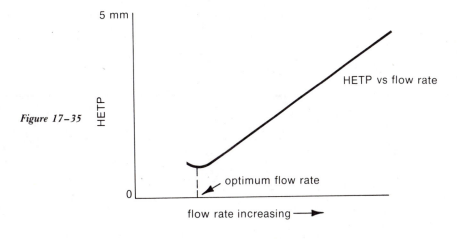

Figure 17–35

Problems

1. Consider the simulated chromatogram (Fig. 17–36) of selected anticonvulsants separated on C_{18} bonded silica. The mobile phase is 10 per cent acetonitrile in water. (Figure 17–36 is also used in problem 2.)

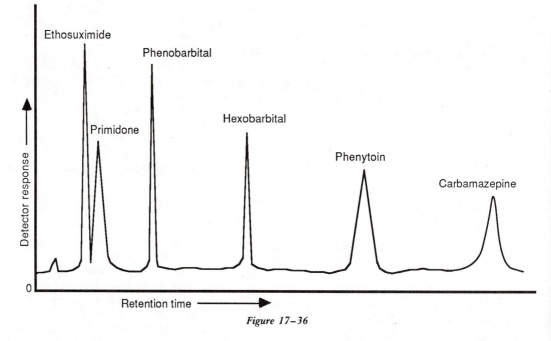

Figure 17–36

The following chromatogram was obtained on a patient's serum. Which anticonvulsants are present? (Conditions for this chromatogram were identical to those for the chromatogram in Figure 17–36.)

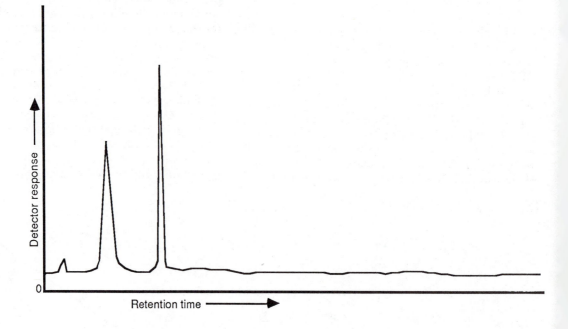

2. The following chromatogram was obtained on a patient's serum. Which anticonvulsants are present? (Conditions for this chromatogram were identical to those for the chromatogram in Figure 17–36.)

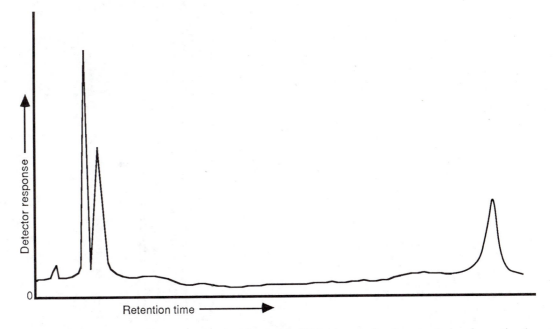

3. Assume that the chromatogram in Figure 17–36 is the result of an analysis of a patient's serum. The peak height for the phenobarbital peak in Figure 17–36 is 56 units. If a phenobarbital standard of 5 mg/dl produces a peak of 93 units under identical operating conditions, what is the concentration of phenobarbital in the patient's serum?

4. Assume that the chromatogram in Figure 17–36 is the result of an analysis of a patient's serum. The peak height for the primidone peak in Figure 17–36 is 40 units. If a primidone standard of 3 mg/dl produces a peak of 98 units under identical operating conditions, what is the primidone level in the patient's serum?

5. The carbamazepine level in the patient's serum must be determined. The patient's serum produces the chromatogram in Figure 17–36; the area under the peak is 29.3 units. If the analysis of five standards produces the following data, what is the patient's serum carbamazepine level?

Solution Analyzed (mg/dl)	Peak Area (mm²)
Blank	0
1	19.7
2	39.9
3	61.0
4	80.5
5	98.7

The common monovalent cations can be separated by ion chromatography. The following chromatogram is an example of such a separation.

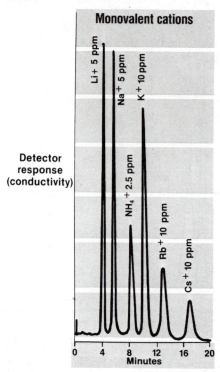

Courtesy of Dionex Corporation, Sunnyvale, Calif. Reprinted with permission.

6. A solution that is 20 ppm in rubidium (Rb^+) and 20 ppm in ammonium (NH_4^+) is chromatographed. If the rubidium peak height is 47 mm and the ammonium peak height is 92 mm, what would be the value of f in Equation 17–3? Assume that rubidium is used as the internal standard in the next question.

7. The Rb^+ concentration of an ammonium unknown is made to be 20 ppm. The unknown is chromatographed. The peak height of the Rb^+ is measured as 47 mm, while the peak height of the NH_4^+ peak is 43 mm. What is the ammonium concentration in the unknown? Assume that the operating conditions are the same in problems 6 and 7.

8. Referring to the preceding chromatogram, assume that sodium ion is to be analyzed using lithium ion as the internal standard. What is f from Equation 17–3?

9. Continuing problem 8, if the lithium concentration of a sodium unknown is made to be 50 ppm, and the lithium peak found on analysis is 26 units, what is the sodium concentration if the sodium peak is 73 units? Assume that the operating conditions are identical in problems 8 and 9.

10. If the separation in Figure 17–36 were done on a 30-cm column, what would be the HETP for carbamazepine?

11. If the small peak that is eluted before ethosuximide is a nonretained compound, what is the relative retention of carbamazepine? (Refer to Fig. 17–36.) Use ethosuximide as the retention time standard.

12. If the separation in Figure 17–36 were done on a 30-cm column, what would be the HETP for the phenobarbital peak?

13. If the first small peak in Figure 17–36 is a nonretained compound, what is the relative retention of phenobarbital? Use ethosuximide as the retention time standard.

18

Ion-Exchange Chromatography

Around the turn of the century, scientists studying soils discovered that the cations in certain soils can be easily traded for other cations. If these soils are placed in a column, such as the one in Figure 17–1 (p. 173), and if a solution containing a different cation is passed through the column, the solution that is eluted contains the cations from the soil, and the cations from the solution are retained in the soil. This phenomenon is called ion exchange, and the soils that exchange ions are called ion exchangers. The ion-exchange properties of these materials led to the development of a number of useful applications, which are discussed in this chapter.

The fact that soils are marginal in their function as ion exchangers has led to the development of synthetic exchangers. The synthetic exchangers are very effective exchange materials. Let's consider the properties of these polymeric organic materials.

ION-EXCHANGE RESINS

The most common synthetic ion exchangers are prepared from copolymers of styrene and divinyl benzene. The styrene in these copolymers is present in the form of long chains. Such a chain is illustrated in Equation 18–1. In these structures N is a large whole number. The divinyl benzene is included in these copolymers to serve as a cross-linking agent. Two styrene chains that are cross-linked are shown in Equation 18–2.

In the most common copolymer, there are 8 molecules of divinyl benzene for every 92 molecules of styrene. Such a copolymer is said to be 8 per cent cross-linked.

It is important to realize that the product illustrated in Equation 18–2 is a two-dimensional representation of a three-dimensional structure. As

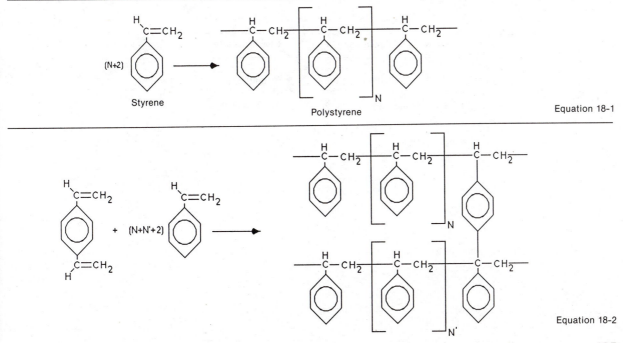

Equation 18-1

Equation 18-2

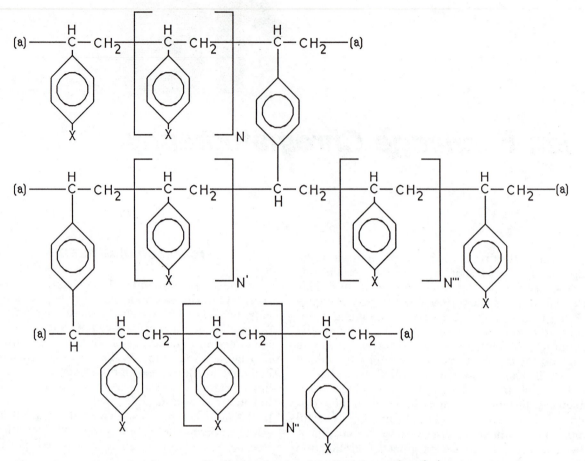

Figure 18–1. Structure of a substituted resin, where X represents the substituent group.

TABLE 18-1. Some Common X Groups

Chemical Structure	Functional Group	Chemical Nature	Type of Exchange
$-SO^-\,H^+$	Sulfonic acid	Strong acid	Cation
$-COO^-\,H^+$	Carboxylic acid	Weak acid	Cation
$-CH_2COO^-\,H^+$	Carboxymethyl (CM)	Weak acid	Cation
$-CH_2\overset{+}{N}(CH_3)_3Cl^-$	Quaternary ammonium	Strong base	Anion
$-CH_2\overset{+}{N}\!\!<^{CH_3}_{CH_3}\!\!-CH_2CH_2OH(Cl^-)$	Quaternary ammonium	Strong base	Anion
$-CH_2NH^+\!\!<^{CH_3}_{CH_3}\ OH^-$	Tertiary ammonium	Weak base	Anion
$-CH_2CH_2NH^+\!\!<^{CH_2CH_3}_{CH_2CH_3}\ OH^-$	Diethylaminoethyl (DEAE)	Weak base	Anion

the copolymer forms, it builds in all three dimensions. The resulting structure is insoluble in all solvents except those that destroy the polymer. The copolymer particles, called resins, are generally spherical. The resin beads come in many different particle sizes.

The 8 per cent cross-linking value is not the only value that is used. Resins with 6, 4, and 2 per cent cross-linking values are available. The smaller the amount of cross-linking, the greater the porosity of the resin.

Once these beads of styrene–divinyl benzene resin have been prepared, it is possible to introduce various chemical groups via standard organic chemical reactions. These groups are added in the para position of the styrene rings. The structure of such a substituted resin is shown in two dimensions in Figure 18–1.

Carefully note the structure in Figure 18–1; one could view this structure as the basic repeating unit of the ion-exchanger resin. There are six unfilled bonds, which are marked by (a). These bonds are the sites of attachment for more of the same repeating units. Also note that N, N', and N'' are large whole numbers that generally are not the same number.

In the styrene–divinyl benzene series of exchangers, four types of substituent groups are used. These groups are described as the strongly acid, the weakly acid, the weakly basic, and the strongly basic. Examples of these groups are provided in Table 18–1.

In order to see how these groups do ion exchange, let's use the sulfonic acid group as an example. Picture each of the X's in Figure 18–1 being replaced by a -SO_3H group. The -SO_3H group is highly ionic. With these groups attached, a resin bead becomes a very large polyanion that is insoluble in all solvents except in those that destroy it. The hydrogen ions are electrostatically bound to the sulfonate groups, but they are exchangeable for other cations. The cations that are electrostatically bonded to a resin are called counter ions. Let's represent this cation exchanger as $ResSO_3H$. When a cation solution is placed in contact with $ResSO_3H$, an equilibrium reaction takes place. Using a sodium solution as an example, the reaction is

$$ResSO_3H + Na^+ \rightleftharpoons ResSO_3Na + H^+$$

This reaction is truly an equilibrium reaction; it is much like the other equilibrium reactions you have studied. Its behavior is predicted by the law of mass action, and it has an equilibrium constant, K. The value of K per se is seldom calculated, but the relative values of K are well known and are called relative selectivities. In Table 18–2 some relative selectivities for a commercial resin are listed. As

TABLE 18–2. Ion Selectivity for AG 50W-X8

Counterion	Relative Selectivity for AG 50W-X8
H^+	1.0
Li^+	0.85
Na^+	1.5
NH_4^+	1.95
K^+	2.5
Rb^+	2.6
Cs^+	2.7
Cu^+	5.3
Ag^+	7.6
Mn^{2+}	2.35
Mg^{2+}	2.5
Fe^{2+}	2.55
Zn^{2+}	2.7
Co^{2+}	2.8
Cu^{2+}	2.9
Cd^{2+}	2.95
Ni^{2+}	3.0
Ca^{2+}	3.9
Sr^{2+}	4.95
Hg^{2+}	7.2
Pb^{2+}	7.5
Ba^{2+}	8.7

Courtesy of Bio-Rad Laboratories, Richmond, Calif. Used with permission.

Table 18–2 shows, the relative selectivity values vary over a considerable range.

Having presented this basic background information, I would like to point out that divinyl benzene copolymers are not the only polymers used as the resin backbone. Some are prepared by the polymerization of acrylic acid or one of its relatives. These reactions are done in the presence of divinyl benzene; the divinyl benzene provides the cross-linking as before. The acrylic acid provides the long chains via the following simplified reaction. The carboxyl group provides the ion-exchange sites.

Acrylic acid / Polyacrylic acid

If anion exchanges are desired, then acrylonitrile is used. Divinyl benzene is still the cross-linking agent. The long chains come from the following reaction:

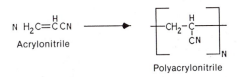

Acrylonitrile / Polyacrylonitrile

The nitrogen of the polynitrile can be converted to a nitrogen-containing anion-exchange group.

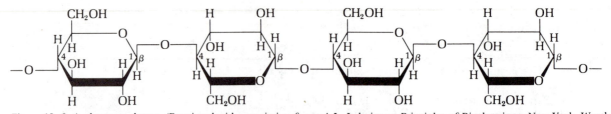

Figure 18–2. A glucose polymer. (Reprinted with permission from: A.L. Lehninger, Principles of Biochemistry, New York, Worth Publishers, Inc., 1982.)

Other exchanges are prepared by introducing carboxyl groups or diethylaminoethyl groups into cellulose, dextran, or agarose. The carboxyl groups are abbreviated CM in the literature; DEAE is the common abbreviation for the diethylaminoethyl group. Cellulose, dextran, and agarose are polymers of glucose that have been cross-linked. These exchangers are very hydrophilic and are used more in research work than in other settings. An example of a glucose polymer is shown in Figure 18–2. The CM or DEAE groups are bonded through the oxygen in the hydroxyl groups of the polymer.

ION-EXCHANGE APPLICATIONS

Using this background, let's consider some of the major uses of cation exchangers. Most uses of ion-exchange resins require that the resin be placed in a column (Fig. 18–3). The column is necessary in order to manipulate the equilibrium reaction of

the resin. The reason for this will be seen in the next paragraph. Let's consider the first type of practical application of cation exchangers.

Ion Interconversions. If we need to convert the cation content of a solution into an equivalent amount of a second cation, we can do so by passing the solution through a column of resin that has the desired cation as its counter ion. For example, if we need to convert the sodium ion content of a solution into protons, we can use $ResSO_3H$.

$$ResSO_3^-H^+ + Na^+ \rightarrow ResSO_3^-Na^+ + H^+$$

Even though the reaction of ions with a resin is an equilibrium reaction, the use of a column forces the reaction completely to the right. This occurs because the hydrogen ions set free in the reaction move down the column. Once they move down the column, they are not available to enter the reverse reaction. This has the effect of forcing the reaction to completion.

For a second example, consider the conversion of the sodium ion content of a solution into potassium ions. This conversion could be done with $ResSO_3K$. The reaction would be

$$ResSO_3^-K^+ + Na^+ \rightarrow ResSO_3^-Na^+ + K^+$$

If a resin has a very great selectivity for the counter ion that it has and a low selectivity for the ion that you want to give it, it can be difficult to force the reaction.

A practical use of this technology is the production of "soft" water. Soft water is water that has been passed through $ResSO_3Na$. The undesirable ions in the water (Ca^{+2} and Mg^{+2}) are retained by the resin as counter ions while an equivalent amount of sodium ions is released.

An additional practical use of this technology is the production of deionized water. Before we can consider this matter, it is necessary to learn about anion-exchange resins. Consider Figure 18–1 again. If the Xs in the structure were replaced by

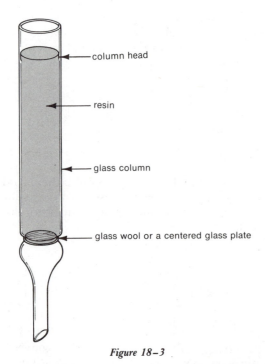

column head

resin

glass column

glass wool or a centered glass plate

Figure 18–3

$$- CH_2\overset{+}{N}(CH_3)_3\,OH^-$$

the resulting resin would be an anion-exchange resin. The anion exchangers behave just like cation exchangers except that they exchange anions instead of cations. Thus, for example, the reaction would be

$$ResCH_2N[CH_3]_3^+OH^- + Cl^- \rightarrow ResCH_2N[CH_3]_3^+Cl^- + OH^-$$

So an anion-exchange resin can be used to convert anions to equivalent amounts of other anions.

With this knowledge of cation-exchange resins and anion-exchange resins as background, let's think about the purification of ionically contaminated water. If you need ionically pure water for your work, and your water source is contaminated by some ionic material, it is possible to use ion exchangers to deionize the water. The ion-exchange column required is one that contains a quantity of $ResSO_3^-H^+$ as well as a quantity of $ResCH_2N(CH_3)_3^+OH^-$. This is a so-called mixed-bed column. The mixed-bed column deionizes the water by converting any cations present to hydrogen ions and any anions present to hydroxide ions. The hydrogen ions react with the hydroxide ions to produce water. The resulting water is highly purified with respect to ionic contaminates.

Removal of Interferences. A second application of ion exchangers is the removal of ions from solution that would interfere with a chemical analysis. For example, the determination of calcium in the presence of phosphate is difficult. If we want to remove the phosphate, we can add the Ca^{+2} solution to a column of resin in the hydrogen form. The Ca^{+2} ions bind to the cation resin and are immobilized in the column. The phosphate, being a negative ion, passes through the column and is converted to phosphoric acid. Distilled water could be added to the column to wash out the last traces of phosphoric acid. During this washing the Ca^{+2} ion would not move. Then, if nitric acid were added, the resin would be converted back to the acid form (remember the law of mass action), and the calcium ions would be set free and eluted as calcium nitrate.

An alternative approach is to use an anion exchanger to retain the phosphate and elute an anion that does not interfere. In either of these two approaches, the interfering phosphate could be removed from the calcium. The solution containing the calcium could then be subjected to analysis. The use of this approach is common in the analysis of urinary porphyrins and plasma ammonia.

The Concentration of Dilute Solutions. A third use of an ion-exchange resin is in the concentration of ions from very dilute solutions. If we had a million liters of 10^{-6} molar gold chloride solution, and we wanted the gold, we could pass the million liters of solution through a cation-exchange column. When the million liters had passed through the cation-exchange column, we could elute the gold with 1 liter of concentrated hydrochloric acid. This would concentrate the gold from 10^{-6} molar to 1 molar. The gold could easily be electroplated from the concentrated solution, whereas it could not be directly recovered from the dilute solution.

This approach is used to clean up spills of radioactive water from nuclear reactor accidents. The ionic radioisotopes can be picked up and concentrated on suitable resins.

Ion Chromatography. Modern ion chromatography has much in common with liquid chromatography (see Chapter 17). The mobile phase in both methods of separation is a liquid. The same approaches to sample introduction are used in both. The data output in each of the separation methods is a chromatogram. The same approaches to qualitative and quantitative analysis are used in both (see *Quantitation* and *Qualitation* in Chapter 17). In some applications of ion chromatography, even the same detectors are used. Knowledge of the material covered in Chapter 17 is necessary for understanding the discussion that follows.

Remembering the similarities between liquid chromatography and ion chromatography, consider the generalized ion chromatograph in Figure 18–4.

The ion chromatograph consists of two liquid reservoirs, two pumps, a mixer, a pulse damper, a sample introduction system, an analytical column, a suppressor column, and a detector system. The two liquid reservoirs, the two pumps, and the mixer are provided in order to be able to change the composition of the mobile phase. These devices are discussed in detail in Chapter 17 under *Mobile Phase Delivery Systems* and *Factors that Affect Separations*. The sample introduction system used in an ion chromatograph is the same as that used in a liquid chromatograph and is discussed in Chapter 17 under *Sample Introduction System*.

The analytical column is a length of heavy-wall stainless steel tubing that contains an ion-exchange packing. The tubing is much like the tubing that was discussed in Chapter 17 under *Columns for Liquid Chromatography*. One common packing material is 10 μm particles of styrene–divinyl benzene ion exchanger. A second common packing is silica that has been modified chemically to make an ion exchanger. The hydroxyl groups on the silica surface can be replaced by -OCH_2COOH groups to produce cation-exchange capacity. Anion-exchange silica can be prepared by replacing the hydroxyl group with -$OCH_2CH_2NH(CH_2CH_3)_2OH$. These silica-based packings can be prepared from 5 μm to 10 μm porous silica spheres; they have mechanical properties similar to those of the silica liquid chromatographic packings.

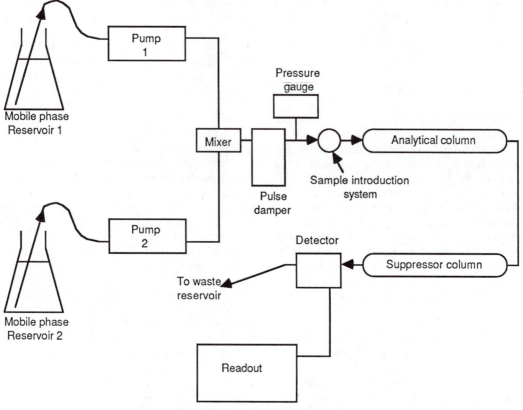

Figure 18–4. An ion chromatograph.

These two types of packing have advantages and disadvantages. The styrene–divinyl benzene exchangers can be used at any pH, whereas the silica packings are restricted to pHs between 2 and about 7 or 8. The silica-based packings are not prone to dimensional changes with variations in ionic strength and solvent composition, but they are sensitive to mechanical shock. Neither flow rates nor mobile phase composition should be changed rapidly. Ion-exchange columns separate things by many mechanisms. Some background on these mechanisms will be provided in this chapter.

The suppressor column may or may not be used in an ion chromatograph; it depends on the detection system. When the system is an electrical conductivity detector, suppressors are frequently used. Let's consider an example. When anions are to be separated on an anion-exchange column, they are frequently driven off the resin with dilute sodium hydroxide solution. This leads to separated anions that are mixed with sodium hydroxide. The high conductivity of the sodium hydroxide solution causes problems for the detector. This situation can be improved by passing the column eluent through a suppressor column that contains a cation-exchange resin in the proton form. The sodium hydroxide interacts with

the resin as follows:

$$ResSO_3^-H^+ + NaOH \rightarrow ResSO_3^-Na^+ + H_2O$$

In this way the NaOH is in effect converted to water. The sample anions, on the other hand, enter the suppressor in the form of sodium salts. These salts undergo the following reaction

$$Na^+X^- + ResSO_3^-H^+ \rightarrow ResSO_3^-Na^+ + H^+ + X^-$$

The HX formed is then detected by the conductivity detector.

In a similar way, the output from a cation-ion exchange column is generally the analyte cations dissolved in an HCl mobile phase. The hydrochloric acid has high conductivity and is a problem for the detector. The HCl can be removed with a suppressor column. An anion-exchange column in the hydroxide form could be used. The fate of the HCl would be

$$ResCH_2N(CH_3)_3^+OH^- + HCl \rightarrow ResCH_2N(CH_3)_3^+Cl^- + H_2O$$

The analyte cations would arrive at the suppressor as chloride salts in solution; they would interact with the suppressor resin as shown below.

$$Cation\ Cl^- + ResCH_2N(CH_3)_3^+OH^- \rightarrow ResCH_2N(CH_3)_3^+Cl^- + Cation\ OH^-$$

The highly ionized cation hydroxide would then be detected.

Suppressor columns must be regenerated from time to time. One unique approach involves the use of ion-exchange membranes. The suppressor regenerates continuously from one side of the membrane even as it is suppressing on the other side of the membrane. Both processes occur simultaneously. The regenerant ions seem to be mobile enough to cross the membrane.

The detectors used in ion chromatography are the common ones used in liquid chromatography plus the highly sensitive electrical conductivity detector. The common liquid chromatographic detectors are discussed in Chapter 17 under *Liquid Chromatographic Detection*.

The typical electrical conductivity detector that is used in ion chromatography consists of a very small tubular flow cell. The cell has an internal volume of about 2 μl and a cell constant, θ, of about 30 cm^{-1}. The design of the cells is shown in Figure 18–5.

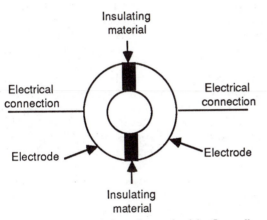

Figure 18–5. An end view of a conductivity flow cell.

The conductivity flow cell is powered with 1000 Hz voltage of less than 3 volts; the value for conductivity of the solution passing through the cell is found by measuring the resulting current. The control of the temperature of the cell and of the associated wiring is very important. These detectors frequently have circuitry that compensates for small temperature changes. They also have electronic provisions for subtracting the background conductivity of the mobile phase. These detectors are very sensitive and find widespread application in ion chromatography.

MECHANISMS OF SEPARATION

Many mechanisms of separation are thought to be operative in ion chromatography. Most of the important clinical applications of ion chromatography seem to have at least two and sometimes three or four of these mechanisms operating simultaneously. This is certainly true in the separation of amino acids and peptides. Because the picture in the clinical examples is so clouded by complexity, I would like to use some examples that are not clinical but are straightforward enough to be understood. The background provided here may be of use in understanding parts of clinical separation.

There seem to be five mechanisms of separation on ion exchangers. Of these, four are important mechanisms of separation in clinical chemistry. They are (1) resin-selectivity phenomenon, (2) ion-exclusion phenomenon, (3) sample-ion equilibria phenomenon, (4) adsorption.

Resin Selectivity. Consider the first mechanism. Picture an ion chromatograph, such as the one in Figure 18–4, that has 5×10^{-3} molar HCl in reservoir 1 and nothing in reservoir 2. The column used in the instrument would contain ResSO$_3$H. Also picture a suppressor column containing ResCH$_2$N(CH$_3$)$_3^+$OH$^-$. The instrument would have an electrical conductance detector.

Let's start the separation by turning on pump 1. With this done, a sample of a chloride solution containing Na$^+$, NH$_4^+$, K$^+$, Rb$^+$, and Cs$^+$ could be injected into the instrument. As discussed previously in this chapter (*Ion Interconversions*), these ions will tend to be picked up by the resin and will be held as counter ions. With HCl being used as the mobile phase, however, there is an ongoing struggle between the cations and the hydrogen ions for the exchange sites. The extent to which each of the cations competes for the sites depends on the relative selectivity of the resin for the cations. An ion with high selectivity will spend more time as a counter ion than will one with low selectivity. When an ion is in a column, it must be either a counter ion or in the flowing mobile phase. Hence, ions with lower selectivity are eluted first, whereas those with higher selectivity spend more time as counter ions and are eluted later. This leads to the separation of different ions. An example of this type of separation is shown in Figure 18–6. The selectivity data presented in Table 18–2 predict that the separation order should be as it is.

Before we consider the next separation mechanism, let's consider the suppressor column that would be used in the previous example. The best suppressor for this separation would be an anion-exchanger resin in the hydroxide-ion form. Such a resin would completely remove the HCl from the mobile phase by the following reaction:

$$ResCH_2N(CH_3)_3^+OH^- + HCl \rightarrow ResCH_2N(CH_3)_3^+Cl^- + H_2O$$

The analyte chlorides would be converted to cation hydroxides. Using KCl as an example, the reaction is

$$ResCH_2N(CH_3)_3^+OH^- + KCl \rightarrow ResCH_2N(CH_3)_3^+Cl^- + KOH$$

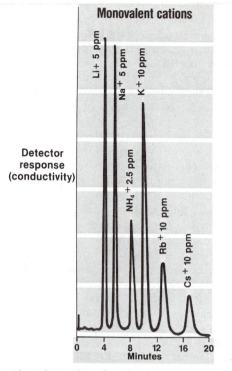

Figure 18-6. Separation of monovalent cations on strongly acid styrene, divinyl benzene ion exchanger. 5×10^{-3} molar HCl. (Courtesy of Dionex Corporation, Sunnyvale, Calif. Reprinted with permission.)

In this way the suppressor removes the highly conductive background caused by the HCl and leaves the analyte in a highly conductive form.

Ion Exclusion. Next let's consider an ion-exclusion type separation. Picture the ion chromatograph in Figure 18-4 set up as it was in the previous example. Picture the cation-exchange column replaced by an anion-ion exclusion column. The mobile phase reservoir would be filled with 10^{-3} molar hydrochloric acid. With the pump turned on and

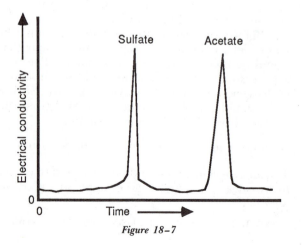

Figure 18-7

the detector baseline adjusted, a sample of sulfuric acid mixed with acetic acid could be injected. If this were done, the chromatogram illustrated in Figure 18-7 would result.

This separation has taken place because of the combination of two phenomena. First, ion-exclusion columns that are designed to separate ions are made from styrene-divinyl benzene packings, which are highly sulfonated. When such a column is used with a mobile phase that has low ionic strength, the negative charge on the resin surface prevents anions from entering the resin beads. On the other hand, molecules in the sample can enter the resin beads and in so doing are temporarily removed from the mobile phase. This slows their passage through the column.

The separation in Figure 18-7 takes place also because of the characteristics of the two sample constituents. The acetic acid in the sample has little tendency to ionize. Thus the molecules of acetic acid are free to enter the resin beads. The sulfuric acid, on the other hand, is highly ionized. The sulfate ions from the sulfuric acid have little tendency to form molecules. Hence the sulfate cannot enter the resin and, therefore, is rapidly eluted.

Before we consider a more complex ion-exclusion separation, let's have a quick look at the suppressor column and its behavior during this anion separation. The common suppressor that is used with HCl mobile phases in this type of separation is one that contains $ResSO_3Ag$. When the HCl interacts with such a suppressor, AgCl is precipitated and the resin is converted to $ResSO_3H$. The anions in the sample do not interact with the suppressor resin. The hydrogen ions that were associated with the anions in the sample are retained by the suppressor. The salts that actually reach the detector are silver sulfate and silver acetate. The salts are sufficiently soluble to remain in solution and hence can be detected by conductance.

Ion-exclusion columns can do separations of weak acids very well. If the mobile phase is selected critically in order to get a lot of variation in the per cent ionization, separations of even closely related acids can be done. Such a separation is illustrated in Figure 18-8. This separation was done on a highly sulfonated ion exclusion-anion column. The sample was injected into a stream of 0.005 molar hydrochloric acid. The HCl causes the weak acids to be protonized but to a somewhat greater or lesser extent depending on their acid ionization constant (Ka) values. The greater the fraction of a given acid in the molecular form, the slower it moves. Consider the Ka values in Table 18-3. Notice that with the exception of formic acid, the weak acids separate in order of decreasing Ka. At a given pH, the maleic acid, because of its larger Ka value, will be more highly ionized than the propionic acid. It was mentioned previously in this section that more than one

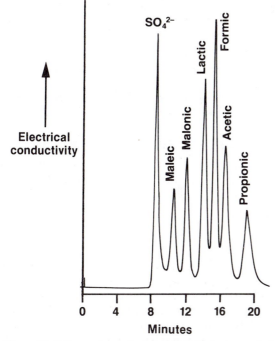

Figure 18–8. Separations of some weak acids by ion exclusion. Mobile phase 5×10^{-3} molar HCl in gradient to 1×10^{-3} molar HCl during the separation. (Courtesy of Dionex Corporation, Sunnyvale, Calif. Reprinted with permission.)

TABLE 18–3. Ka Values for Weak Acids

Compound	Acid Ionization Constant (Ka_1)
Maleic acid	1.00×10^{-2}
Malonic acid	1.43×10^{-3}
Lactic acid	1.32×10^{-4}
Formic acid	1.77×10^{-4}
Acetic acid	1.75×10^{-5}
Propionic acid	1.34×10^{-5}

reservoirs were used to change from pH 3.1 to pH 5.3 and then on to pH 7.4. This variation in pH is used to successively force the amino acids into their anion forms. They enter the column in their zwitterion forms.

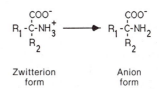

The zwitterions behave as if they were neutral molecules. As neutral molecules, they can enter the ion-exclusion resin. As pH increases, they are converted into anions and are forced from the resin. Once they are forced from the resin, they leave the column very rapidly.

In our example, amino acids have been reacted with a fluorogen in a post-column reactor, and they have been detected by fluorescence. No suppressor is required.

Sample-Ion Equilibria. The next type of separation is based on the equilibria of the sample ions. This type of separation was done many years ago

mechanism of separation can operate at a time. The lactic acid–formic acid reversal is an example of such a phenomenon.

These separations serve as the basis of some of the amino acid analyzers that are common in medical research. Such a separation is illustrated in Figure 18–9. In this separation, the two pumps and

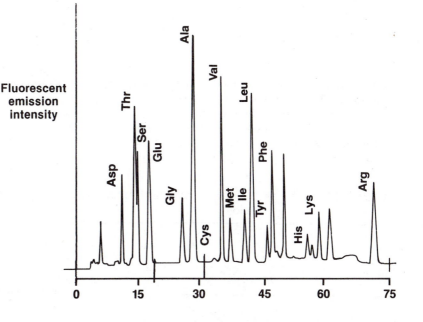

Figure 18–9. Notice that the mobile phase changes at 19 minutes and at 32 minutes. (Courtesy of Dionex Corporation, Sunnyvale, Calif. Reprinted with permission.)

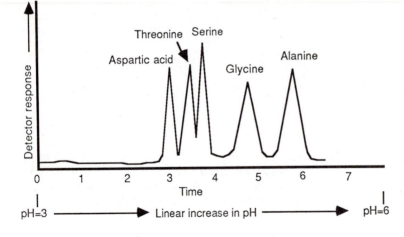

Figure 18–10. Computer-simulated chromatogram for five amino acids.

by Moore and Stein. Moore and Stein separated amino acids on Dowex 50X4; this resin is a styrene–divinyl benzene resin with sulfonate ion-exchange sites. In their work, amino acids were injected on column at low pH. The amino acids were in their cation form and were picked up by the resin as counter ions.

$$
\begin{array}{c}
COOH \\
| \\
R_1 - C - NH_3^+ \\
| \\
R_2
\end{array}
$$

Amino acid in
cation
form

The pH of the mobile phase was then caused to increase slowly. This increase in pH caused the carboxyl group of the most acidic amino acid to revert to the zwitterion form.

$$
\begin{array}{c}
COO^- \\
| \\
R_1 - C - NH_3^+ \\
| \\
R_2
\end{array}
$$

Amino acid in
zwitterion
form

The zwitterions were perceived by the resin as neutral molecules. In this way, the most acidic amino acid was able to leave the column.

As the pH continued to increase, the next most acidic amino acid was released and so on. The resulting chromatogram for five selected amino acids

is shown in Figure 18–10. This separation follows the order of increasing pK_1 values (Table 18–4).

In separations in which the concentration of an equilibrium ion is varied, the difference in the equilibrium constants frequently affects the separations. This phenomenon is a factor in the separation of amino acids by ion exclusion as well as in the previous example.

Adsorption. The fourth mechanism of separation is the mechanism of adsorption. Some sample molecules adsorb to the divinyl benzene–styrene copolymer. This happens even if no ion-exchange groups are present. It can also happen if ion-exchange groups are present. If an adsorbed sample is treated with a displacing agent, the various sample constituents can be driven from the resin surface. This is thought to be the mechanism of the separation illustrated in Figure 18–11.

TABLE 18–4. The pK_1' Values for Selected Amino Acids

Aspartic acid	$pK_1' = 1.88$
Threonine	$pK_1' = 2.15$
Serine	$pK_1' = 2.21$
Glycine	$pK_1' = 2.34$
Alanine	$pK_1' = 2.35$

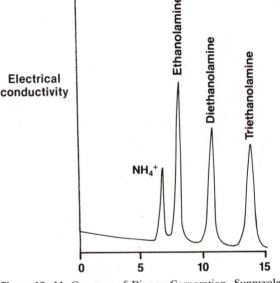

Figure 18–11. Courtesy of Dionex Corporation, Sunnyvale, Calif. Reprinted with permission.

ODDS AND ENDS

When working with ion-exchange resins in non–high performance ion chromatography applications, it is not uncommon for a worker to prepare ion-exchange columns for his or her own use. It is important to realize that dry resins soak up a lot of water and increase in volume. A bead may double or triple its size by swelling in water. For this reason, resins are always presoaked until the swelling has

TABLE 18–5. Analytical Grade Resin Substitutions

Type and Exchange Groups	Bio-Rad Analytical Grade Ion Exchange Resins	Dow Chem. Company "Dowex"	Diamond Shamrock "Duolite"	Rohm & Haas Co. "Amberlite"	Ionac Chemical Company	Permutit Company (England)	Mitsubishi Chem. Co. "Diaion"	Bayer "Lewatit"[a] / Permutit, Inc. Germany[b]	Fisher "Rexyn"[c] / Merck[d]	Akzo Chem. Co. "Imac"[e] / Montedison "Kastel"[f]	Wolfen Dye Factories "Wofatit"
Type 1, Strongly Basic polystyrene gel-type resins $\varnothing\text{-CH}_2\text{N}^+(\text{CH}_3)_3\text{Cl}^-$	AG 1-X2 AG 1-X4 AG 1-X8	1-X2 1-X4 1-X8 (SBR) SBR-P 11	A-143 A-101D ES-131 A-109 A-104	IRA 401 IRA 402 IRA 400 IRA 420 IRA 430 IRP-67, IRN 78	A-540 A-548 ASB-1 A-440 A-546 A-935	Zerolit FF (lightly crosslinked) Zerolit FF Permutit S-1	SA11A, SA11B SA10A, SA10B SA100	M 500[a] M 5020[a] M 5080[a] ESB[b]	201[c] III[c]	S 5-40[e] S 5-50[e] A 500[f]	
Type 1, Strongly Basic non-polystyrene gel-type resins $\varnothing\text{-CH}_2\text{N}^+(\text{CH}_3)_3\text{Cl}^-$		ES-132		IRA 458				MN[a]			L 165 L 150
Type 2, Strongly Basic polystyrene gel-type resins $\varnothing\text{-CH}_2\text{N}^+(\text{CH}_3)_2 (\text{C}_2\text{H}_4\text{OH})\text{Cl}^-$	AG 2-X8	2-X4 2-X8 (SAR)	A-102D	IRA 410	A-550 ASB-2	S-2	SA21A, SA21B SA20A, SA20B	M 600[a] ES[b]		S 5-42[e] A 300[f]	
Strongly Basic, Macroporous resins, type 1 $\varnothing\text{-CH}_2\text{N}^+(\text{CH}_3)_3\text{Cl}^-$	AG MP-1	MSA-1	A-161	IRA 900 IRA 904 IRA 938 IRA 958 Amberlyst A-26, A-27	AFP-100 A 641		PA 304 PA 306, PA 308 PA 310, PA 312 PA 318, PA 320	MP 5080[a]		A 500 P[f]	
Strongly Basic, Macroporous resins, type 2 $\varnothing\text{-CH}_2\text{N}^+(\text{CH}_3)_2 (\text{C}_2\text{H}_4\text{OH})\text{Cl}^-$		MSA-2	A-162	IRA 910	A-642 A-651		PA 404 PA 406, PA 408 PA 410, PA 412 PA 414, PA 416 PA 418, PA 420			A 300 P[f]	
Intermediate Base resins $\text{R-N}^+(\text{CH}_3)_2\text{Cl}^-$ and $\text{R-N}^+(\text{CH}_3)_2 (\text{C}_2\text{H}_4\text{OH})\text{Cl}^-$	Bio-Rex 5		A-30B	IRA 47	A-305 MS-170	F			208[c] 205[c]		
Weakly basic gel-type polystyrene, phenolic or polyamine $\varnothing\text{-CH}_2\text{N}^+(\text{R})_2\text{Cl}^-$	AG 3-X4A	WGR WGR-2	A-6 A-4F ES-375 A-340	IRA 45 IR 4B IRA 47 IRA 68 IRA 60, IRP 58	A-375 A-260	G	WA 10 WA 11	MIH[a] E[b]	207[c] 203[c] 206[c] II[d]	A13, 17, 19[e] A 27[e] A 101[f]	MD
Macroporous intermediate and weak base exchangers $\varnothing\text{-CH}_2\text{N}^+(\text{R})_2\text{Cl}^-$		MWA-1	A-7 ES 308 A-374 ES 368 A-378 ES 366 A-561	IRA-35 IRA-93 IRA-94 IRA-99 Amberlyst A-21	MG-1 A-328 AFP 329		WA 20 WA 21 WA 30	MP 7080[a]		A 20[e]	
Strongly acidic polystyrene, gel type $\varnothing\text{-SO}_3^-\text{H}^+$	AG 50W-X2 AG 50W-X4 AG 50W-X8 AG 50W-X10 AG 50W-X12 AG 50W-X16	50W-X2 50W-X4 50W-X8 HCRW-2 50W-X10 HCR, HCR-S	C-20 C-225X10 C-20X12	IR-116, IR-118 IR-120, IRN-77 IRN-218 IRN-163 IRN-169 IR-122 IR-124, IR-130 IR-140, IR-169	C-298 C-249, CF C-240, 242 C-250, 251 C-253, 255 C-256, 257 C-258, 299	Zeocarb 225 (X4) Zeocarb 225	SK 102, SK 103 SK 104, SK 106 SK 1A, SK 1B SK 110, SK 112 SK 116	PN[a] S 100[a] S 1080[b] RS[b]	101[c] I[d]	C-22[e] C-12[e] C-300[f]	KPS 200
Strongly acidic Non-polystyrene, $\varnothing\text{-SO}_3^-\text{H}^+$						Zeocarb 215					F, P
Macroporous Strong Acid Cation Exchangers $\varnothing\text{-SO}_3^-\text{H}^+$	AG MP-50	MSC-1	C-25D C-26 C-3	200 252 Amberlyst 15	CFP 110 CFS		PK 204, PK 208 PK 212, PK 216 PK 220, PK 224 PK 228			C 8P[e] C 16P[e]	
Weak acid cation exchange resins $\text{R-COO}^-\text{Na}^+$	Bio-Rex 70	CCR-2 MWC-1	C-433 C-464	IRC-84 IRC-50, CG-50 DP-1, IRC-72 IRP-64	CC CCN	Zeocarb 226	WK 10 WK 11	CNO[a] C[b] CP 3050[b]	102[c] IV[d]	C 101[f] Z-5[e]	CN CP 300
Weakly Acidic chelating resin $\varnothing\text{-CH}_2\text{N} \begin{smallmatrix} \text{CH}_2\text{COO}^-\text{H}^+ \\ \text{CH}_2\text{COO}^-\text{H}^+ \end{smallmatrix}$	Chelex 100	A-1	ES-466				CR 10				
Mixed Bed Resins											
$\varnothing\text{-SO}_3^-\text{H}^+$ & $\varnothing\text{-CH}_2\text{N}^+(\text{CH}_3)_3\text{OH}^-$	AG 501-X8	MB-3	GPM-331 G	IRN-150 MB-1	NM-60, NM-40 NM-65, M-747 MI-747, NM-42		SMN-1		300[c] V[d]		
$\varnothing\text{-SO}_3\text{H}^+$ & $\varnothing\text{-CH}_2\text{N}^+(\text{CH}_3)_3\text{OH}^-$ indicator dye	AG 501-X8 (D)								I-300[c]		

Courtesy of Bio-Rad Laboratories, Richmond, Calif. Reprinted with permission.

stopped; then they are placed in the column. If the dry resin were added to a column, and water were then added, the column might explode from the pressure.

Resins, as they are sold, are contaminated with various ions and small organic molecules. The resins are prepared for use by repeated washing and repeated cycling from the acid to metal ion form for cation resins, or from the hydroxide to anion form for anion resins. After the resins are cleaned up, they are stored in the appropriate form in the swollen state. When one wishes to make a column,

one makes a slurry of a proper amount of resin and pours it into the column. After the resin has settled and excess liquid has drained off, the column is ready for use. *Never let the column go dry during its useful life*. This will ruin the column. A small plug of glass wool or clean, fine sand is often placed on top of the resin in the column to prevent it from being disturbed by the addition of liquid to the column.

Most companies sell many different forms and types of resins. The resins vary as to the type of ionic group present, the particle size, and the de-

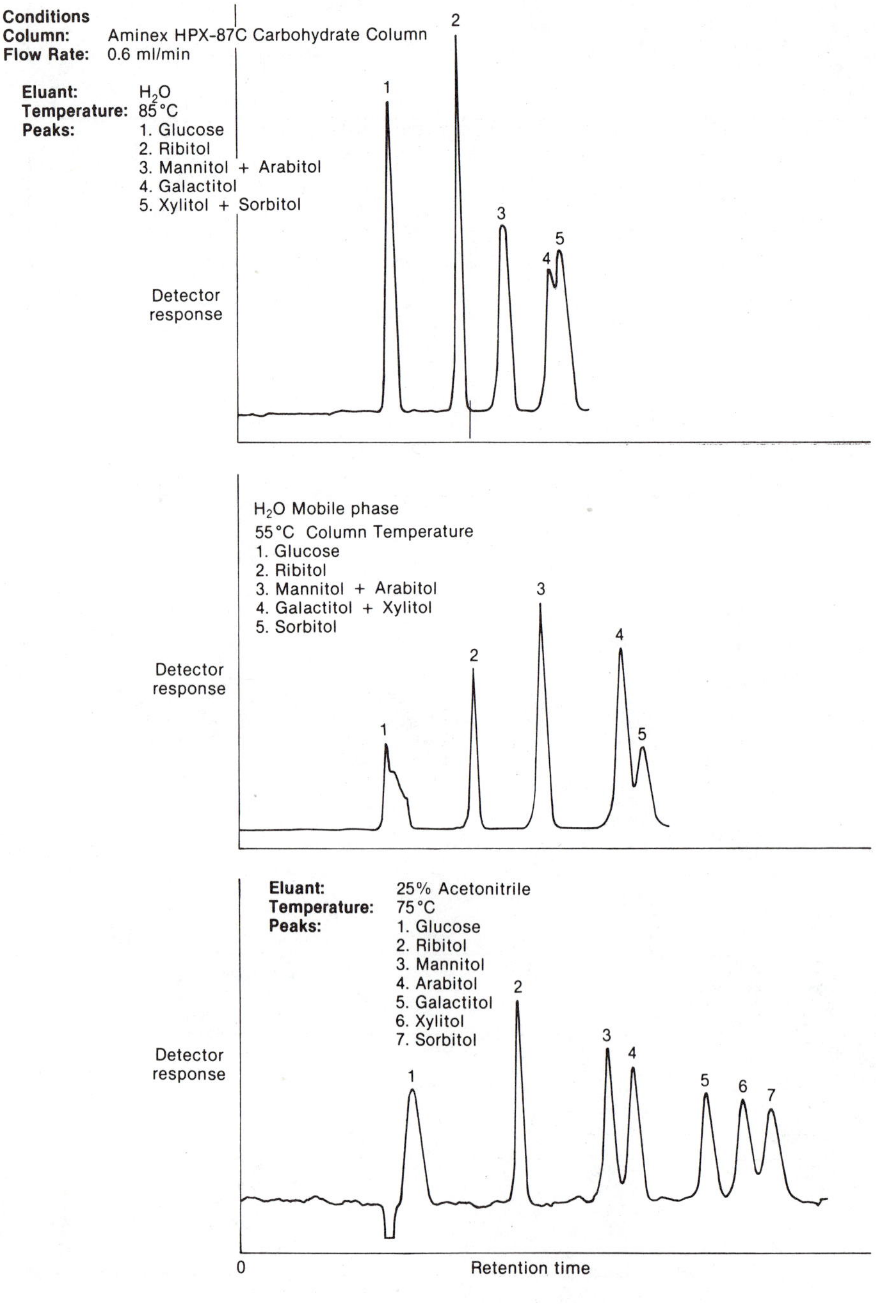

Figure 18–12. Courtesy of Bio-Rad Laboratories, Richmond, Calif. Reprinted with permission.

gree of cross-linking. Resins prepared from mixtures containing larger amounts of divinyl benzene are more likely cross-linked and swell less. Highly cross-linked resins swell only a little; hence the pores in the beads are only a few angstroms. In lightly cross-linked resins, the swelling is very great; the pores in such resins are sometimes greater than 100 angstroms. Resins also vary with respect to their ion-exchange capacity. The capacity refers to the number of equivalents of ions that can be exchanged.

Chemically, the styrene–divinyl benzene resins are quite stable; only strong oxidizing and reducing agents cause decomposition. Most of these resins are stable up to 100°C. Strongly basic anion-exchange resins in the hydroxide form start to decompose above 60°C.

Unfortunately, resins are sold under trade names. In Table 18–5 the resins from different companies along with their trade names are listed.

In addition to organic ion-exchange resins, a few inorganic ion-exchange solids exist. These are generally sodium-aluminum silicates. Although they were discovered first, they are not used as much as organic resins at the present time. Almost all medical applications are done with organic resins. Some liquid ion-exchange compounds are also known. The liquid exchangers are used in some potentiometric electrodes. The valinomycin membrane electrode that is used to measure potassium activity is a good example.

In ion chromatography remember that column temperature can affect separations. The presence of organic solvents can have an effect as well, particularly in high-performance ion chromatography. This fact is illustrated by the chromatograms in Figure 18–12. These chromatograms have the same time axis. The same flow rate and the same column were used for all three separations.

Gas Chromatography

Gas chromatography and liquid chromatography have much in common. Both techniques use a system to supply mobile phase. Both use the same types of sample introduction systems. Both techniques use chromatographic columns to make separations, and both have detection systems. Both techniques produce chromatograms. Qualitative analysis and quantitative analysis are performed in the same way under both systems. Because of these similarities, Chapter 17 should be considered a prerequisite of the study of this chapter.

The major difference between liquid chromatography and gas chromatography is that gas chromatography uses a gaseous mobile phase. The gaseous mobile phase is called a carrier gas. The physical properties of gases are much different from those of liquids. Because of this fact, there are major differences in the instruments used in these two forms of chromatography.

INSTRUMENTATION

The flow path of a quality gas chromatograph is illustrated in Figure 19–1. Let's consider each of the components used in these instruments.

Mobile Phase Supply System

The mobile phase, or carrier gas, is a fairly pure gas that is purchased in a highly compressed state. The choice of gas is dependent on the instrument's detector. The cylinder generally arrives in the laboratory at 2500 psi and contains 250 cubic feet of gas at standard temperature and pressure (STP).

The two-stage pressure regulator (Fig. 19–2) helps control the flow rate of the carrier gas. The pressure gauge on the regulator nearest the cylinder indicates the pressure of the gas in the cylinder. The gauge farthest from the cylinder indicates the pressure of delivery to the system. It is best if the regulator has a stainless steel diaphragm.

The carrier gas purifier, or scrubber, can be one of several devices, but at the very least it should be a water vapor scrubber. Most gas chromatographic columns do not get along well with water. Its presence adversely affects separation quality.

The next item in the diagram is the reference path. The reference path is present on most chromatographs. Once in a while a chromatograph does not have a reference path, or the path may be present but unused. The reference path is provided as

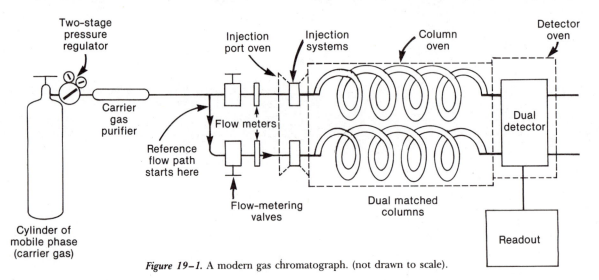

Figure 19–1. A modern gas chromatograph. (not drawn to scale).

Figure 19–2. A two-stage pressure regulator. (Reprinted with permission of Supelco, Inc., Bellefonte, PA 16823. Supelco Catalog 22, p. 92.)

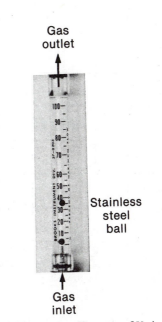

Figure 19–3. Rotameters. (Courtesy of Varian Instrument Division of Varian Techtron Pty. Ltd., Australia. Reprinted with permission.)

an instrument noise reduction system. More will be said about this later.

Next in the flow path is a pair of flow-metering valves. There is one in each of the two paths. These valves, in conjunction with the regulator, are provided to keep the flow rate of the carrier gas constant. The constant flow rate is needed in order to get reproducible retention times for the sample components.

The next device in the flow path is a set of flow meters. These may or may not be present, but they are often provided to help the instrument operator reproduce flow rates on a day-to-day basis. The most common flow meter is a rotameter (Fig. 19–3). The gas passing through the device levitates the ball. The higher the ball, the higher the flow rate. The rotameter does not have a nice correlation between flow rate and scale number; it is nonlinear and is pressure and gas dependent. For specifics, see the literature accompanying your instrument. Two sample calibrations from an instrument in our laboratory are shown in Figures 19–4 and 19–5. Note the nonlinearity, the pressure dependence, and the effect of the different gases.

Sample Introduction System

The chromatograph must have some system for the introduction of samples into the column. Generally, this system consists of a hypodermic syringe

Figure 19–4. Courtesy of Hewlett-Packard Co., Avondale Div., Avondale Penn. Reprinted with permission.

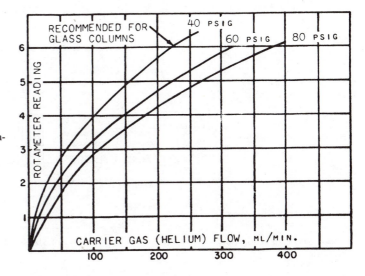

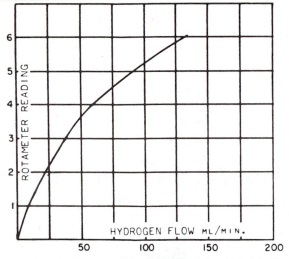

Figure 19–5. Courtesy of Hewlett-Packard Co., Avondale Div., Avondale, Penn. Reprinted with permission.

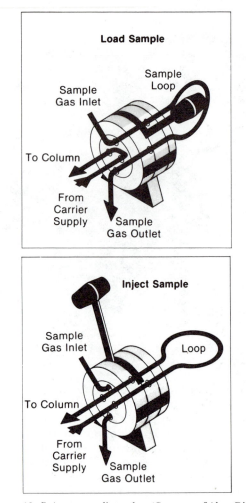

and a silicone rubber septum (Fig. 19–6). The syringe needle is introduced through the septum into the central tube. The needle length is such that it almost reaches the column. The injection port is generally heated in order to vaporize liquid samples. The oven is generally heated to about 30°C hotter than the highest boiling sample constituent. This guarantees that small liquid samples vaporize instantaneously and are rapidly carried into the column.

When gaseous samples at pressures other than atmospheric must be chromatographed, a gas-sampling valve should be used (Fig. 19–7). Its operation is well illustrated in the figure; note the different positions of the handle.

Figure 19–7. A gas-sampling valve. (Courtesy of Altex Division, Beckman Instruments, Inc., San Ramon, Calif. Adapted with permission.)

Gas Chromatographic Columns

The next devices in the flow path are the chromatographic columns, of which there are two major types—packed columns and capillary columns (Fig. 19–8). A packed column is prepared by filling a length of tubing with a packing material. The most commonly used tubing diameter is 1/8 inch outside diameter; other diameters, larger and smaller, are

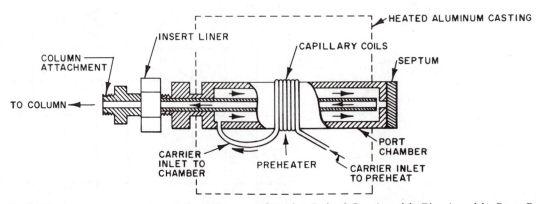

Figure 19–6. Injection port—cross-sectional view. (Courtesy of Hewlett-Packard Co., Avondale Div., Avondale, Penn. Reprinted with permission.)

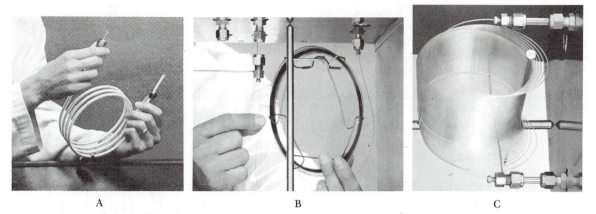

Figure 19–8. Columns; *A,* packed glass column (Supelco Catalog 21). *B,* Silica capillary column (Supelco Catalog 22). *C,* Wide-bore capillary column (Supelco Catalog 22, p. 3). (Reprinted with permission of Supelco, Inc., Bellefonte, PA 16823.)

used. The most common tubing materials are stainless steel and glass. The column tubing is coiled to a suitable diameter for installation in the column oven. In the case of stainless steel, coiling is done after packing; in the case of glass, it is done before packing.

The packing is prepared by coating small particles of so-called support with a suitable low-volatility liquid. The typical supports (Fig. 19–9) are 100 to 120 mesh* particles of fluxed and chemically treated diatomaceous earth. These particles have a surface area of about 1 square meter per gram. The surface of the support is covered by a thin layer of stationary phase. The stationary phase is a high-boiling liquid; it possesses little vapor pres-

---

*Mesh refers to the number of holes per square inch in a sieve.

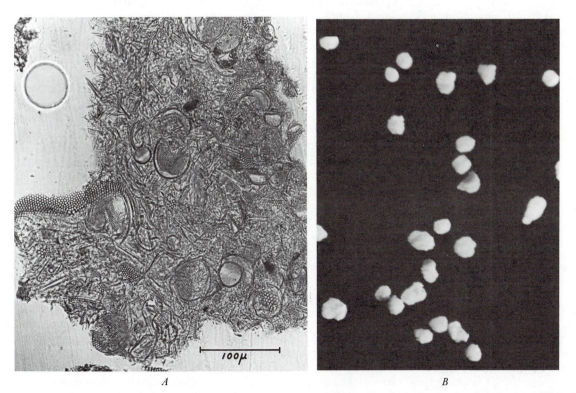

Figure 19–9. A, Photomicrograph of a typical support. (Reprinted with permission of Supelco, Inc., Bellefonte, PA 16823.) *B,* A typical gas chromatographic support, Chromosorb W, 60/80 mesh, magnified 400 times.

sure at the temperature of use. A common stationary phase is OV-17. It is a phenyl methyl silicone polymer that contains 50 per cent w/w phenyl groups and can be used with columns reaching temperatures as high as 350°C. OV-17 has the following structure:

Above 350°C, the vapor pressure of OV-17 becomes excessive, and significant amounts of the silicone will vaporize from the column. The 350°C value is the temperature limit of the stationary phase; all stationary phases have a temperature limit. Gas chromatographic columns can be operated at temperatures above their temperature limits for a brief period of time. This practice does produce drift on the chromatogram baseline. The use of matched columns and dual detectors will keep the drift minimal. Keep in mind that operation above the temperature limit does, in time, destroy the column.

Generally, capillary columns are prepared from either glass or fused silica. They generally are 0.8 mm outside diameter and 0.25 mm inside diameter. Common lengths are 30 and 60 meters. Capillary columns are quite different from packed columns in that there is no support present except for the inside of the capillary tube itself. If one were to examine the cross-section of a capillary column, one would find that the stationary phase covers the tubing walls (Fig. 19–10). Capillary columns are some-

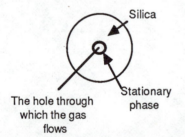

Figure 19–10. Cross-sectional view of a capillary column. The stationary phase film thickness is 0.1 to 1 μm.

times called wall-coated, open tubular (WCOT) columns. The film thickness is controlled in the manufacturing process. Some stationary phases can be bonded chemically to the capillary walls; other are held in the column by surface tension. These columns operate at volume flow rates of about 2 ml per minute. They are extremely high-resolution columns. A sample chromatogram that was produced from a capillary column separation is shown

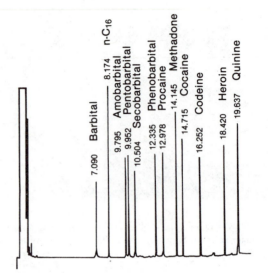

Figure 19–11. Sample chromatogram produced from a capillary column separation. Temperature programming to 280°C, at 10°C per minute 25m 0.31 mm I.D., 0.52 μm film, 5% phenyl methyl silicone, cross linked. Approximately 10 ng of each compound. (Courtesy of Hewlett-Packard, Palo Alto, Calif. Adapted with permission.)

in Figure 19–11. Notice that the peaks are very narrow. A second type of capillary column is the support-coated, open tubular (SCOT) columns. These columns have celite support fused into the capillary walls. In other respects they are similar to the wall-coated, open tubular columns.

In conclusion, the gas chromatographic column is a device in which the sample constituents are allowed to partition themselves between the carrier gas and the thin layer of stationary phase. Compounds that do not like to dissolve in the stationary phase move through the column rapidly. The sample constituents that do like to dissolve in the stationary phase are retarded but eventually are eluted from the column. Thus, the retention time varies according to the relative attraction that a sample component has for the stationary phase.

Detectors

The only part of the instrument left to discuss is the detector. There are about ten different detectors that one can buy for a gas chromatograph. As far as clinical laboratories are concerned, I have seen only three different types of detectors. The first type is the thermal conductivity detector that is used in the identification of anaerobic bacteria. The second is the hydrogen flame ionization detector that is so common in toxicology. The third type of detector is the mass spectrometer that is used in some high-volume toxicology laboratories.

Thermal Conductivity Detectors. Thermal conductivity detectors measure the concentration of a gas mixture by measuring in effect the thermal conductivity of the mixture. There exists a linear relationship between the thermal conductivity of a binary mixture of gases and the composition of that mixture. This variation in thermal conductivity is particularly useful if hydrogen or helium is used as the carrier gas. These two gases have very high thermal conductivity compared with that of other gases (Table 19–1). In the United States, helium is

TABLE 19–1. Relative Thermal Conductivity

Gas	Thermal Conductivity
Hydrogen	50
Helium	40
Nitrogen	8
Carbon dioxide	5
Ethane	8
Isopentane	5
Benzene	4
Ethanol	5
Carbon tetrachloride	2

the carrier gas of choice; hydrogen is flammable. Using helium as a carrier gas produces a great change in thermal conductivity when a sample constituent is mixed with the carrier (Fig. 19–12). With

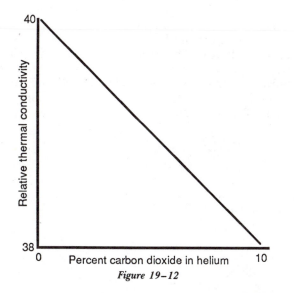

Figure 19–12

this knowledge of the variation of thermal conductivity with concentration, let us look at the detector.

The thermal conductivity detector makes use of four tungsten filaments like the one shown in Figure 19–13. When a voltage is applied to the rigid metal supports, the resulting current causes the thin tungsten filament to heat up. The gas flowing through the filament chamber tends to cool the filament. The filament is bathed most of the time in pure carrier gas; this condition produces a given equilibrium filament temperature. When helium is

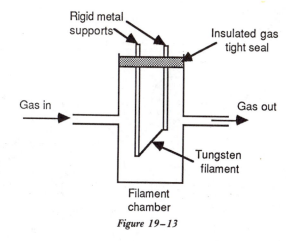

Figure 19–13

used as the carrier gas, the equilibrium temperature tends to be lower because helium has a high thermal conductivity. When a sample constituent mixed with helium is eluted from the column, the mixture has a lower thermal conductivity. The mixture does not cool the filament as well than that of pure helium. This allows the filament to heat up. This variation in filament temperature is directly proportional to the concentration of the sample constituent in the carrier gas.

It is not necessary to actually measure the temperature of the filament. As discussed in Chapter 1, the resistance of a conductor is related to the temperature of the conductor.

$$R = Ro(1 + \alpha T)$$

where
R is the resistance at temperature T°C
Ro is the resistance at 0°C
α is the temperature coefficient of resistivity
So, by measuring the resistance of the filament we can find the concentration of the sample constituent that is mixed with the carrier gas.

In the GC detector, four filaments are used— two as reference filaments, two as filaments exposed to the gas coming from the column. The detector is illustrated in Figure 19–14 and is really a differential detector. The filaments in chambers 1 and 4 are exposed to the gases coming from the column. The filaments in chambers 2 and 3 are exposed to pure carrier gas. These filaments then become the four resistances in a Wheatstone bridge (Fig. 19–15). When a sample is eluted from the column, filaments 1 and 4 are cooled less well. This causes their resistance to increase and also causes a voltage to appear across points a and b. The voltage across a and b is directly proportional to the instantaneous concentration of the sample constituent that is passing through the filament chambers 1 and 4.

Chromatograms that are produced by a thermal conductivity detector are really plots of the voltage

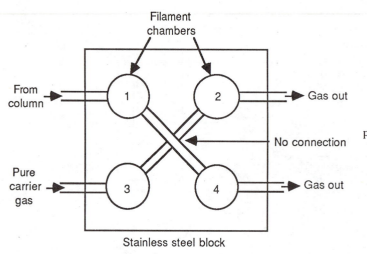

Figure 19–14. The pure carrier gas is frequently supplied through a reference flow path.

across points a and b (Fig. 19–15) versus time. The chromatograms can be used for quantitative analysis and qualitative analysis as outlined in Chapter 17.

The thermal conductivity detector is the least sensitive of the common GC detectors, but it is not expensive and is fairly stable. It is widely used in clinical laboratories on instruments that measure the volatile fatty acids that are produced by anaerobic bacteria. The sample chromatogram in Figure 19–16 is of a volatile fatty acid standard. The chromatograms of the extracts of bacterial metabolites show characteristic patterns of these acids. Some examples are shown in Figure 19–17. These patterns are used to identify the anaerobe that has been cultured from a patient.

Flame Ionization Detectors (FID). A second common GC detector used in clinical analysis is the hydrogen flame ionization detector. It is very popular in toxicology work, since it is highly sensitive to organic compounds. Its sensitivity to organic compounds is about 10^{-12} g/sec. This is about 10^6 times more sensitive than a thermal conductivity detector. Consider Figure 19–18.

The detector consists of a flame jet mounted in a detector housing. Hydrogen is mixed with the gases coming from the column. The mixture flows to the tip of the flame jet, where it is burned. The oxidant is air that is entrained from the sides of the jet; the air flows up between the jet and the concentric collector. The flame jet is the negative electrical connection for an external circuit. The flame burns continuously at the head of the jet. The flame gases flow up through the collector and out through vent holes at the top of the collector. A dust cap is

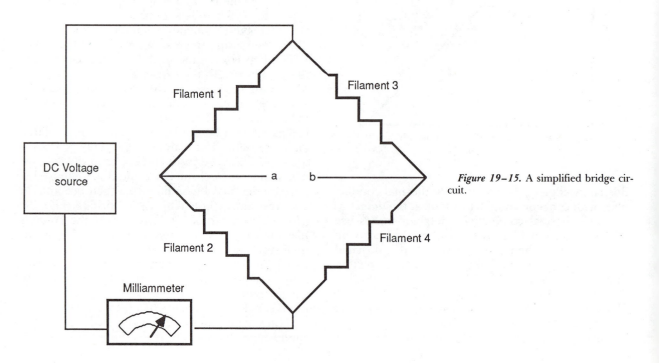

Figure 19–15. A simplified bridge circuit.

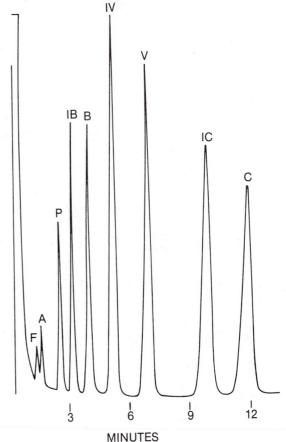

Figure 19–16. Volatile fatty acid (VFA) standard on a 15% SP-1220/1% H₃PO₄ column at 145°C. Flow rate 70 ml/minute; support 100/120 mesh; chromosorb W. *VFA Code:* F, formic acid; A, acetic acid; P, propionic acid; IB, isobutyric acid; B, butyric acid; IV, isovaleric acid; V, valeric acid; IC, isocaproic acid; C, caproic acid. (Reprinted with permission from: K.J. Hauser and R.J. Zabransky, J. Clin. Microbiol. *2*(1):1, 1975.)

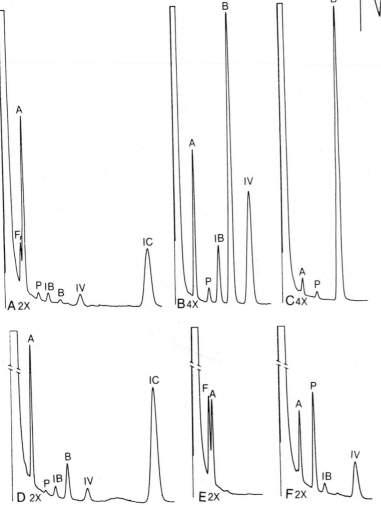

Figure 19–17. Chromatograms of extracts of anaerobic bacteria on SP-1220 at 145°C and 70 cc/minute gas flow rate. *A, C. sordellii; B, C. hastiforme; C, F. necrophorum; D, P. anaerobius; E, C. ramosum; F, B. fragilis subsp. thetaiotaomicron.* VFA: F, formic; A, acetic; P, propionic; IB, isobutyric; B, butyric; IV, isovaleric; V, valeric; IC, isocaproic; C, caproic. (Reprinted with permission from: K.J. Hauser and R.J. Zabransky, J. Clin. Microbiol., *2*(1):1, 1975.)

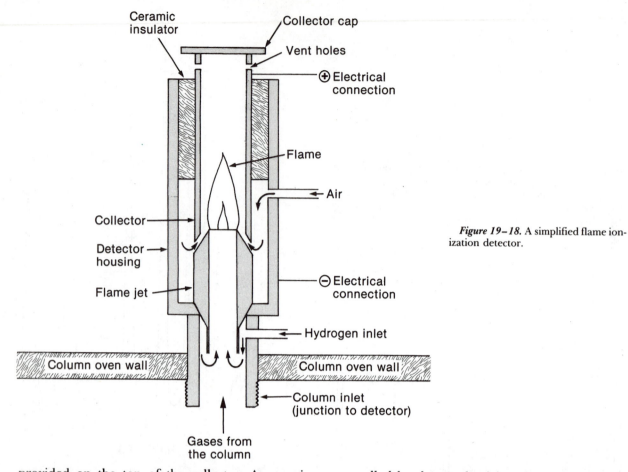

Ceramic
insulator

Collector cap

Vent holes

⊕ Electrical
connection

Flame

Air

Collector

Detector
housing

Flame jet

⊖ Electrical
connection

Hydrogen inlet

Column oven wall   Column oven wall

Column inlet
(junction to detector)

Gases from
the column

Figure 19–18. A simplified flame ionization detector.

provided on the top of the collector. A ceramic electrical insulator supports the collector inside the housing. The collector is the positive electrical connection for the external power supply and readout electronics. The purpose of this design is to allow the measurement of the electrical conductivity of the flame. The electrical layout is shown in Figure 19–19. The flow of current through the flame is controlled by the conductivity of the flame. Thus the current in the series circuit that is made up by the power supply, the electrometer resistor, and the flame is controlled by the conductivity of the flame. The voltage drop across the electrometer resistor is a function of the current. The field effect transistor voltmeter (FETVM) measures the voltage and provides the result to the recorder or data system.

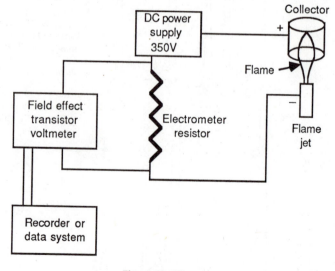

Figure 19–19

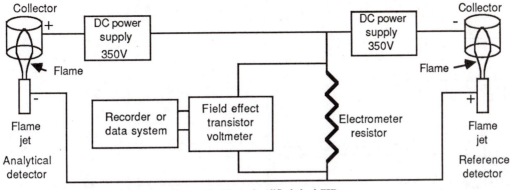

Figure 19–20. A simplified dual FID.

In this way, the conductivity of the flame ultimately is measured as a proportional voltage.

Next consider the things that affect the conductivity of the flame. When only nitrogen is coming from the column and passing through the flame, the flame has a very low conductance. Nitrogen does not ionize well under these conditions, nor does water. Thus there is nothing to carry the current through the flame. On the other hand, when a carbon compound comes from the column and enters the flame, the conductivity increases. Carbon compounds that are combusted under these conditions produce many ionic products. The amount of ions produced depends on the type and the amount of compound present. The conductivity of the flame is directly proportional to the number of ions in the flame. The number of ions in the flame is directly proportional to the mass of compound in the flame, at least for a given compound. Thus the chromatograms that result are plots of voltage as seen by the FETVM versus time. The voltage that is seen is directly proportional to the mass of compound in the flame at a given instant. Because the flame conductivity is so greatly affected by the amount of carbon-containing compound in the flame, the detector works very well for organic compounds.

When a chromatographic separation requires program temperature operation, it is necessary to use matched columns and two flame ionization detectors. If this is done, the baseline instability that results from temperature programming can be overcome. This is done in such a way that the column stationary phase that bleeds into the mobile phase affects both detectors equally but in an opposite way. (Fig. 19–20). Because the current that is produced by stationary phase bleed travels in opposite directions, it cancels out in the electrometer resistor. The net result is that the readout does not reflect the drift in the base line. This is a very common detector arrangement on better instruments.

The flame ionization detector requires a source of hydrogen and clean compressed air. The detector response to a fixed amount of compound is a function of the flow rate of all of the gases involved. For example, the FID in our laboratory has the following characteristics. When a fixed sample size is injected under differing air flow rates, the curve illustrated in Figure 19–21 is produced. On our

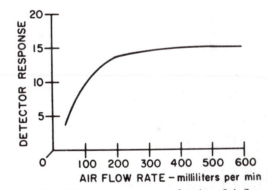

Figure 19–21. Detector response as a function of air flow rate. (Courtesy of Hewlett-Packard Co., Avondale Div., Avondale, Penn. Reprinted with permission.)

instrument, for proper detector operation, the air flow rate must be approximately 500 ml/min.

The flow of hydrogen affects the detector response as well. For our instrument, the curve illustrated in Figure 19–22 was obtained by injecting a constant amount of sample at differing hydrogen flow rates. For the greatest sensitivity, our instrument must be operated at a hydrogen flow rate of about 40 cc/min.

Note that the flow rate of the carrier gas has an effect on the FID. On our instrument, flow rates below 50 ml/min are undesirable. If flow rates below this value must be used, makeup gas can be added through the hydrogen inlet. The makeup gas is always the same gas as the carrier gas.

The flow rates that are discussed in the preceding three paragraphs are for a specific type of instrument. The flow rates for your instrument will

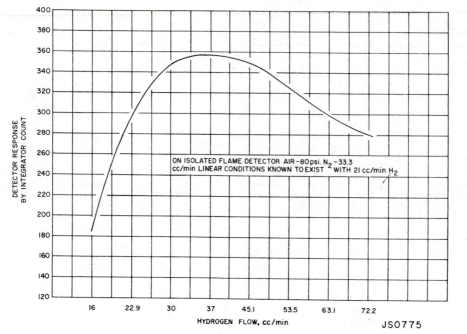

Figure 19–22. Response change for a change in hydrogen flow. (Courtesy of Hewlett-Packard Co., Avondale Div., Avondale, Penn. Reprinted with permission.)

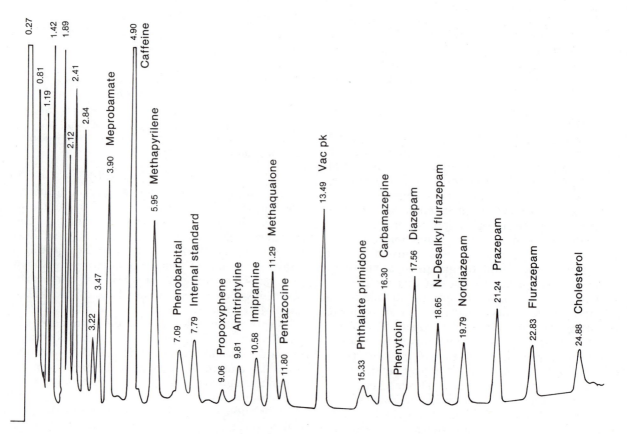

Figure 19–23. Courtesy of David Squiliace, Mayo Clinic, Rochester, Minn.

TABLE 19-2. Peak Identification

Retention Time	Compound
1.19	Barbital
1.42	Methyprylon
1.89	Butalbital
2.12	Amobarbital
2.41	Pentobarbital
2.84	Secobarbital
3.22	Internal standard I
3.47	Diphenhydramine

be different. See your instrument manual for informed operation.

The FID is a good detector for organic compounds. The more oxygen or halogen in the organic compound, the poorer the sensitivity of the detectors. For example, formic acid (HCOOH) and carbon tetrachloride are almost undetectable. The detector is widely used on instruments utilized in toxicology. Figure 19–23 provides an example of a chromatogram that was prepared to standardize an instrument for such a drug screen. This drug screen is used for qualitative analysis only. The peak identification for early peaks shown in Figure 19–23 is provided in Table 19–2.

Also consider the chromatogram in Figure 19–24. This chromatogram is the standardization chromatogram for the typical blood-alcohol procedures that are used in forensic toxicology. In these procedures n-propanol is added to blood as an internal standard, and the resulting solution is injected into the instrument. A small precolumn that contains only some support material is used to trap the nonvolatile components of the blood. The volatile components then move into the column. The results are calculated just as they are for liquid chromatography.

Quadrupole Mass Spectrometers. One of the most elegant detectors that one can use on a gas chromatograph is the quadrupole mass spectrometer. The mass spectrometer is elegant because in addition to supplying a chromatogram, it provides

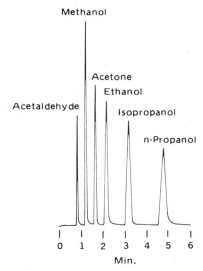

Figure 19–24. Simultaneous, isothermal identification of blood alcohols. Clinical packing 1-1766, 60/80 Carbopack B/5% Carbowax 2OM, 6 ft. × 2mm ID, glass column. Column temperature: 85°C, Flow rate: 20 ml/minute, He, Det.: FID, at $16 \times 10^{+10}$ sensitivity, Sample: 1 μl aqueous volatiles standard, Conc.: 0.05–0.10% each component. (Reprinted with permission of Supelco, Inc., Bellefonte, PA 16823. Supelco Catalog 21, p. 10, fig. 1.)

the mass spectra of the compounds eluted. The mass spectrum of a compound can be used for its qualitative identification. Because the chromatographic retention time and the mass spectrum of a compound are independent of each other, the probability of obtaining an incorrect identification of an analyte is almost zero. This identification is approached by determining the retention times and mass spectra of bona fide samples of suspected analytes. The retention times and spectra of samples of unknown composition can then be compared with the data determined with bona fide samples. Before we discuss this matter in detail, let's consider the detector and its principle of operation.

A simplified quadrupole mass spectrometer (Fig. 19–25) consists of an ion source, a quadrupole

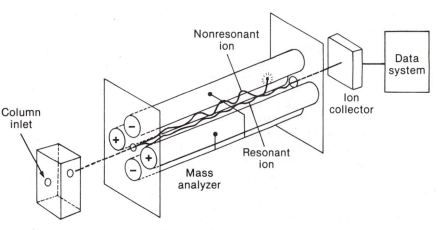

Figure 19–25. Adapted with permission from: D. Lichtman, Research/Development, *15* (2):52, 1964.

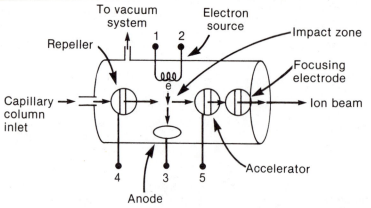

Figure 19-26. A simplified ion source.

mass analyzer, an ion collector, a data system, and a vacuum system. Let's consider the parts of the detector. The ion source generally is a device that causes the sample molecules to be bombarded with high-energy electrons. These electrons cause the sample molecules to become either ionized or fragmented and ionized. The second role that the ion source plays is that of ejecting the ionized sample into the mass analyzer. Consider Figure 19–26.

The source of electrons is a heated tungsten filament; it is heated by electrical power that is supplied at contacts 1 and 2. The electrons that are boiled off the filament are accelerated toward the anode. This takes place because a source of DC voltage is applied between connections 1 and 3. Higher anode voltages produce electrons with more energy. The choice of the energy level of the bombarding electrons depends on the nature of the experiment. In gas chromatography–mass spectrometry work, fairly high-energy electrons are used in order to achieve fragmentation of the sample molecules.

The positive ions so produced are repelled by the repeller and accelerated by the accelerator. This takes place because a DC voltage of about 15 volts is applied between connections 4 and 5. The repeller is made positive, and the accelerator is made negative. The resulting ion beam is allowed to drift into the quadrupole mass analyzer.

The quadrupole mass analyzer is a device made from four metal rods with hyperbolic faces. The rods are mounted in an array and connected to an electrical circuit (Fig. 19–27). The application of these voltages to the four rods produces an alternating electrical field in the space between the rods. When an ion from the ion source is injected into the analyzer, the ion travels an oscillatory path. The oscillatory path is characterized by increasing amplitude. In Figure 19–28, this path is illustrated for an ion "too heavy," an ion "too light," and an ion "just right."

The ion "too light" picks up too much amplitude and crashes into one of the rods. The ion "too

heavy" cannot respond fast enough to the alternating fields and misses the exit slit. The ion that is "just right" in mass passes through the analyzer and out the exit slit. The condition of being able to transverse the analyzer and escape through the exit slit is called being in resonance. The range of ion masses to be studied can be caused to be resonant by varying the frequency of the radio frequency power supply. One of the common instruments on the market can scan from 10 to 800 atomic mass units (amu) in less than 1 second.

The ion collector shown in Figure 19–25 is, for all practical purposes, a photomultiplier tube. The device is called an electron multiplier. The positive ions that leave the exit slit of the mass analyzer fall onto the first dynode of the electron multiplier. The impacting ion knocks photoelectrons out of the first dynode surface, and they are attracted to the second dynode, and so on. (Refer to the discussion of the photomultiplier tube in Chapter 4.) The ion current produced by the electron multiplier is amplified and sent to the data system.

The data system is a digital computer that does many things. First, it controls the frequency of the radio frequency power supply (Fig. 19–27). Be-

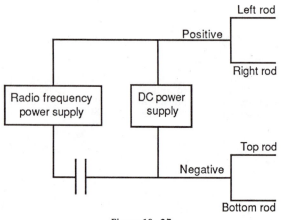

Figure 19-27

ION DETECTOR

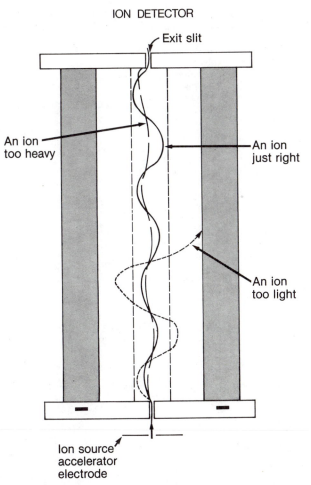

Figure 19–28. Quadrupole mass analyzer—top view. Top positive rod not shown.

cause it does so, the computer knows the mass of the resonant ion at any given time. During the mass sweep, the ion current is stored versus time and is given a spectrum number. The total current that is delivered during the sweep is also measured and stored. For example, assume that the sweep time is 1 second and that the sweep range is 10 to 800 amu. If a chromatographic separation takes 10 minutes, the data system will contain 600 mass spectra and 600 total ion current values. On command, the computer will produce a plot of the total ion current versus time. This provides the total ion current chromatogram. Such a chromatogram is illustrated in Figure 19–29. Once the operator locates a peak of interest, the computer can be asked for the mass spectrum of the molecules that were in the mass spectrometer at that point during the separation. If the separation has provided a pure compound to the ion source, then the mass spectrum of the pure compound will be produced by the computer. Most instruments have a library of spectra of many known compounds. If you ask, the computer will compare

the spectrum of an analyte with the spectra in the library. It will then tell you what the unknown compound is. Some examples of mass spectra are illustrated in Figures 19–30 to 19–38.

Often a specific ion is known to be produced by only one compound. When that is the case, the computer can be asked to display the ion current of that particular ion versus time. The other data do not need to be displayed. The better instruments can provide ion currents for up to 20 preselected ions versus time. These instruments can scan a narrow range of ion masses very quickly. Scan rates up to 2000 amu/sec are possible. The system can scan a 100-amu range 20 times per second. The data system must be large in order to hold that much data.

The ion source, the mass analyzer, and the electron multiplier must all be operated under high-vacuum conditions. When capillary columns are used, the 1 to 2 ml/min carrier gas flow rate can be introduced directly into the ion source. The vacuum pumps can handle that much gas and still maintain the low pressure that is necessary.

With packed columns, a jet separator (Fig. 19–39) is commonly used to strip off most of the carrier gas. The packed columns are used with carrier gas flow rates of about 40 ml/min. The pumps cannot keep up with that much gas. Because the use of capillary columns is supplanting packed columns in mass spectrometry work, you may never have to use such a separator.

In order to illustrate an application of GC-MS, I am presenting some general information from a drug screening procedure. In the procedure, 2 ml of serum, 2 ml of 1 molar acetic acid, and an internal standard are extracted with 6 ml of 80 per cent chloroform, 20 per cent ethyl acetate. The resulting solution is dried with sodium sulfate and evaporated to dryness at 40°C under a flow of nitrogen. The residue is redissolved in 100 μl of methanol. Two μl of the methanol solution is injected into a GC-mass spectrometer with a 30-meter capillary column. A simplified total ion chromatogram for the drug screen standard is illustrated in Figure 19–29. The drugs and their retention times are listed in Table 19–3.

The mass spectra of several drugs are provided in Figures 19–30 through 19–38. Spectra such as these can be used to interpret the mass spectrum of an unknown analyte. More often, the computer will consider the mass spectrum produced by an unknown analyte and will search the spectra library for a good fit. The computer uses the patterns of the mass spectral peaks; notice that no two are alike. In this way, the GC-MS system is used to identify the drugs that are found in the sera of patients.

Quantitation of a given drug is best done by using the total ion chromatogram. The area under the peak for the drug of interest is used to find the

Text continued on page 226

TABLE 19–3. Simplified GC-MS Readout for the Drug Screen Standard

Retention Time	Area Under Peak	Amount	Name
1.04	—	—	Solvent
3.26	201	50	Ethchlorvynol (Placidyl)
5.63	694	50	Salicylate
9.03	731	10	Methyprylon
9.36	296	10	Benzocaine
10.13	483	10	Ibuprofen
11.28	572	10	Butalbital
11.56	559	0.559	—
12.17	248	50	Acetaminophen
12.67	598	10	Pentobarbital
13.91	382	10	Meprobamate
14.30	415	10	Caffeine
14.71	848	10	Glutethimide
15.43	174	2	Phencyclidine
15.73	132	2	Mephobarbital
16.83	309	10	Phenobarbital
19.02	931	10	Methaqualone
19.26	145	2	Propoxyphene
19.57	169	2	Imipramine
20.07	715	15	Cyheptamide
20.55	656	10	Phenytoin
21.21	146	2	Diazepam
21.72	386	10	Chlorpromazine (Thorazine), unresolved chlordiazepoxide hydrochloride (Librium)
23.57	167	10	Breakdown product of librium
24.58	501	100	Cholesterol
25.02	116	2	Thioridazine

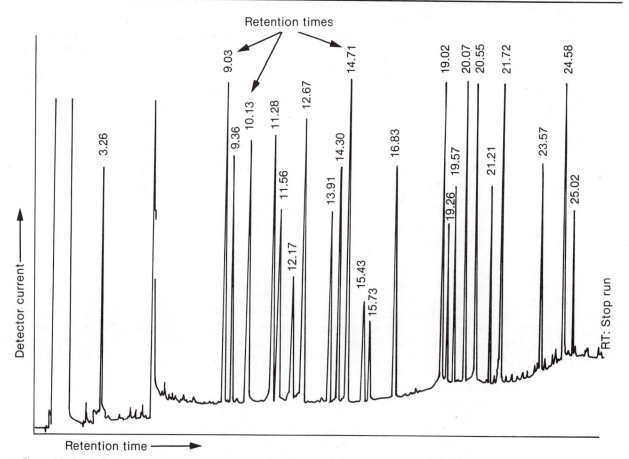

Figure 19–29. Column temperature 80°C programmed to 320°C (nonlinear program). Injection port temperature 250°C. Column 30 meter capillary. Chart speed 1 cm/min. (Courtesy of David Squiliace, Mayo Clinic, Rochester, Minn.)

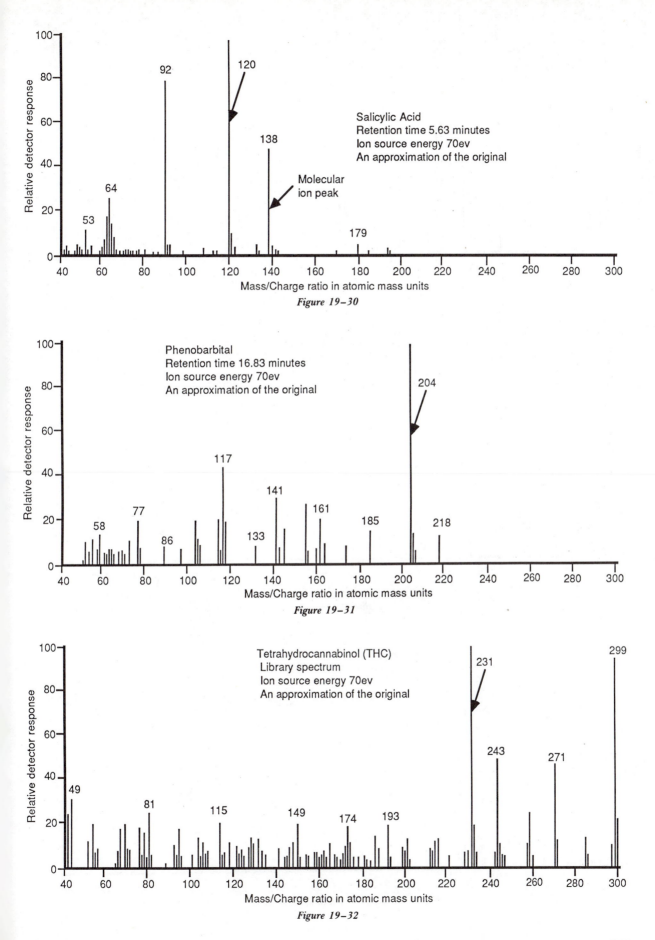

Figure 19–30

Figure 19–31

Figure 19–32

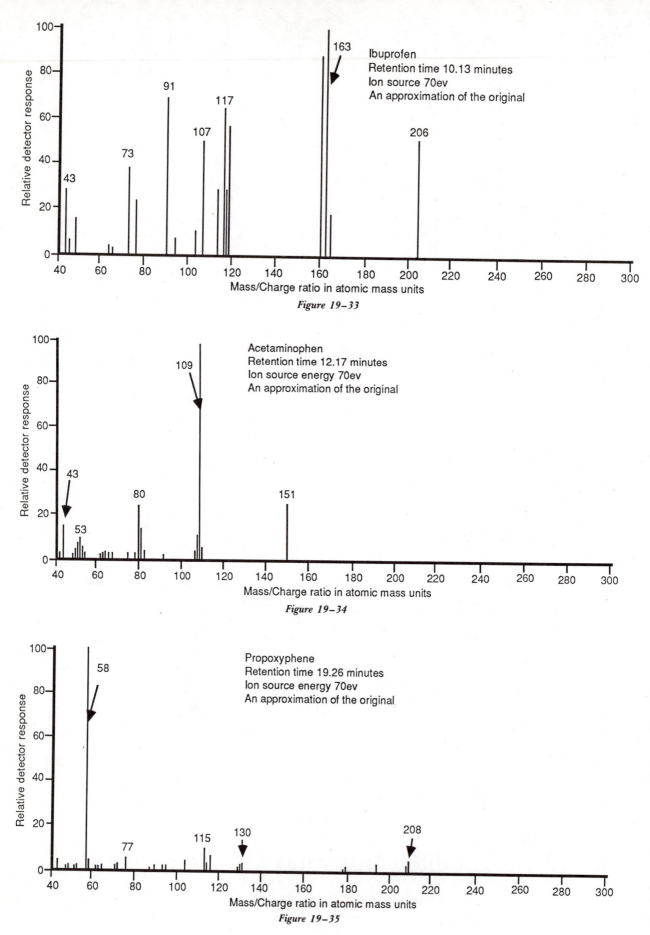

Figure 19–33

Figure 19–34

Figure 19–35

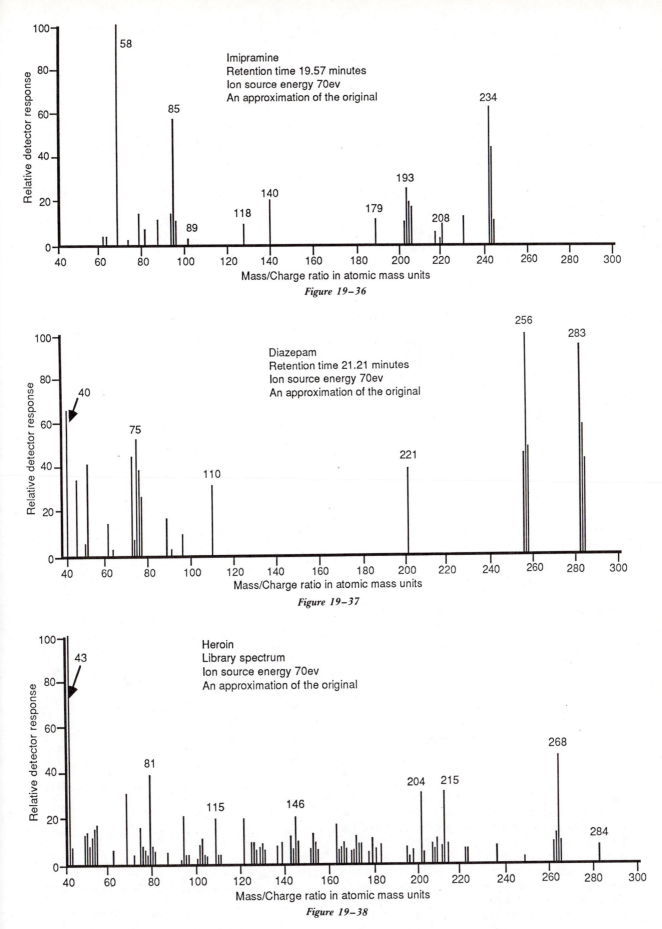

Figure 19–36

Figure 19–37

Figure 19–38

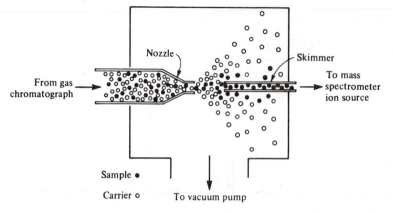

Figure 19–39. A jet separator. (From Instrumental Methods of Analysis, 6th Ed., by H. Willard, L. Merritt, J. Dean, & F. Settle, Copyright © 1981 by Litton Educational Publishing, Inc. Reprinted by permission of Wadsworth, Inc.)

concentration. The internal standard method is used (see Chapter 17).

Mass Spectra and Their Origin. You will recall from the previous discussion that the ion source bombards the sample molecules with high-energy electrons. The high-energy electrons that are bombarding the sample interact with it in several ways. Consider salicylic acid, for example.

When salicylic acid is bombarded with high-energy electrons, the first reaction that takes place is the formation of a molecular ion.

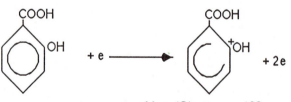

Mass/Charge = 138 amu

The molecular ion, $C_7H_6O_3{}^+$, has a mass-to-charge ratio of 138 amu and is somewhat unstable. The parent ion frequently possesses so much energy that it fragments (see next column). These ions are all seen in the mass spectrum of salicylic acid (see Fig. 19–30). Notice that the mass spectrum has small peaks at mass/charge ratio 169, 179, and 194. These peaks are due to collisions between smaller ions that lead to production of ions that are bigger than the molecular ions. These various fragments along with the parent ion are the positive ions that

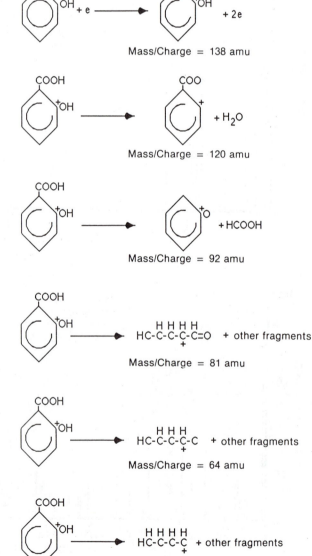

lead to the mass spectrum of a compound. The fragments provide a lot of information about the molecular structure of a compound.

FACTORS THAT AFFECT SEPARATIONS

In gas chromatographic separations, the selection of the stationary phase has a profound effect on the outcome of a given separation. This fact can be best illustrated by the chromatograms in Figures 19–40 and 19–41. These separations are of 3 μl

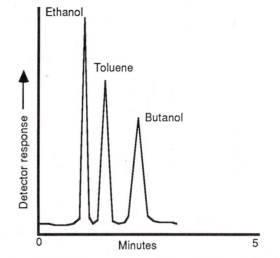

Figure 19–41. Column: 2 meters; Carbowax 20M on GasChrom Q. Column temperature 80°C; attenuation 32.

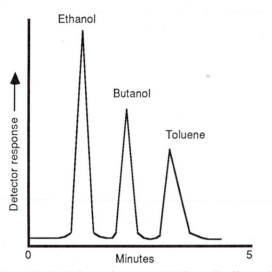

Figure 19–40. Column: 2 meters; OV-17 on GasChrom Q. Column temperature 190°C; attenuation 32.

of a mixture containing ethanol, butanol, and toluene. All chromatographic conditions not listed are identical and ordinary. The column temperature in Figure 19–40 had to be higher than 80°C in order to get the retention times down to minutes instead of hours. Because the stationary phases in the two columns have different polarities, the sample molecules interact differently. This leads to the change in the elution order.

The proper choice of a GC stationary phase is somewhat empirical. The companies that sell columns are happy to suggest which stationary phases to use. There are also published tables of data concerning which stationary phases should work for different separations.

The choice of the carrier gas in GC work depends on the detector. The mobile phase gas has little effect on the separations.

The gross effect of increasing the carrier-gas flow rate is to shorten retention times. The shape of the chromatographic peaks may or may not be improved by increasing the flow rate. This topic is covered in the next section.

The gross effect of increasing the column temperature is to shorten retention times. The resolution between low-boiling sample constituents is generally made worse by increasing temperature. An example of this is shown in Figures 19–42 and 19–43. The same sample is chromatographed under conditions of different temperature. All the other chromatographic variables are the same in these two separations. Notice that at 180°C the resolution between compounds 1 and 2 is lost and that the resolution between compounds 3 and 4 is also adversely affected. Also notice that the chromatogram that was obtained at 180°C was done in 6 minutes.

On many better gas chromatographs it is possible to change the temperature of the column during the analysis. This is called programmed temperature operation and can be used so that the column is cooler at the start of the separation and hotter at the end.

When programmed temperature operation is used to separate the mixture used in Figure 19–41, a really nice separation is possible (Fig. 19–44). The disadvantage of program temperature operation is that the column must be cooled to the starting temperature before a second analysis can be begun.

Peak Shape and Resolution. In addition to the gross effects of stationary phase selection, column flow rate, and column temperature, there are some subtle factors that affect separations. These subtle considerations are the effects of the operating variables on the chromatographic peaks. The HETP values (height equivalent of a theoretical plate) are somewhat affected by most of the operating conditions of the chromatograph. The concept of HETP is discussed in detail in Chapter 17. Famil-

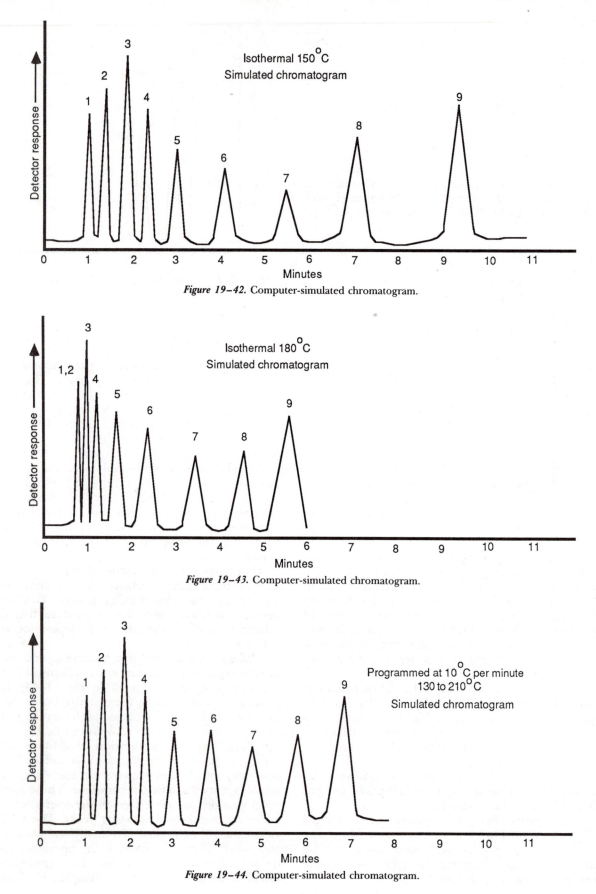

Figure 19–42. Computer-simulated chromatogram.

Figure 19–43. Computer-simulated chromatogram.

Figure 19–44. Computer-simulated chromatogram.

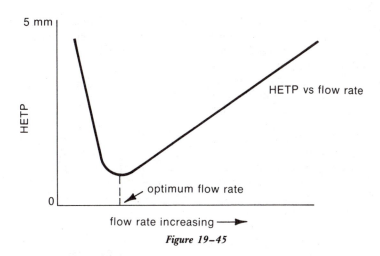

Figure 19–45

iarity with this concept is assumed in the following discussion. If necessary, review the presentation of HETP in Chapter 17 before proceeding in this chapter.

Briefly, HETP is affected by many things. First, consider the effect of column flow rate (Fig. 19–45). The curve tells us that an optimum flow rate exists for the model compound. The optimum flow rate for the model compound will be close to the optimum flow rate for many compounds.

The amount of stationary phase in the column has an effect on HETP. Consider Figure 19–46.

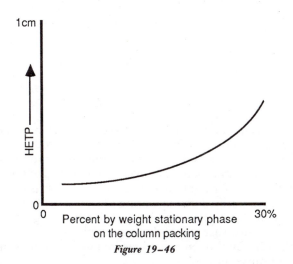

Figure 19–46

This curve tells us that the smaller the percentage of stationary phase in the packing, the better the peak shape. Note, however, that the percentage of stationary phase should not go too low; the packing must have enough stationary phase to handle the sample. If large samples are used, larger loadings of stationary phase are needed. An inappropriately selected stationary phase can lead to large values

of HETP. Use selection guides to select the proper stationary phase.

The temperature of the column has an effect on the HETP of the peaks. Within reason, higher temperatures of separation favor somewhat smaller HETPs. Decreases in retention time at higher temperatures do limit this approach to improved separations. The stationary phase temperature limit may also limit the temperature of operation.

The smaller the sample size, the lower the HETP. The need for sufficient sample to allow detection is, of course, a limitation to this approach to improved separations.

Rapid sample introduction is necessary for the production of good, narrow peaks. The analyst can produce peaks with large HETP values by injecting the sample slowly.

Injection port temperatures that are not well above the boiling point of the highest boiling constituent can also lead to peaks with large HETP values. As a rule of thumb, the injection port should be about 30°C hotter than the boiling point of the highest boiling sample component. This provides enough heat to produce instantaneous vaporization of the sample.

The detector temperature needs to be well above the boiling point of the highest boiling constituent. If it is not, some condensation in the detector is possible, which would lead to larger HETP values.

If all these matters are given some consideration, then the instrument can provide the best possible separations.

ODDS AND ENDS

When working in gas chromatography, it is necessary to handle cylinders of compressed gas. Read about the safe handling of compressed gas. The literature can be obtained from your gas supplier.

Beware of the effect of water on the separating ability of your columns. HETP values increase during the life of a column. The rate of increase is much accelerated by the use of wet samples and wet carrier gas. Use a scrubber to get rid of the water in the carrier gas. If possible, dry the samples that are to be injected. Various desiccants have been used; I like anhydrous sodium sulfate.

If possible, use pressure regulators with stainless steel diaphragms. In work at high sensitivity, these regulators provide better baselines. The difference can be seen in Figure 19–47.

Before
Baseline of an on-column injection of a hydrocarbon mixture. This chromatogram shows extra column effects caused by pressure regulator with a neoprene type diagram.

After
Baseline of the same sample and column after installing a Moisture Trap. Oven temp. 70°C (program 4°C/min.)

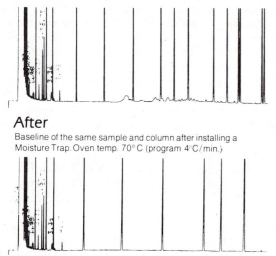

Figure 19–47. Courtesy of The Anspec Company, Inc, Ann Arbor, Mich. Reprinted with permission.

Some samples do not get along well with metal columns. Glass columns or glass-lined metal columns may improve the sensitivity in these cases.

Do not forget to replace injection-port septa. They should be good for between 100 and 200 injections. A leaking septum really hurts sensitivity. Septa have temperature limits; use the correct material for the temperature used in your work. The higher the temperature, the fewer the injections before leaking begins.

When injecting samples with a nonvolatile content, it is important to provide a short precolumn (guard column). The precolumn contains support material so that the nonvolatile substances are collected in this short, disposable segment. This practice protects the analytical column from premature

degeneration. The effect of nonvolatile substances in the column inlet is shown in Figure 19–48.

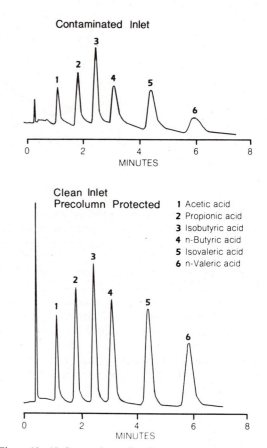

Figure 19–48. Comparison of analyses using contaminated versus clean inlet; 0.1% VFA in water. (Courtesy of The Anspec Company, Inc, Ann Arbor, Mich. Reprinted with permission.)

Frequently it is necessary to condition a column for a few hours before it can provide reproducible results. This is best done by heating the column to its temperature limit and maintaining it at that temperature overnight. It is best not to hook up the column to the detector during this process. Carrier gas should be flowing slowly through the column. Make sure that the hydrogen supply is *shut off* to the FID. Hydrogen could flow into the column oven and cause an explosion.

Columns in storage should be capped to prevent water vapor from being absorbed.

Soap bubble flow meters are inexpensive and adequate for many flow rate measurements.

Capillary columns are prone to oxidation problems. Oxygen should be scrubbed from the carrier gas. This is especially true when the temperature of operation is higher than 50°C.

Problems   The problems in Chapter 17 cover the concepts in Chapter 19, so I am providing only two additional problems.

1. Consider the following mass spectrum:

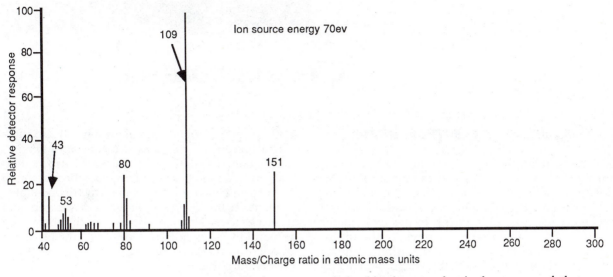

Limiting your search to the compounds in this chapter, what is the compound that was analyzed in the figure?

2. Consider the following mass spectrum:

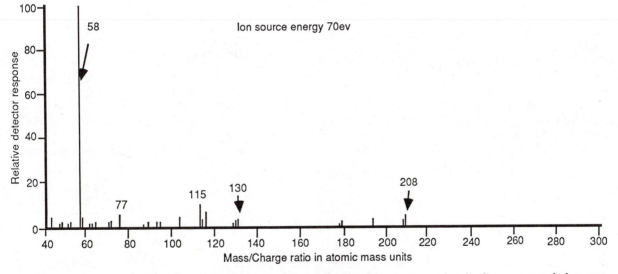

Limiting your search to the compounds in this chapter, what is the compound that was analyzed in the figure?

Electrophoresis

PRINCIPLES OF ELECTROPHORESIS

Electrophoresis is a separation method based on the fact that charged particles migrate through solution under the influence of an electrical field. The principles of electrophoresis are best illustrated by a very idealistic experiment. In this experiment we must *assume* that ions in solution do not diffuse from regions of higher concentration into regions of lower concentration. With this assumption made, let's consider the glass and platinum cell in Figure 20–1. If the cell were filled with distilled water, and if a thin band of a colored anionic compound were layered in as illustrated, the cell would be ready for electrophoresis. Once the DC power supply was turned on, one would observe that the sample of colored anions would begin to migrate toward the anode. One would find that the rate of migration would be constant. The colorless cations that were added with the colored anions would, of course, migrate in the opposite direction.

Keeping in mind the idea that ions can be made to migrate through fluids, let's consider the forces that act on an ion in an electrophoretic cell. This will lead to an understanding of why electrophoresis can separate different types of ions. It will also show why a specific ion migrates at a constant rate.

In our study of these forces, we will use a spherical ion as a model. First picture a spherical ion in an environment that is free from any frictional phenomena. If the ion were placed in a vacuum, and if an electrical field were imposed, the ion would respond by being accelerated toward the electrode of opposite charge. The equation that predicts this accelerating force is

$$F = QE \qquad \text{Equation 20–1}$$

where F is the accelerating force that is caused by the electrical field acting on the ion;
Q is the charge on the spherical ion;
E is the electrical field strength expressed in volts/cm.

Also recall from physics that

$$F = MA \qquad \text{Equation 20–2}$$

where M is the mass of the spherical ion;
A is the acceleration that the ion undergoes.

Combining Equations 20–1 and 20–2, we obtain

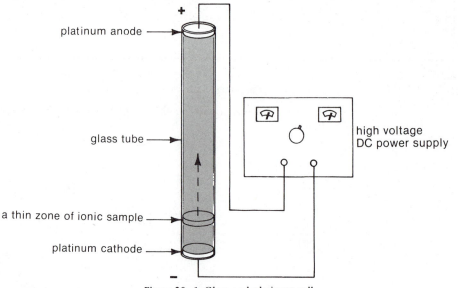

Figure 20–1. Glass and platinum cell.

platinum anode

glass tube

a thin zone of ionic sample

platinum cathode

high voltage
DC power supply

$$QE = MA \qquad \text{Equation } 20\text{--}3$$

or

$$A = \frac{QE}{M}$$

Thus the acceleration that the ion undergoes is directly proportional to the charge and inversely proportional to the mass of the ion.

The effect of this accelerating force on the velocity of the ion can be best shown by example. Consider an ion that has one unit of charge and a molecular weight of 100. If such an ion were placed in an electrical field of 100 volts/cm, the ion would experience a force that would produce an acceleration of 1 cm per second per second. The velocity of the ion at the end of the first second would be 1 cm/sec. The velocity at the end of the second second would be 2 cm/sec, and at the end of the third second it would be 3 cm/sec, and so on.

Now let's move closer to reality. The electrophoresis cell in Figure 20–1 contains water. Water is a fluid and as such would be expected to exert a frictional retarding force on our ion. When a spherical body moves through a fluid, the frictional force can be predicted by Stokes' law. The law states

$$Fs = 6\pi\, v\eta V \qquad \text{Equation } 20\text{--}4$$

where Fs is the frictional force
v is the radius of the spherical particle
η is the viscosity of the fluid medium
V is the particle velocity

Thus the faster the ion's velocity, the greater the frictional retarding force.

The net result of these two opposing forces is that an ion placed in an electrical field will be rapidly accelerated to a certain velocity, that is, the velocity necessary to generate a frictional force that just equals the accelerating force. When this velocity has been achieved, the ion has reached its terminal velocity and can go no faster. This statement is valid for a specified field strength and for the specified solvent. Thus the molecular parameters that affect the value of the terminal velocity are seen in Equations 20–1 and 20–4. In Equation 20–1 we see that

the charge on the ion is significant, whereas in Equation 20–4 we see that the ion radius is significant. In a given experimental situation, all the other terms in these equations are fixed. The greater the ionic charge, the faster the terminal velocity; the smaller the ion radius or ion size, the greater the terminal velocity.

Picture a sample that consists of three ionic compounds of equal charge but of different molecular size. When the sample is subjected to electrophoresis, the ions with greatest molecular size will migrate with smallest terminal velocity. The ions with the smallest molecular size will migrate with the largest terminal velocity, and the ions of intermediate size will migrate with an intermediate velocity. The net result is that the three compounds will separate. This is the physical basis of electrophoretic separations.

PRACTICAL ELECTROPHORESIS

Now let us return to reality and see how electrophoresis is actually done. In our experiment we assumed that there were no forces that would cause the sample to mix with the solvent. Actually, mixing would occur. So, in order to minimize this problem, the chemist causes the sample to migrate through a support material that decreases the amount of diffusion. Many support materials have been used—chromatography paper, starch gel, dextran gel, cellulose acetate, polyacrylamide gel, and agarose gel, for example. Currently the most commonly used materials are cellulose acetate and agarose gel. Cellulose acetate is the more common of the two. The use of cellulose acetate electrophoresis will be covered first.

Cellulose acetate is a plastic material that can be made in the form of thin strips. The strips have pores and channels so that ions can migrate through their structure.

In electrophoresis experiments, the control of pH is generally very important. For this reason, a buffer solution of the proper pH is used instead of just water. The strips of cellulose acetate are soaked in the buffer before use and then are mounted in a cell that looks much like the one in Figure 20–2.

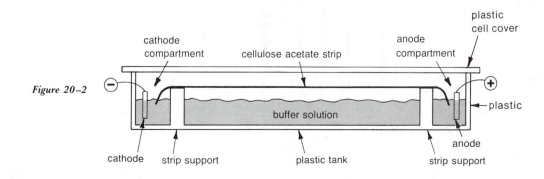

Figure 20–2

cathode compartment · cellulose acetate strip · anode compartment · plastic cell cover · plastic · buffer solution · cathode · strip support · plastic tank · anode · strip support

The buffer in the anode and cathode compartments is the same as the buffer in the center of the tank, but the compartments are insulated electrically from the center tank and from each other by the plastic strip supports. The anode and cathode are generally made of platinum wire.

A cellulose acetate strip is depicted in Figure 20–3. The sample is applied to the presoaked strip

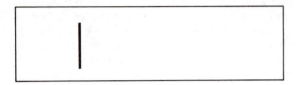

Figure 20–3. A cellulose acetate strip.

with an applicator and appears as shown in the figure. The edges of the strip are not used because it is common for the band to trail along the edge. If the sample carries a negative charge, the strip is placed in the cell so that the sample is nearest the cathode. Once the strip is in place, the voltage is turned on and the sample is separated.

Some Experimental Considerations

First, we must consider the buildup of heat in the strip. This takes place according to the equation

$$H = \frac{I^2}{q^2 K}$$

where H is heat in watts/cm³
      I is current in amps
      q is the cross-sectional area of the strip in cm²
      K is the specific conductance of the solution in ohm⁻¹ cm⁻¹

Heat buildup can be a big problem. If the strip gets too hot, buffer solution will evaporate off the strip, or the sample may be destroyed. In order to prevent this, the current is kept low, and the buffer may be refrigerated before or during use. The heating

limitation imposed on current restricts the applied voltage; the lower the voltage, the slower the separation will be. Equipment used at a very high voltage is often provided with a built-in refrigeration unit.

A second problem is the electroendosmotic effect. When water is placed in a capillary tube and subjected to an electrical field, the water moves slowly in the direction of the negative electrode. This same effect is found with some of the electrophoretic support materials. If the data obtained are to be exactly reproducible from laboratory to laboratory, this effect must be counterbalanced. In order to assess the movement of the sample that is caused by its being carried with the solvent by the electroendosmotic effect, a nonpolar, nonionic compound can be added to the sample. If the nonionic compound moves from the point of sample application, this movement can be subtracted from the sample movement. A nonionic, nonpolar compound is used because, by itself, it would not move under the influence of an electrical field.

Quantitation and Qualitation

Next consider how electrophoresis is used in quantitation and qualitation of charged particles. We will use the separation of serum proteins as an example. If serum is placed on a cellulose acetate strip (see Fig. 20–3) and electrophoresis is performed, the proteins will be separated into several fractions. If the proteins are stained and the white background of the strip is cleared to colorless, the strip would look something like Figure 20–4. The most common method of qualitation is to run a normal serum on a second strip and compare the relative positions. This approach is much like the use of R_f values in paper or thin layer chromatographic methods.

Quantitation is generally done with a special visible photometer called a densitometer Fig. 20–5. Densitometers are photometers that have a mechanism for mounting and moving an electrophoretic strip past the light beam.

The absorbances of the various regions of the

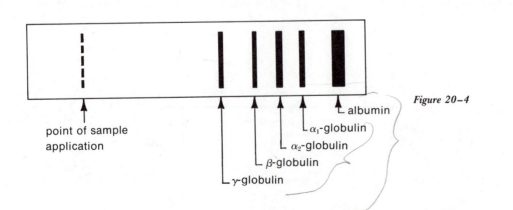

Figure 20–4

point of sample application

γ-globulin
β-globulin
α₂-globulin
α₁-globulin
albumin

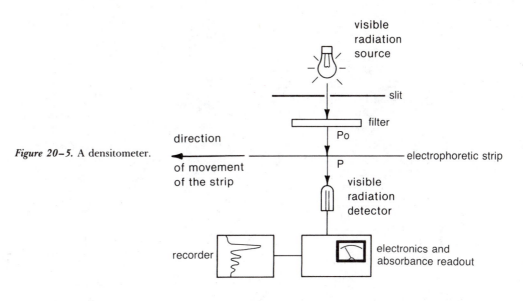

Figure 20–5. A densitometer.

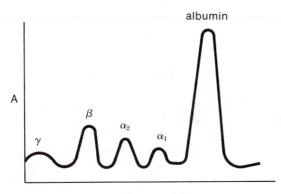

Figure 20–6. Readout from a densitometer.

electrophoretic strip are measured. If a densitometer is linked with a recorder, a plot of absorbance versus position on the strip is obtained. Figure 20–6 shows the readout from such a densitometer; the

separation is that of normal serum. The area under the various peaks is, of course, proportional to the amount of the protein that is present. The relative area of each peak is used to make the medical diagnosis.

GEL ELECTROPHORESIS

The first stabilizing medium used in electrophoresis was paper. Paper suffered from a large electroendosmotic effect and was inconvenient for densitometry because of its opaque nature. Cellulose acetate suffered less from the electroendosmotic effect and could be converted after use into

a transparent film. Because of these properties, cellulose acetate became the standard stabilizing medium of the clinical laboratory. Cellulose acetate was adequate for the separation of serum proteins into five groups of proteins.

In recent years it has become apparent that each of the five groups of proteins that is found by cellulose acetate electrophoresis is really a complex mixture of proteins. Because of the need to further separate these mixtures, separation scientists started to investigate various other stabilizing media. The materials that have proved most useful are agarose gel and polyacrylamide gel.

Agarose gel is a natural linear polymer of the polysaccharides galactose and 3,6-anhydrogalactose. The material is derived from the agar of the seaweed *Gelidium amansii*. The gel is prepared by dissolving the powdered polymer in boiling water. The resulting solution can be spread on plastic sheets. The solution jells at about 45°C and produces a thin film of transparent, fairly rigid gel. The gel has very large pores and has little tendency to separate materials on the basis of gel permeation.

Experimentally these gel plates are used much like cellulose acetate strips. The electrophoretic separation of blood serum proteins is of somewhat higher resolution than it would be if done on cellulose acetate. The electrophoretic patterns can be quantitated or qualitated by several means. Staining can be used; this is the practice that is common in cellulose acetate electrophoresis. The use of immunochemical techniques is a common way to identify the various proteins. One technique uses specific antisera to form precipitin bands from specific proteins of interest.

The use of agarose gel electrophoresis is becoming more common in routine clinical analysis. An example of a serum protein separation is shown

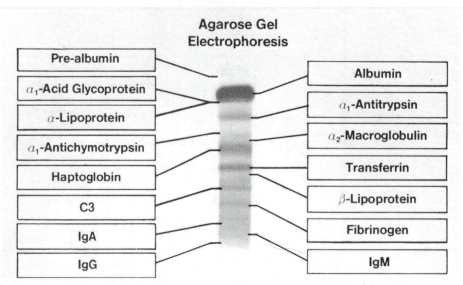

Figure 20–7. This agarose gel electrophoresis pattern of normal human plasma indicates 15 major proteins. (From: High Resolution Electrophoresis, Copyright 1985, Helena Laboratories Corporation, Beaumont, Texas. Reprinted with permission.)

in Figure 20–7. Notice that there are 15 bands as opposed to the five found on cellulose acetate strips.

In addition to agarose gel, there is some interest in polyacrylamide gel as a stabilizing medium in electrophoresis. Polyacrylamide gel is made by the polymerization of acrylamide and N,N-methylenebisacrylamide. The second compound is a cross-linking agent. The more cross-linking agent, the smaller the pores in the resulting gel. The resulting gel is reasonably rigid and is fairly transparent, but, unlike agarose gel, the pore size in polyacrylamide gel is small enough to be active in a gel permeation mode of separation. Polyacrylamide gel electrophoresis can resolve serum proteins into about 20 bands compared with the five bands on cellulose acetate strips.

The location of proteins that have been separated on polyacrylamide gel is frequently found by immunochemical means. The use of the gel requires some care, as the monomers are toxic and absorbed through the skin. Polyacrylamide gel electrophoresis has not been widely accepted in the clinical laboratory because the results are very hard to interpret. There is some indication that it might soon be refined enough to enter routine clinical use.

APPLICATIONS

Electrophoresis finds considerable application in clinical analysis. Serum proteins, lipoproteins, glycoproteins, and some enzymes are routinely determined in this way. Hemoglobin and haptoglobin are also studied through electrophoresis. Some isoenzymes are separated from each other by the technique.

Osmometry

OSMOTIC PRESSURE

Osmotic pressure is best defined in terms of a hypothetical experiment that involves the apparatus in Figure 21–1. The apparatus consists of a perfect

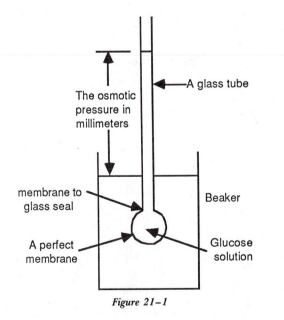

Figure 21–1

membrane filled with a glucose solution that is tightly connected to a glass tube. This assembly is then placed in a beaker of distilled water. The membrane allows water molecules to pass through its walls but does not allow the passage of other materials. If this experiment could be done, one would find that distilled water passes through the membrane and increases the solution volume to the point that the solution rises in the tube. This process would continue until the weight of the solution in the tube is great enough to counteract the propensity of the glucose solution to attract more

water. The height of the water in the tube relative to the level of distilled water in the beaker is called the osmotic pressure.

Osmotic pressure has been studied in the laboratory by the use of various less-than-perfect membranes. In the course of these studies, it has been found that the approximate osmotic pressure of a solution depends only on the number of particles of solute per kilogram of solvent. Let me provide some examples. The osmotic pressure of a solution that contains 0.1 mol of glucose dissolved in 1 kg of water would be 2.24 m of water. This is to say that the height of the water in the glass tube in Figure 21–1 would be 2.24 m. As a second example, the osmotic pressure of a solution that contained 0.1 mol of sodium chloride dissolved in 1 kg of water would be 4.48 m of water. This would be so because of the dissociation of 0.1 mol of NaCl into 0.1 mol of Na^+ and 0.1 mol of Cl^-. This dissociation would lead to the production of 0.2 mol of particles. As a third example, picture a solution that contains 0.1 mol of glucose and 0.1 mol of sodium chloride all dissolved in 1 kg of water. The osmotic pressure of this solution would be 3×2.24 m, or 6.72 m, of distilled water.

One can see from these examples that the osmotic pressure, at least to a first approximation, is dependent only on the number of moles of particles per kilogram of solvent. Even the ionic charge does not have an effect. The molal* osmotic pressure constant is 22.4 meters of water per molal. There are secondary effects that cause the osmotic pressure constant to be somewhat different for different compounds, but a value of 22.4 meters of water per molal is a good average value.

If one wants to consider the nonideal behavior

---

*The molality of a solution is the number of moles of solute per kilogram of solvent. The molality of a solution is sometimes called its osmolality. This proliferation of terms is unfortunate.

of electrolytes and its effect on osmotic pressure, it can be done by the use of the following equation:

The osmotic pressure expressed in meters of water $= \sum_{\substack{\text{over all solutes} \\ \text{in the solution}}} (\phi NC)\ 22.4\ \text{m/molal}$

where   C is the molality of a given solute in mol per kg of solvent

N is the number of particles that dissociate from a molecule of the solute

ϕ is a factor called the osmotic coefficient that reflects nonideal behavior of the solute

ϕ is concentration dependent and parallels activity coefficients for ionic solutes. ϕ for glucose is 1, for NaCl at 0.1 molal is 0.93*

For example, if we use this equation on the third example in the previous discussion, we obtain

$$\text{osmotic pressure} = [(1)(1)(0.1\ \text{molal}) +$$

$$(0.93)(2)(0.1\ \text{molal})]\ \frac{22.4\ \text{m}\ H_2O}{\text{molal}}$$

osmotic pressure = 6.41 meters of water

The osmotic pressure is important in the body because it is one of the chief factors that regulates the equilibrium between the cellular and extracellular fluids. The osmotic pressure of blood serum and that of urine are both of diagnostic value and are routinely, if indirectly, measured.

A property of a solution that does not depend on what the solute is but only on how many particles there are in solution per kg of solvent is called a colligative property. There are other colligative properties associated with the dissolution of solutes in a solvent. These are the depression of the solvent freezing point, the elevation of the solvent boiling point, and the lowering of the solvent vapor pressure. These other colligative properties follow the same type of pattern variation as the osmotic pressure.

In summary, the osmotic pressure of a solution is a nonspecific measure of the total number of particles in the solution. There are secondary effects that are caused by nonideal solute-solute and solute-solvent interactions. These secondary effects impose no serious problems in the use of osmotic pressure data. The value of the osmotic pressure of blood serum and of urine is diagnostically significant.

THE MEASUREMENT OF OSMOLALITY

The direct determination of osmotic pressure is, for a number of reasons, experimentally difficult.

---

*Wolf, A. V., Aqueous Solutions and Body Fluids, New York, Hoeber Medical Division of Harper and Row, 1966.

Because of this fact, the direct measurement of the osmotic pressure is never done in a clinical laboratory. Instead, one of the other colligative properties is used to measure the total molality of the serum or urine. This value is called the osmolality and is directly proportional to the osmotic pressure. When a health professional speaks of the osmolality of a sample, he or she is thinking in terms of the effect of the osmotic pressure. The value of the osmotic pressure is simple being expressed in terms of the total apparent molality.

With this background in mind, let us consider the actual measurements that are used. First, there are instruments on the American market that measure the total molality (osmolality) by the depression of the vapor pressure of water. These instruments are not common enough to merit a discussion at this time. Second, there are a number of instruments that measure the total molality (osmolality) by the depression of the freezing point of water.

The measurement of serum or urine osmolality by freezing point depression is based on the fact that the freezing point of water is decreased by 1.858°C per molal of ideal particles. The freezing temperature can be measured very accurately, so it is not surprising that this approach has found widespread use.

The Freezing Point Osmometer

Consider Figure 21–2. The sample is placed in the glass cell; a sample volume of about 250 μl is common. The stirring wire and the thermistor

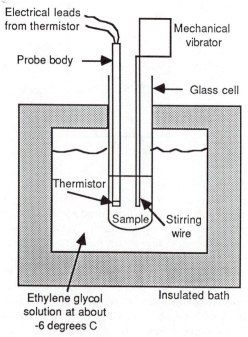

Figure 21–2. A simplified freezing point osmometer.

probe are introduced into the cell. The ethylene glycol solution is maintained at about −6°C by a refrigeration unit.

Under these conditions, the subfreezing coolant solution starts to cool the sample. This can be seen graphically in Figure 21–3; the cooling seg-

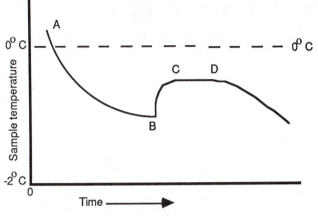

Figure 21–3. A time-temperature curve.

ment of the illustration is from point A to point B. By point B on the curve, the sample is supercooled. This means that its temperature is below its equilibrium freezing point. At point B, the vibrator is turned on, and the agitation causes the sample to freeze rapidly. This freezing process pumps the heat of fusion into the sample. These calories cause the sample to increase in temperature to the point that liquid sample and solid sample are in a slushy coexistence. This state has been achieved by point C on the curve. This state is, of course, the freezing point of the sample. The temperature of the sample stays at the freezing point for a minute or so, but eventually the cooling bath removes enough calories to lower the temperature below the freezing point. The difference between 0°C and the tem-

perature of the CD plateau in Figure 21–3 is the depression of the freezing point of the solvent.

The temperature at the freezing point is measured by the thermistor probe, which is discussed in Appendix C. The circuit that could be used to measure the temperature is depicted in Figure C–31 p. 327. The newer instruments are designed to give the osmolality by direct readout, which is done by running a high standard and a low standard. These values are used to calibrate the microprocessor in the electronics. The electrical resistance of the thermistor when measured during the CD plateau (Fig. 21–3) is then used to calculate and display the sample osmolality electrically.

ODDS AND ENDS

The temperature of the cooling bath is fairly important. If it is too warm, the sample freezes slowly, which causes the composition of the solid to be different from that of the liquid. This leads to an incorrect result. If the bath is too cool, time is insufficient for the plateau to become well defined.

The material used to standardize the instrument is sodium chloride. Solutions that are 100 millimolal (milliosmolal) in total ions and 500 millimolal (milliosmolal) in total ions are used as the low and high standards for serum measurements. A 900 milliosmolal standard may be substituted for the 500 milliosmolal standard for measurements of urine.

The sample should be free of particulates, as these interfere with obtaining a proper cooling curve. Their presence does not allow sufficient supercooling to occur before freezing takes place. The result is that the solid and liquid in the cell have different compositions. As discussed previously, this condition leads to an incorrect value.

The cell needs to be cleaned and dried carefully between determinations.

Problems

1. The osmolality of a sample is measured by the freezing point depression method. If the freezing point of the solution is depressed 0.832° C relative to distilled water, what is the osmolality of the sample?

2. The osmolality of a sample is measured by the freezing point depression method. If the freezing point of the solution is depressed 0.617° C relative to distilled water, what is the osmolality of the sample?

Nuclear Radiation and Its Measurement

The scope of this chapter is not as broad as that of some in this book. It is not meant to provide a broad background on the topic of radionuclides. It is particularly important to note that the topics of radiation safety and health physics are left to more specialized works (see **References**). This chapter is written from the narrow perspective of current applications of nuclear radiation in the routine clinical laboratory and is meant to provide background on the instrumentation that is used in that setting.

The nuclei of atoms are made up of neutrons and protons. All atoms of a given element have the same number of protons in their nuclei. This number of protons is the atomic number of the specific element. It is possible to have nuclei with the same atomic number but differing atomic mass number. This occurs because the nuclei have differing numbers of neutrons. Atoms with the same atomic number but differing atomic mass numbers are said to be isotopes of the same element. Using typical symbolization, some examples of isotopes of the same element are $^{12}_{6}C$, $^{13}_{6}C$, $^{14}_{6}C$, and $^{1}_{1}H$, $^{2}_{1}H$, $^{3}_{1}H$. The number of neutrons in the nucleus is equal to the atomic mass number of the isotope minus the atomic number. Most naturally occurring elements are mixtures of isotopes. These isotopes are called natural isotopes. The term nuclide is a synonym for the term isotope.

NUCLEAR STABILITY AND RADIOACTIVITY

When the ratio of neutrons to protons in an isotope is just right, the nucleus is stable and is not prone to emit nuclear radiation. On the other hand, if the ratio is too high or too low, the nucleus is prone to the emission of nuclear radiation. Such an isotope is said to be radioactive and is called a radionuclide. If the ratio of neutrons to protons is too low for a heavy isotope, the nucleus may emit an alpha particle. An alpha is a helium nucleus. Consider the following example of an alpha emission.

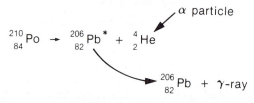

Reaction 22–1

The product of the reaction has 82 protons and is an isotope of lead. The ratio of neutrons to protons in the lead isotope is in keeping with stability for atomic number 82. This is not the case for the polonium isotope. Note that the lead isotope that is formed is originally in a nuclear excited state. When the nuclear excited state reverts to the ground state, a photon of energy is emitted. The emitted photon is a gamma ray. Gamma rays are short-wavelength electromagnetic radiation that originate in the nucleus. The emission of a gamma ray frequently follows an alpha emission.

When the ratio of neutrons to protons is too high for an isotope, the isotope frequently converts a neutron into a proton and an electron. The proton stays in the nucleus, but the electron is ejected. This high-energy electron is called a beta particle. An example of a beta emission follows:

$$^{14}_{6}C \rightarrow {}^{14}_{7}N + {}^{0}_{-1}\beta \qquad \text{Reaction 22–2}$$

where $^{0}_{-1}\beta$ is a beta particle.

The ratio of neutrons to protons is too high in carbon-14 but is a value that leads to stability in nitrogen-14. A second example of a beta emitter is tritium.

$$^{3}_{1}H \rightarrow {}^{3}_{2}He + {}^{0}_{-1}\beta \qquad \text{Reaction 22–3}$$

In both cases the daughter produced is a stable isotope in the ground state. There are examples, however, in which excited-state daughters are produced. In these cases, gamma-ray emission will follow a beta emission.

Most naturally occurring radioisotopes are alpha or beta emitters; many are also gamma-ray emitters. It is interesting to note, however, that most of the isotopes used in routine clinical chemistry are not naturally occurring isotopes. The most commonly used isotope is iodine-125, which is generated in a nuclear reactor by the following reactions.

$$^{124}_{54}Xe + ^{1}_{0}N \rightarrow ^{125}_{54}Xe + \gamma \qquad \text{Reaction 22-4}$$

$$^{125}_{54}Xe + ^{0}_{-1}e \xrightarrow[\text{capture}]{\text{electron}} ^{125}_{53}I + \gamma + \text{x-rays}$$

$$\text{Reaction 22-5}$$

Because the xenon-125 has a neutron-to-proton ratio that is too low for stability, it captures an extra nuclear-orbiting electron and converts the electron and one of its protons into a neutron

$$^{1}_{1}P + ^{0}_{-1}e \rightarrow ^{1}_{0}N$$

where $^{1}_{1}P$ is a proton
$^{1}_{0}N$ is a neutron

This process leads to an excited-state iodine-125. When the iodine-125 reverts to the ground state, a gamma ray is emitted. The hole that is left when the electron is captured is filled by outer orbital electrons; this process is accompanied by x-ray emission. In addition to all this, there are other energy-emitting processes that may take place in an electron capture event. One is the product of a high-energy electron, called Auger electron. Auger electrons are detected as if they were beta particles. If this information is of interest to you, see the references listed at the end of this chapter or any comprehensive nuclear physics book.

The product of these nuclear reactions is the radioisotope iodine-125. Like xenon-125, iodine-125 has a neutron-to-proton ratio that is too low. It takes the same approach to stability and captures one of its inner-shell electrons and produces gamma rays, x-rays, and other energetic emissions. The decay routine is as follows

$$^{125}_{53}I + ^{0}_{-1}e \xrightarrow[\text{capture}]{\text{electron}} ^{125}_{52}Te + \gamma\text{-ray} + \text{x-rays} + \text{Auger}$$
$$\text{electrons}$$

$$\text{Reaction 22-6}$$

A second manmade isotope used in clinical chemistry is cobalt-57. Cobalt-57 is an electron-capture isotope as well. Consider its decay scheme.

$$^{57}_{27}Co + ^{0}_{-1}e \rightarrow ^{57}_{26}Fe + \gamma\text{-ray} + \text{x-rays} + \text{Auger}$$
$$\text{electrons}$$

$$\text{Reaction 22-7}$$

In summary, radioisotopes may emit several types of radiation. The emission types are alpha particles, beta particles, gamma rays, x-rays, Auger electrons, and others. Alpha emitters have no routine clinical applications; they will not be discussed further. There are, however, "routine" applications of some beta emitters. Tritium ($^{3}_{1}H$) and carbon-14 ($^{14}_{6}C$) do find their way into some clinical laboratories. These materials are used for some radioimmunoassay procedures. The use of these beta emitters does present some special problems because of their low penetrating ability. The decay schemes of these beta emitters are shown in Reactions 22-2 and 22-3.

The most important isotope in use in clinical applications is iodine-125. Its decay scheme is shown in Reaction 22-6. Cobalt-57 is used in the clinical laboratory for vitamin B_{12} measurements. Its decay scheme is shown in Reaction 22-7. These isotopes are electron-capture isotopes; they emit gamma rays, x-rays, and other energetic emissions.

THE RATE OF RADIOACTIVE DECAY

For sources that contain only one radioisotope, it has been found that the rate of radio emission is directly proportional to the number of radioactive atoms present in the source. This finding can be stated mathematically.

$$\text{Rate of nuclear emission} = A = \lambda N \qquad \text{Equation 22-1}$$

where λ is the decay constant
N is the number of radioactive atoms
A is the activity in disintegrations per unit time

In the process of radioactive decay, the number of radioactive atoms is constantly declining with time. Because of this fact, it is at times necessary to calculate the number of radioactive atoms present at a given time, t. This can be done if the number of radioactive atoms is known at time equals zero. The equation that predicts the number of radioactive atoms is

$$N = N_0\, e^{-\lambda t} \qquad \text{Equation 22-2}$$

where N_0 is the number of atoms at time equals zero
t is time in appropriate units
e is 2.718

It is not uncommon to modify Equation 22-2 by substituting Equation 22-1 into it. In that case,

Equation 22–2 becomes

$$\frac{A}{\lambda} = \frac{A_0}{\lambda} e^{-\lambda t}$$

or

$$A = A_0 e^{-\lambda t} \qquad \text{Equation 22–3}$$

This equation relates the disintegration rate at time equals zero to the disintegration rate at times equals t. Consider an example problem.

A tritium ($_1^3$H) source has an activity of 643 counts/minute on January 9, 1985. What will be the source's activity on January 9, 1989? Assume that λ is 5.63×10^{-2} years.

$$A = A_0 e^{-\lambda t}$$

$$\ln A = \ln A_0 + (-\lambda t)$$

$$\ln A = 6.47 - (5.63 \times 10^{-2} \text{ years})(4 \text{ years})$$

$$A = 513 \text{ counts/minute}$$

An indication of how rapidly an isotope will decay is the time required for half of the original atoms to disintegrate. This time is called the half-life of the isotope. The half-life of an isotope is generally found graphically by plotting the log of the count rate of the isotope versus time. The time required to realize a 50 per cent reduction in count rate is equal to half-life. The half-lives of the common clinical isotopes are given in Table 22–1.

The concept of the half-life leads to a useful equation for calculating the decay constant, λ. The half-life of an isotope is the time necessary for

$$A = A_0/2$$

If this fact is substituted into Equation 22–3 we can solve for λ.

$$A = A_0/2 = A_0 e^{-\lambda t_{1/2}}$$

$$\frac{1}{2} = e^{-\lambda t_{1/2}}$$

$$2.303 \log \frac{1}{2} = 2.303 \log e^{-\lambda t_{1/2}}$$

$$-0.693 = -\lambda t_{1/2}$$

$$t_{1/2} = 0.693/\lambda \quad \text{or} \quad \lambda = 0.693/t_{1/2}$$

TABLE 22–1. Common Isotopes Used in Clinical Chemistry

Isotope	Half-Life	Emission Type	Energy (Kev)
^{3}H	12.3 years	β	18
^{14}C	5730.0 years	β	156
^{57}Co	270.0 days	γ-rays, x-rays, and Auger electrons	122* 136*
^{125}I	60.0 days	γ-rays, x-rays, and Auger electrons	36*

*γ-ray energy

The principal unit of measure of a radioisotope is the Curie (Ci). A curie of a radioisotope is the amount necessary to produce 3.7×10^{10} disintegrations per second. The mass of a radioisotope that is needed in order to have a curie of an isotope varies greatly. It can be as little as a few milligrams of elements with short half-lives, or it can be more than 1000 kilograms for isotopes with long half-lives.

The curie is prefixed with metric modifiers in order to get the correct-sized unit. We often hear the terms millicurie and microcurie in the everyday laboratory applications.

THE INTERACTION OF BETA PARTICLES WITH MATTER

When a beta particle is allowed to interact with matter, it does so in one or more of three ways. In the process of interacting with matter, the beta particle loses its energy until it is an electron at thermal equilibrium with its surroundings.

The most common interaction (Fig. 22–1) takes place when the beta particles pass close to an orbiting electron. When this happens, the electrical fields interact, and the orbiting electron is ejected from the atom. A less energetic beta particle then travels on to its next interaction. In this way, a beta particle frequently dissipates its energy by leaving

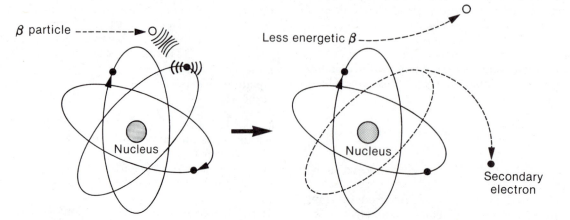

Figure 22–1. Redrawn from: J.A. Sorenson and M.E. Phelps, *Physics in Nuclear Medicine,* New York, Grune & Stratton, 1980.

a string of ions and electrons in its wake. These electrons are the particles that generally make it possible to detect the beta particles.

The second type of interaction between beta particles and matter takes place when the beta particle does not come close enough to ionize the atom, but does come close enough to interact. In this case, the electrical field of the beta particle causes a valence electron in an atom or molecule to be promoted to an electronic excited state. When the atom or molecule reverts to the ground state, a photon of energy is emitted. The energy emitted is generally visible or ultraviolet radiation.

The third type of interaction, (Fig. 22–2) that occurs between atoms and beta particles takes place

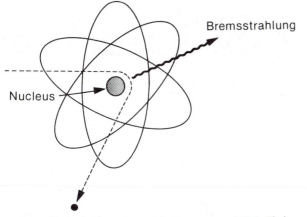

Figure 22–2. Redrawn from: J.A. Sorenson and M.E. Phelps, Physics in Nuclear Medicine, New York, Grune & Stratton, 1980.

when the electrical field of the beta particle and the electrical field of the nucleus interact. This interaction leads to a drastic change in direction of propagation of the beta particle and to a considerable loss of energy for the beta particle. The energy that is lost is emitted from the interaction site as a photon of electromagnetic radiation. The photon is called bremsstrahlung radiation and is, for practical purposes, an x-ray.

Beta detectors make use of the electrons and ions produced when a beta particle interacts with matter. These detectors are discussed later in this chapter.

THE INTERACTION OF GAMMA RAYS WITH MATTER

For the purposes of this discussion, the term gamma ray includes all high-energy photons. This includes x-rays, bremsstrahlung radiation, and any other nonparticulate, high-energy photons.

When a gamma ray passes through matter, it interacts with the matter in one or more of the following ways. First, if it interacts with an inner-orbit electron, the gamma ray is absorbed, and the electron is ejected from the atom (Fig. 22–3). The

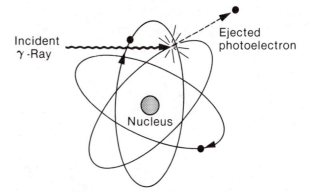

Figure 22–3. Redrawn from: J.A. Sorenson and M.E. Phelps, Physics in Nuclear Medicine, New York, Grune & Stratton, 1980.

ejected electron is called a photoelectron; it possesses the energy of the gamma ray except for the energy used in removing the electron from the atom. At this point, the photoelectron can do the things that were discussed in the previous section. It is not much different from a beta particle. When the photoelectron departs from its inner orbit, an electron from one of the outer orbits must fall in to replace it. This gives rise to an x-ray, which can, in turn, interact with matter. The net result of this process is that many electrons are knocked out of their former locations.

If the gamma ray interacts with a valence electron, the interaction reduces the energy of the gamma ray and results in the ejection of an electron from the atom (Fig. 22–4). The gamma ray continues to interact with matter. The ejected electron is called a Compton electron, and like a beta particle, it proceeds to interact with matter. The net result of Compton scattering is that a number of electrons are knocked out of their former locations.

A third possible interaction is pair production. When a gamma ray with at least 1.022 MeV of energy interacts with an atom, the gamma ray can be absorbed, and the resulting energy can eject a

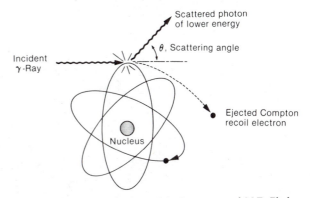

Figure 22–4. Redrawn from: J.A. Sorenson and M.E. Phelps, Physics in Nuclear Medicine, New York, Grune & Stratton, 1980.

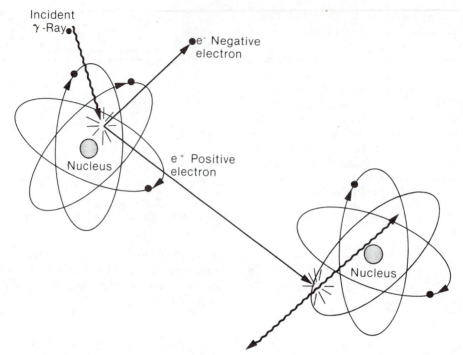

0.511-MeV Annihilation photons

Figure 22–5. Redrawn from: J.A. Sorenson and M.E. Phelps, Physics in Nuclear Medicine, New York, Grune & Stratton, 1980.

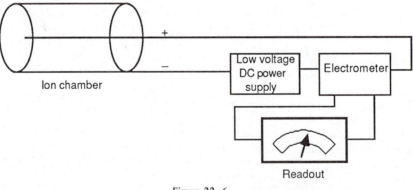

Figure 22–6

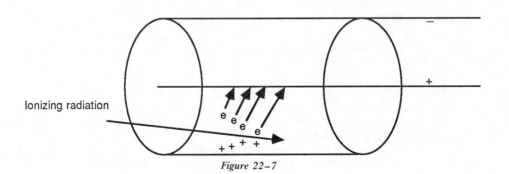

Figure 22–7

beta particle and a positron. The positron can then react with an electron. The result of this reaction is the annihilation of both particles with the ejection of two 0.511-MeV annihilation photons (Fig. 22–5). The beta particle can interact with matter as discussed previously. The annihilation photons are just two more gamma rays, and they can interact with matter. The net effect of this is that a number of electrons are knocked out of their former locations.

In summary, the energy of gamma rays and other high-energy photons is dissipated in matter by the generation of ions and electrons. The electrons produced are generally the agents detected with gamma ray detectors. The nature of this process is discussed later in this chapter.

Figure 22–8. Courtesy of Warrington Inc., Pflugerville, Texas. Reprinted with permission.

ION-CHAMBER COUNTERS

The main effect of ionizing radiation as it interacts with matter is the production of electrons and positive ions. Several types of radiation detectors make use of these ions and electrons. The first is the ion chamber. A simple ion chamber and its associated circuitry are illustrated in Figure 22–6.

The ion chamber consists of a metal cylinder with a wire electrode supported down the center. The cylinder may or may not be sealed. The cylinder is charged negatively by a DC power supply. The wire is charged positively. The applied voltage is relatively low. In some instruments, the power source is a battery. The electrometer (discussed in Chapter 19 under *Detectors*) is a very sensitive current-measuring device. The cylinder is filled with a gas, such as air at atmospheric pressure. When ionizing radiation enters the detector, it causes gas molecules to dissociate into positive ions and electrons. These charged particles are collected by the electrodes and provide the basis for an electrical current (Fig. 22–7). For a specified type of radiation, the higher the level of radiation, the greater the current.

Ion-chamber counters are not very sensitive to gamma rays because the ionization produced per unit distance traveled is low for gamma rays. The ionization efficiency of ion-chamber detectors is generally less than 1 per cent for gamma rays. Nonetheless, these detectors are used in survey meters for radiation-safety purposes. The popular Cutie Pie survey meter (Fig. 22–8) manufactured by Warrington Laboratories, Inc., is an example of such a battery-operated ion chamber. Ion-chamber detectors work well for the detection of beta-emitting gases. This is the basis of operation for the BAC-TAC instruments (Fig. 22–9) manufactured by Johnston Laboratories, Inc. The BACTAC instruments are used to determine whether bacteria have grown in a culture broth. The broth contains carbon-14–labeled nutrients. If the bacteria can utilize the provided nutrient, the metabolic product of C-14 carbon dioxide is produced and collects in the head space of the vial. After a suitable incubation period, the head-space gases are swept into an ion chamber for counting. If the count rate is 20 or 30 per cent of full scale on the instrument, it is assumed that the bacteria could utilize the nutrient. By selecting appropriately labeled nutrients, one can confirm the presence or absence of various bacteria in the inoculum. The instrument is adjusted to read full scale when 25 nCi (nanocuries) of $^{14}CO_2$ are in the ion chamber. Thus the production of five or six nCi of the gas is used to confirm the presence of a particular bacterium.

THE GEIGER-MÜLLER COUNTER

When scientists made an effort to improve the sensitivity of the ion-chamber detector, it became apparent that more amplification of the current produced was needed. An easy way to get this amplification is to seal the ion chamber and place a counting gas in it. The gas used is generally argon that contains some chlorine. The gas-filled chamber is powered by a high-voltage DC power supply (Fig.

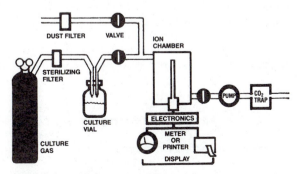

Figure 22–9. Courtesy of Johnston Laboratories, Inc., Cockeyville, Md. Reprinted with permission.

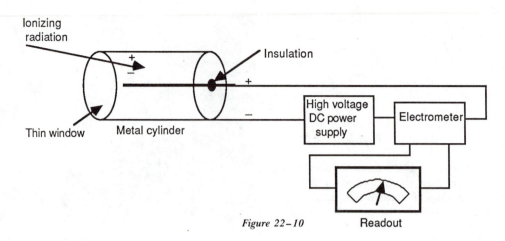

Figure 22-10   Readout

22-10). The center electrode is a metal wire. The cylinder is metal or glass with a metallic coating on the internal surface. The detector is called a Geiger-Müller tube.

The gas in the Geiger-Müller (GM) tube interacts with radiation in a manner similar to that described for the ion-chamber counter. The electrons and argon cations produced are accelerated by the high voltage. These highly energetic ions start ionizing other argon atoms, and then the new ions are themselves accelerated. The process is repeated many times. The net effect is an amplification of about 10^{10}. For every ion pair formed by the radiation, about 10^{10} ion pairs are collected by the electrodes. This gas-amplification factor makes the GM detector about ten times more sensitive than an ion-chamber detector.

The chlorine is included in the counting gas as a quenching agent. This gas is used to absorb excess energy at the end of a counting event. If this were not done, the energy of the argon cations hitting the cathode could activate the tube again. These secondary ionizations must be avoided.

The GM counter is not as sensitive to gamma rays as one would hope. The tube also has a certain amount of down time. The avalanche of current set off by the radiation requires some time to clear. The tube cannot count a second radio emission during this time. The GM tube cannot cope with count rates over about 15,000 CPM. If rates higher than this are allowed to interact with the detector, it misses some of the radiation.

The GM counter is used primarily for survey work. A battery-powered survey meter is illustrated in Figure 22-11.

SCINTILLATION COUNTERS

Liquid Scintillation Counters

When it is necessary to count the beta emitters, tritium, or carbon-14, the detection system of choice is a liquid scintillation counter. The emitting sample is mixed with a compatible solvent that contains a suitable fluorescent compound. The fluorescent compound commonly used is 2,5-diphenyloxazole (POP). A second fluorescent compound is added as a wavelength modifier. The common one used is 1,4-di-[2,5-phenyloxazole]benzene (POPOP). Sometimes other additives are used. The mixture of counting solution and sample is placed in a vial, and the vial is placed between two photomultiplier tubes (Fig. 22-12). (Photomultiplier tubes are discussed in Chapter 4.)

When a beta particle is emitted, it interacts mainly with the solvent. Through one mechanism or another, electrons are ejected from the solvent molecules. These electrons possess considerable energy. These energetic electrons interact with the 2,5-diphenyloxazole, POP. The POP takes the energy of the electrons and converts some of it into visible radiant energy. The emission spectrum of POP is shown in Figure 22-13.

Figure 22-11. Battery-powered survey meter. (Courtesy of Warrington Inc., Pflugerville, Texas. Reprinted with permission.)

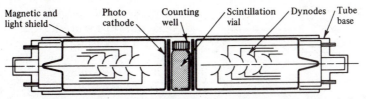

Figure 22–12. Diagram of a typical beta scintillation counter showing only the counting well and photomultiplier tube detectors. (From: Instrumental Methods of Analysis, 6th Ed., by H. Willard, L. Merritt, J. Dean, & F. Settle, Copyright © 1981 by Litton Educational Publishing, Inc. Reprinted by permission of Wadsworth, Inc.)

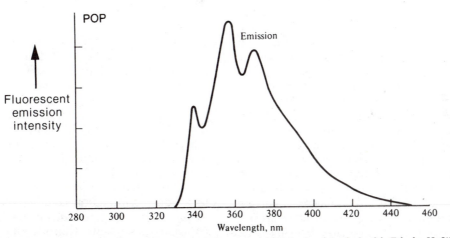

Figure 22–13. The emission characteristics of POP. (From: Instrumental Methods of Analysis, 6th Ed., by H. Willard, L. Merritt, J. Dean, & F. Settle, Copyright © 1981 by Litton Educational Publishing, Inc. Reprinted by permission of Wadsworth, Inc.)

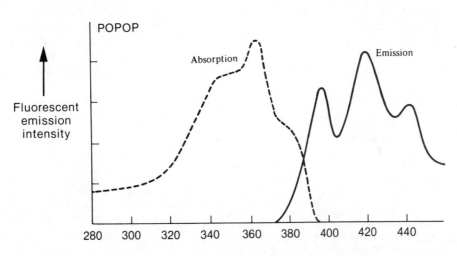

Figure 22–14. The absorption and emission characteristics of POPOP. (From: Instrumental Methods of Analysis, 6th Ed., by H. Willard, L. Merritt, J. Dean, & F. Settle, Copyright © 1981 by Litton Educational Publishing, Inc. Reprinted by permission of Wadsworth, Inc.)

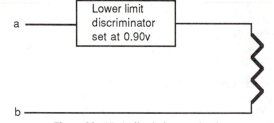

Figure 22–15. A discriminator circuit.

The radiant energy that is put out by POP is not well suited for detection by a photomultiplier tube. Because of this, a *wavelength shifter*, POPOP, is added. The POPOP absorbs the energy that is emitted by POP and reemits some of the energy at a longer wavelength (Fig. 22–14). The longer wavelength emission of POPOP is detected more efficiently by the photomultiplier tubes. It is interesting to note that the POPOP emits its energy in all directions. When one of the photomultiplier tubes sees a POPOP emission, so does the other. If both tubes fail to see an emission, it is not counted. This reduces the false counts that are introduced by electrical problems with the photomultiplier tubes. In this way, the energy that a beta particle loses in the scintillation solution is seen by both photomultiplier tubes as a flash of light. The photomultiplier tubes convert the flash of light into an electrical current. The magnitude of the current is directly proportional to the energy of the original beta particles. If these current pulses are passed through an electrical resistor, they are converted to voltage pulses. The voltage pulses can then be amplified and sent to a discriminator circuit or circuits for selection and counting. This is done in order to gain more selectivity. A discriminator circuit can reject pulses that are too high or too low in energy. In this way, it is possible to avoid the counts that come from an isotope that is not of interest.

Let's consider some of the properties of discriminator circuits. A discriminator circuit will conduct current only when the applied voltage exceeds

a present value. Such a circuit is shown in a block diagram in Figure 22–15. When a voltage of less than 0.90 v is applied across terminals a and b, no current flows through the resistor. On the other hand, if the voltage applied across terminals a and b is greater than 0.90 v, current flows through the resistor.

If two discriminator circuits are combined with an anticoincidence circuit, the resulting device is called a single-channel pulse-height analyzer. Such a device is shown as a block diagram in Figure 22–16. If a voltage pulse that is less than 0.90 v is applied to terminals a and b, neither of the discriminators will conduct current. If neither of them conducts current, then the anticoincidence detector sees nothing and does not direct the scaler to count. If a voltage pulse greater than 1 v is applied, both discriminators will conduct current. If both conduct current, the anticoincidence detector will detect that coincidence has occurred and will again direct the scaler not to count. If a voltage pulse that is between 0.90 v and 1 v is applied to terminals a and b, the lower limit discriminator will conduct, but the upper limit discriminator will not conduct. This causes the anticoincidence circuit to signal the scaler to count. The voltage range accepted is the *window* range and is electrically adjustable.

When the circuit in Figure 22–16 is connected to a liquid scintillation counter and its associated amplifier, an instrument that is suitable for counting beta particles results. Consider the block diagram in Figure 22–17. This instrument can be used with the upper limit discriminator turned off. If this is done, all radiation of energy greater than the lower limit will be counted. This is called integral counting. The instrument more commonly is used with both discriminators set to appropriate values. This type of counting is called differential counting.

It is possible to build instruments that have many discriminators. The popular multichannel analyzers have 400 sets of discriminators and can be used to obtain the beta spectrum of an emitter. Not all the beta particles emitted by an isotope have

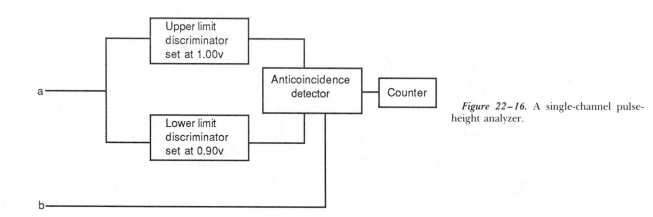

Figure 22–16. A single-channel pulse-height analyzer.

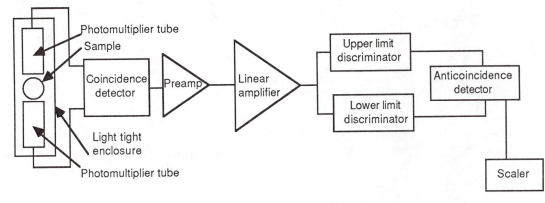

Figure 22–17

the same energy. There is, in fact, much variation in the energy of beta particles coming from the same source. Beta spectrometers are of interest in research. The beta spectrum of carbon-14 is included here to illustrate the point that multichannel analyzers make possible the determination of spectra of beta emissions (Fig. 22–18).

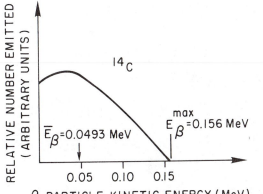

Figure 22–18. Energy spectrum (number emitted versus energy) for β particles emitted by ^{14}C. Maximum β^- particle energy E_β^{max} is Q, the transition energy. Average energy $\overline{E}_\beta$ is 0.0493 MeV, about 1/3 E_β^{max}. (From: J.A. Sorenson and M.E. Phelps, Physics in Nuclear Medicine, New York, Grune & Stratton, 1980. Reprinted by permission of Grune & Stratton, Inc. and the author.)

Liquid scintillation counting is troubled by the phenomenon of quenching (see Chapter 7). Quenching is caused when the sample matrix in some way affects the conversion of nuclear energy to light energy. Most of the problem is with the POP and POPOP system. Various materials including dissolved oxygen can affect the quantum efficiency of the fluorescent processes. The presence of radiant energy photon absorber in the sample can cause quenching. In all of these cases, the scaler does not see as high a count as it should. Scientists try to correct for quenching by various standardization methods. These methods are: (1) the internal standard method, (2) the external standard method, and (3) the channels-ratio method. In each of these methods, a sample of known count rate is placed in the instrument, and the experimentally measured count rate is compared with it. The difference between the known and measured count rates is used to calculate a correction factor. The resulting correction factor is used to compensate for quenching. For a detailed discussion of these methods, consult the references at the end of this chapter.

Solid Scintillation Counters

The common gamma ray counters used in routine clinical chemistry make use of solid, single-crystal scintillators. The most common scintillation crystals are prepared from sodium iodide that contains 1 per cent thallium. These crystals have a density of 3.7 g/cm^3 and have an effective atomic number of about 50. Hence they provide a reasonable opportunity for a gamma ray to interact with the crystal. The energy conversion efficiency is about 10% for a 100-keV gamma ray. This yield of radiant energy is relatively high, while the duration of a pulse is relatively short. Sodium iodide is hydroscopic, so it is necessary to seal the crystal into an aluminum foil envelope that fits the crystal tightly. It is fairly common to drill a well into the scintillation crystal. The sample can then be inserted into the well, and in this way the scintillation crystal surrounds the sample.

The scintillation crystal is bonded to an end-window photomultiplier tube (see Chapter 4). Figure 22–19 shows several NaI/PM assemblies.

Upon entering a scintillation crystal, the gamma ray interacts with the NaI by producing free electrons. These electrons are produced by one or more of the mechanisms discussed earlier in this chapter. The energy of the electrons produced is directly proportional to the energy of the gamma ray. These electrons are collected in the vicinity of the thallium atoms, which are called luminescent centers. This is followed by the emission of a pulse

Figure 22–19. NaI(T1) crystal and PM tube assemblies. (Courtesy of Bicron Corporation, Newbury, Ohio. Reprinted with permission.)

Figure 22–20. Arrangement of NaI(T1) crystal and PM tube in a typical detector assembly. (From: J.A. Sorenson and M.E. Phelps, Physics in Nuclear Medicine, New York, Grune & Stratton, 1980. Reprinted by permission of Grune & Stratton, Inc. and the author.)

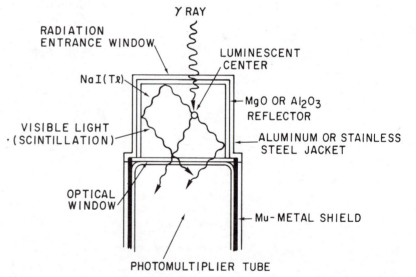

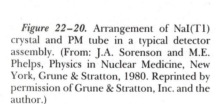

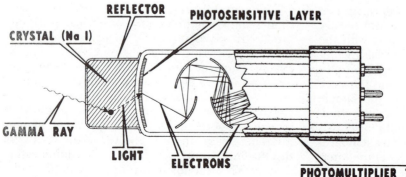

Figure 22–21. Schematic diagram of a solid scintillation detector. (Reprinted from: N.W. Tietz, Fundamentals of Clinical Chemistry, 2nd Ed., Philadelphia, W.B. Saunders Company, 1976.)

of radiant energy. The intensity of the radiant-energy pulse is directly proportional to the energy of the gamma ray that was involved in its formation. This process is illustrated in Figure 22–20. The photomultiplier tube converts the radiant energy to an electrical current that is proportional to the intensity of radiant energy. As the radiant energy is proportional to the energy of the gamma ray, the output of the photomultiplier current is directly proportional to the gamma energy. The combination of the scintillation crystal and the photomultiplier tube is shown in Figure 22–21.

The scintillation crystal and photomultiplier tube are connected to a preamplifier and associated circuitry, as was shown in Figure 22–8. In routine clinical work, the instrument used is a single-channel analyzer (SCA) with its pulse-height analyzer adjusted to look only at the energy of the isotope of interest. Multichannel gamma spectrometers do exist. They are used in research. Multichannel instruments can be called gamma spectrometers. They are used to obtain gamma spectra. The gamma spectrum of iodine-125 (Fig. 22–22) is included as a point of interest.

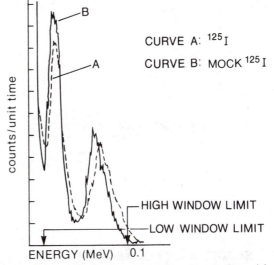

Figure 22–22. Mock ^{125}I is an intimate mixture of americium-241 and iodine-129. It provides an energy spectrum that reliably simulates the spectrum of iodine-125 in a well-type detector. As such, it may be used as a long-lived standard to calibrate instruments such as well scintillation spectrometers in which measurements are to be made involving iodine-125. (Courtesy of The Nucleus, Inc., Oak Ridge, Tenn. Reprinted with permission.)

Conclusion—Scintillation Counters

In conclusion, liquid and solid scintillation counters are used to measure the amounts of beta and gamma emitters, respectively. These measurements are important in the clinical laboratory because they are used to follow the course of immunochemical reactions. The immunochemical technique that uses a radioisotope is called radioimmunoassay, RIA. Immunochemistry and RIA are discussed in detail in Appendix A. The importance of RIA in clinical analysis has been very far-reaching.

REFERENCES

Early P.J., Razzak M.A., and Sodee D.B.: Textbook of Nuclear Medicine Technology, 3rd Ed. St. Louis, C.V. Mosby Company, 1979.
Sorenson J.E. and Phelps M.E.: Physics in Nuclear Medicine. New York, Grune & Stratton, 1980.

Problems

1. A sample contains only one radionuclide; the radionuclide has a long half-life. A 1.000-g sample produces a count rate of 4,739 dpm. A 1.000-g standard of similar composition and geometry is known to contain 3.0 mg of the radionuclide. The standard produces a count rate of 10,000 dpm; what is the mass of radionuclide in the sample?

   Standard   10,000 dpm
   Unknown   4,739 dpm

2. If a sample has as its only radioactive species a radionuclide with a long half-life, and if a 1.000-g standard is known to contain 3.0 mg of the radionuclide, what is the amount of the same isotope in a 1.000-g sample of similar composition?

   Standard   12,394 dpm
   Unknown   9,327 dpm

Instrumentation in Hematology:
Blood Cell Counters

BLOOD: A LITTLE BACKGROUND[1]

Blood is a highly complex mixture of suspended solids in liquid. The liquid is called blood plasma—a very complex true solution. The solids are sometimes called formed elements and include red blood cells (erythrocytes), white blood cells (leukocytes), and platelets. Red blood cells are generally 6.8 to 7.5 μm in diameter, with an average of 7.2 μm. They have a cellular volume of about 90 fl* and contain a high concentration of hemoglobin. There are about 5×10^{12} red cells per liter of blood. The hemoglobin is an iron-containing compound that is red in color and reversibly adsorbs oxygen. The major role of red blood cells is to transport oxygen from the lungs to all body tissues.

White blood cells, generally larger than red blood cells, range in size from 6 to 18 μm and include many different types. The major classifications are lymphocytes, monocytes, neutrophils, eosinophils, and basophils. There are about 7×10^9 white blood cells per liter of blood. Their major role is to control infection.

Platelets average 2 to 4 μm in diameter and tend to be disc shaped. They have a cellular volume of 4 to 7 fl, and there are about 3×10^{11} platelets per liter of blood. Their major role is to maintain hemostasis. When a vein or an artery is damaged, platelets help form a plug and close the wound, thus preventing excessive bleeding.

A number of hematological measurements are made on blood samples. The *hemoglobin* concentration is frequently determined by lysing the red blood cells in a known volume of blood. The resulting homogeneous red solution is investigated by visible spectroscopy. The resulting hemoglobin level is expressed in grams of hemoglobin per liter.

*Femtoliter (fl): femto equals 10^{-15} of the basic unit.

The determination of the packed cell volume is another analytical measurement frequently made. This is done by centrifuging whole blood that has been stabilized with an anticoagulant. This is done in a tube of uniform diameter. The height of the packed cells in the tube is divided by the height of the blood. The resulting value is greater than 0 but less than 1 and is called the *hematocrit*.

Cell counters have been designed that identify and count red blood cells, white blood cells, and platelets. (The red blood cell count is abbreviated RBC; the white blood cell count, WBC.) In some cases, the instruments can estimate the volume of the cells counted. When this is possible, the mean cell volume (MCV) is calculated.

Clinical scientists often speak of the *erythrocyte indices*. The erythrocyte indices are

1. The mean red cell volume (MCV) expressed in femtoliters

$$MCV = \frac{hematocrit}{red\ blood\ cell\ count\ per\ liter}$$

2. The mean red cell hemoglobin (MCH) expressed in picograms per cell

$$MCH = \frac{hemoglobin\ in\ grams\ per\ liter}{red\ blood\ cell\ count}$$

3. The mean red cell hemoglobin concentration (MCHC)

$$MCHC = \frac{hemoglobin\ in\ grams\ per\ liter}{hematocrit}$$

The exact meaning of a differential white blood cell count varies depending upon the laboratory. In every case, however, the differential count is an attempt to subtype the white blood cells. The count may be as simple as a three-part differential or it may be as complex as a five-part differential.

The commonly requested complete blood count (CBC) is defined as a red blood cell count, a white blood cell count, a hematocrit and hemoglobin determination, a differential white blood cell count, the mean cell volume, the mean cell hemoglobin mass, the mean cell hemoglobin concentration, and sometimes a platelet count.

BLOOD CELL COUNTERS

Over the years there has been considerable interest in the development of instruments that can identify and count different types of blood cells. The current instruments available vary considerably in their capabilities and complexities. The simplest instruments can identify and count red blood cells and white blood cells as well as platelets, if programmed to do so. At the other extreme are flow cytometers that can identify and count the different subtypes of white blood cells. Some flow cytometers can even sort the different subtypes into different collection vessels.

When one considers the blood cell counters that are found in non–research oriented laboratories, there are only two approaches whereby cells are identified and counted: the electrical resistance approach (developed first) and the optical approach.

Cell Counters Based on Electrical Resistance Measurements

According to Gulati, Hyun, and Ashton,[2] the blood cell counters manufactured by Coulter Electronics, Baker Instruments, TOA Medical Electronics, Sequoia-Turner, Clay-Adams, Mallinckrodt, and Angel Engineering are based on electrical resistance (impedance)* measurements.

It is important to realize that the following discussion of resistance-based counters is illustrated with examples drawn from the literature of specific manufacturers. The selection of these illustrations was based on their ready availability and clarity for teaching the principles of blood cell counting. Their use does not imply endorsement of any instrument or manufacturer.

Cell counters based on resistance measurements make use of the relatively high electrical resistance of blood cells. This high cellular resistance is exploited by carefully diluting the blood with a highly conductive (low resistance) diluent and then passing the resulting suspension through a unique flow cell. The flow cell has a small orifice, or aperture, that causes most of the blood cells to pass

*Impedance is the alternating current equivalent of resistance. This topic is discussed in Appendix C. To a first approximation, impedance can be thought of as resistance.

through in single file. It also has platinum electrodes for the measurement of electrical resistance. The resistance of the flow cell is monitored during the time that a known volume of blood is passed through the aperture. For the purpose of teaching, I have illustrated the principle of operation of this type of counter with a very crude system (Fig. 23–1).

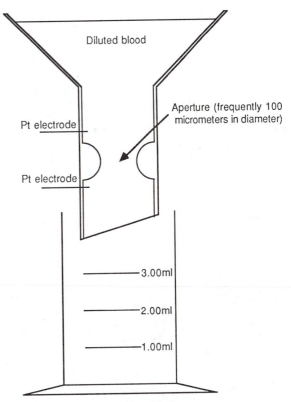

Diluted blood

Aperture (frequently 100 micrometers in diameter)

Pt electrode

Pt electrode

—3.00ml

—2.00ml

—1.00ml

Figure 23–1

The system consists of three major parts: (1) a reservoir that holds the diluted blood sample; (2) the highly specialized electrical conductivity cell with an aperture or narrowing between a pair of platinum electrodes; (3) and the graduated cylinder that measures the volume of diluted blood under examination.

With this general overview in mind, let us consider how the cell count is obtained. Because of the diluteness of the blood cell suspension and the smallness of the aperture, most of the blood cells pass through the aperture one at a time. When a blood cell passes through the aperture, the electrical resistance across the aperture increases. This occurs because the electrical resistance of the blood cell is much higher than the resistance of the electrolyte solute that it displaces. The resistance measured is the sum of the resistance of the blood cell and the undisplaced diluent. The reason for this fact can be

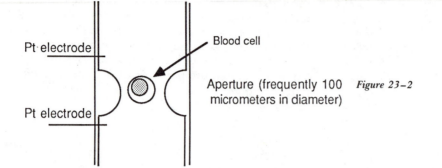

Pt electrode

Blood cell

Aperture (frequently 100 micrometers in diameter)

Figure 23–2

seen in Figure 23–2. All of the volume that the blood cell occupies as it passes through the aperture is largely unavailable for the conduction of current. This results in an increase in the resistance through the aperture. When the blood cell has passed through the aperture and the aperture is filled with the highly conducting diluent, the resistance returns to its baseline value. Each spike in resistance that occurs while the diluted blood passes through the aperture is counted. The actual count is done by an electronic event counter. The blood count, which is expressed in numbers of cells per liter, is calculated from the volume of diluted blood counted, the number of counts recorded, and the amount that the blood was diluted in sample preparation.

In order to obtain the very best results, one

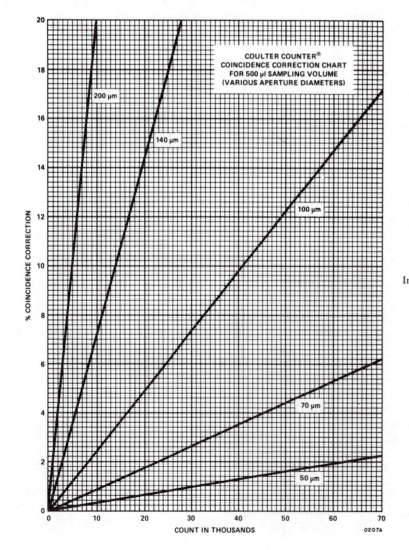

Figure 23–3. Courtesy of Coulter Electronics, Inc., Hialeah, Fla. Reprinted with permission.

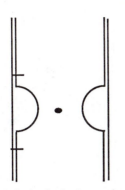

Figure 23–4. A platelet in the conductivity cell.

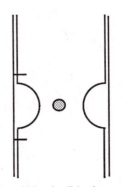

Figure 23–6. A red blood cell in the conductivity cell.

must take into account the two or more blood cells that sometimes simultaneously pass through the aperture. The frequency of this occurrence is related to the number of cells present and the size of the aperture. The numerical value for the correction is predictable and can be calculated from statistical considerations. The value for the correction can be found from the graph in Figure 23–3, which was obtained by analyzing various blood samples of known cell count. The graph is supplied courtesy of Coulter Electronics and applies to their ZBI counter. Be sure to consult the literature supplied with your counter.

The electronic requirements for this type of cell counter are relatively straightforward. The system should have a low-voltage AC power supply of great stability. There must also be a simple circuit for noting momentary increases in resistance across the aperture and a scaler for totalling them. Generally, a conductance circuit is used. Many of these instruments also use circuits to count only certain sizes of resistance changes. It is interesting to note that this same system can be used to differentiate among cells of different sizes. Cells with different sizes have different cellular volumes. In turn, cells with different volumes displace differing amounts of electrolyte solution from the aperture as they pass through. This means that a small cell, like a platelet, will generate a small increase in aperture

resistance, but a white blood cell, being much larger, will produce a large increase in resistance. A red blood cell, which is of intermediate size, will produce intermediate-sized spikes in resistance. This relationship is illustrated in Figures 23–4, 23–5, and 23–6, respectively.

Because these three classes of cells produce different sizes of electrical impulses, it is possible to build counters that can differentiate among different classes of cells. Electronic discriminator circuits are used to sort the impulses by size and to cause the scaler to respond only to cells of a certain class. Some instruments also use these data to calculate the mean cell volume.

Identification and Counting of Red Blood Cells, White Blood Cells, and Platelets on a Specific Instrument

In an effort to reinforce your knowledge of cell counting by resistance measurements, I will discuss a specific instrument—the Coulter model ZBI. The ZBI is the most common instrument used to calibrate or standardize the highly automated instruments that perform the majority of the workload in the laboratory.[2] The ZBI is designed to peform red blood cell counts and white blood cell counts using a 100-μm aperture tube; it can count platelets if a 70-μm aperture tube is installed. The instrument has several subsystems that play a specific role in the counting process. These subsystems will be illustrated as the discussion progresses.

Figure 23–7 is a highly simplified illustration of the conductivity cell and part of the vacuum system of the Coulter ZBI. The conductivity cell consists of a pair of platinum electrodes that are separated by the aperture tube. The aperture has a 100μm opening; other sizes are also available. The model ZBI's system of two electrodes and an aperture tube is the same arrangement as that illustrated in Figure 23–2. The vacuum system is used to move the diluted blood through the aperture. The nature of this system is illustrated in Figure 23–8. The illustration shows the cell with a some-

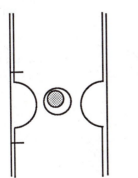

Figure 23–5. A white blood cell in the conductivity cell.

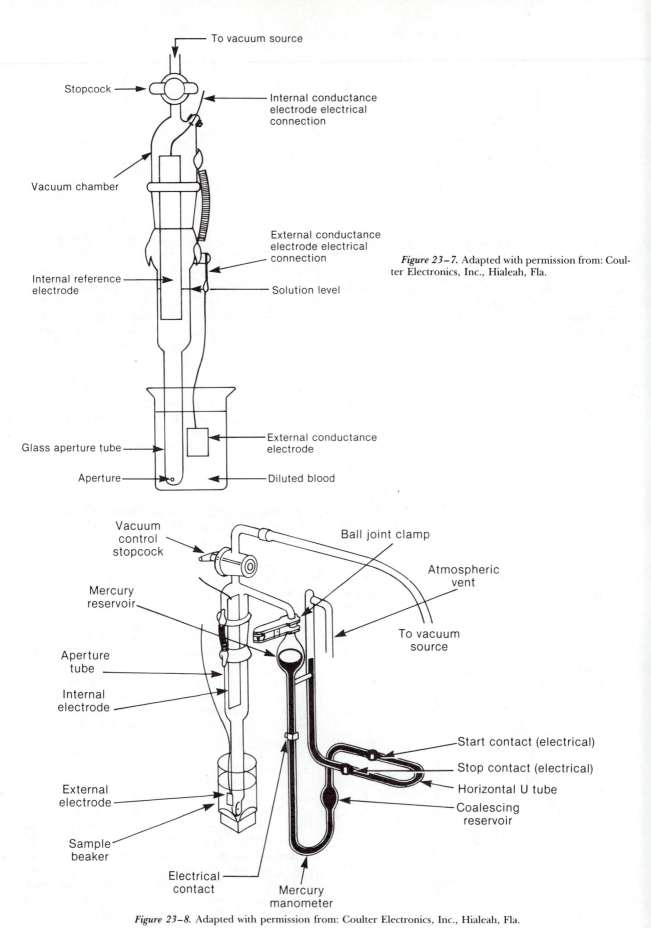

Figure 23–7. Adapted with permission from: Coulter Electronics, Inc., Hialeah, Fla.

Figure 23–8. Adapted with permission from: Coulter Electronics, Inc., Hialeah, Fla.

what simplified vacuum and sample metering system. When the vacuum control stopcock is opened to the vacuum source, the vacuum does two things. First, it draws some diluted blood through the aperture into the aperture tube. This ensures that the internal conductance electrode makes contact with the solution. Second, the vacuum draws mercury from the manometer into the storage bulb located under the ball joint clamp. When enough mercury has been moved so that the horizontal "U" tube is empty, the counter is ready for use.

Once the vacuum control stopcock is closed, the mercury in the mercury reservoir starts to move back through the manometer. As it does so, it reduces the pressure of the air in the aperture tube. The resulting low pressure draws more diluted blood through the aperture. As this process continues, the mercury reaches the start contact in the horizontal U tube. The counter then starts counting the blood cells that come through the aperture. The counting continues until the mercury reaches the stop contact or the horizontal U tube; it is automatically stopped when the mercury reaches the stop contact. The horizontal U tube generally contains 500 μl of mercury between the start and stop contacts. Thus, the instrument reads the cell count for 500 μl of blood. The number is not corrected for coincident cell passage events. This must be done using correction values, as illustrated in Figure 23–3.

In addition to the systems already shown, there are several others on the counter (Fig. 23–9). First,

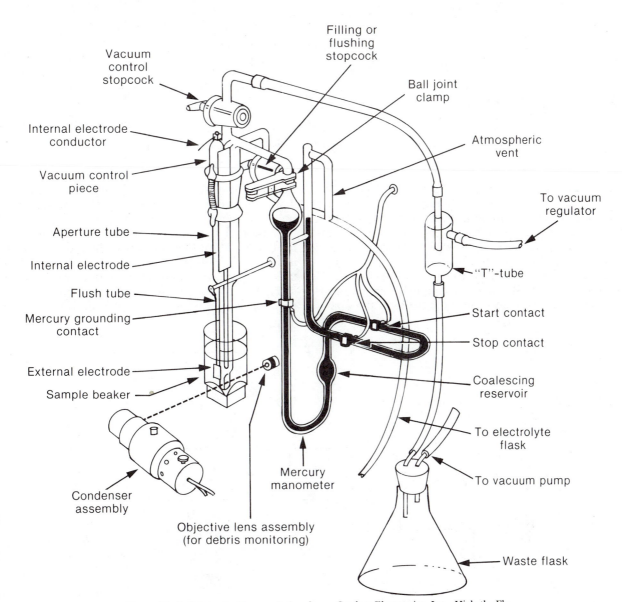

Figure 23–9. Adapted with permission from: Coulter Electronics, Inc., Hialeah, Fla.

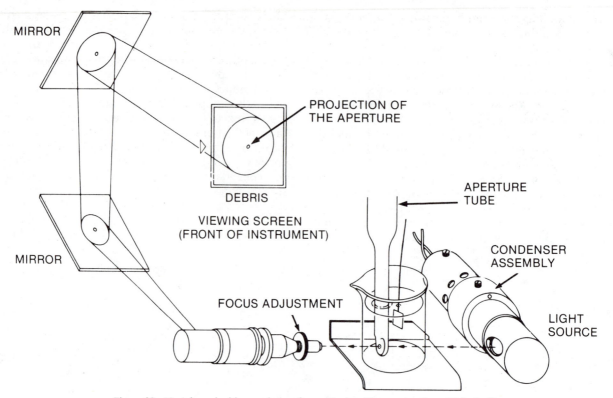

Figure 23–10. Adapted with permission from: Coulter Electronics, Inc., Hialeah, Fla.

there is a system used to flush the aperture tube. The system consists of the filling or flushing stopcock and a line running from the electrolyte flask. The aperture tube is flushed by opening the vacuum control and the flushing stopcock. This sends a torrent of solution down the flushing tube, out through the vacuum line, and on to the waste flask. When a less violent flush will do, a sample beaker of rinse solution can be placed on the outside of the aper-

ture tube, and the vacuum control can be opened. This draws rinse solution through the aperture and on out through the vacuum line.

Another system is used to monitor the aperture for debris. This is necessary because if the aperture is blocked by debris, inaccurate data will result. Details of the debris monitor's operation are provided in Figure 23–10. The image of the aperture is actually projected on a ground glass screen. If there

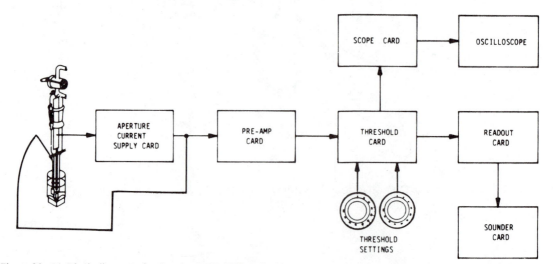

Figure 23–11. Block diagram of a Counter ZBI. (Adapted with permission from: Coulter Electronics, Inc., Hialeah, Fla.)

is an indication of a problem, the instrument operator can observe the aperture.

Now that you have seen the cell and its associated systems, let's consider some of the electronics of the instrument (Figure 23–11). First are the discriminator circuits, which Coulter calls threshold circuits. Figure 23–11 illustrates the threshold adjustments for the high and low discriminator. Second, a built-in oscilloscope is provided so that the shape of the resistance spikes can be observed. These changes in resistance are converted to proportional voltages for oscilloscopic observations.

The instrument counts red blood cells in a 1 to 50,000 suspension. The diluent is isotonic with red blood cells. White blood cells are also present in this solution, but the threshold setting prevents them from being counted. They are counted in a 1 to 500 suspension. The diluent is not isotonic with red blood cells; hence, they lyse. The nucleated cells do not lyse; thus, they can be counted.

The State of the Art: Electrical Resistance Instruments

Highly automated blood cell counters that rely on electrical resistance measurements have been developed and marketed. By incorporating a number of small design improvements, it is now possible to differentiate the white cell population into lymphocytes, mononuclear cells, and granulocytes. These instruments are under computer control and can provide data that were not available in the precomputer days. The most notable example of this is the trend toward the presentation of histograms. A histogram is a graph of the relative number of cells versus the cell size. An example of a normal white blood cell histogram is shown in Figure 23–12.[3]

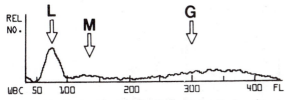

Figure 23–12. A white blood cell histogram (L: lymphocytes, M: monocytes, G: granulocytes). (Reprinted with permission from: R. Keller, Lab. Med., *15* (11):743, 1984.)

Cell Counters Based on Optical Measurements

The blood cell counters that are manufactured by Ortho Diagnostic Systems and Technicon Instruments count blood cells on the basis of their ability to "scatter" electromagnetic radiation. There is some question about the mechanism of this scattering process. Several explanations have been of-

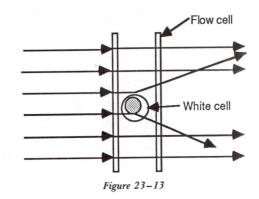

Figure 23–13

fered: radiant energy is simply reflecting off the cells; the radiant energy is being transmitted through the cells and then refracted as it passes on out the other side; or the radiant energy is being truly diffracted by the cell. No matter which explanation is true, it is certain that considerable low-angle "scattering" from the cells takes place (Fig. 23–13).

This low-angle scattering can be exploited for counting purposes. All that is required is a low-angle radiation-scattering photometer and a flow cell that forces the blood cells to pass through the beam of radiation one cell at a time. Consider the low-angle radiation-scattering photometers that are illustrated in Figures 23–14 and 23–15.

The optical paths of the two illustrated photometers have a lot in common with other instruments on the market. The source of electromagnetic radiation generally is a visible source. Tungsten sources as well as lasers have been used. The entrance slit in the photometer illustrated in Figure 23–14 allows for the selection of a tiny part of the source radiation. Lens 1 focuses the diverging radiation on the flow cell and provides the flow cell with a highly columnated region of illumination. The flow cells may be made in many different forms, but generally speaking all manufacturers have designed their instruments so diluted blood is passed through the flow cell. In their design of flow cells, many manufacturers have used laminar flow sheaths in an effort to position the blood cells at the center of a larger tube. The flow cell on the Technicon H6000 (Fig. 23–16) is a good example of this approach. It passes two solutions through the flow cell simultaneously. The diluent occupies most of the cell volume and flows through as a concentric sheath, while the diluted blood passes through the center of the concentric sheath. This results in a very small active volume. The manufacturer states that the active volume on the H6000 is 20 μm in diameter—an amazing dimension considering that the tube has an internal diameter of 250 μm. Because of the very small active volume, platelets and red blood cells can be counted in the presence of each other. White blood cells are counted after lys-

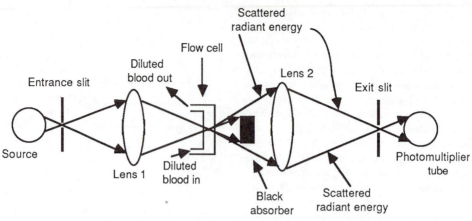

Figure 23–14. Low-angle radiation-scattering photometer.

ing the red blood cells. The black absorber, or blocker, is just large enough to absorb all of the nonscattered radiation striking it. Most of the scattered radiation, however, misses the black absorber. It is collected by lens 2 and focused on the photomultiplier tube. When a blood cell is not present in the active volume, the photomultiplier tube does not receive radiation. When a blood cell is present, the photomultiplier tube receives the scattered radiation. The intensity and duration of the photomultiplier tube output is characteristic of the cell that passed through the active volume. Based on this information it is possible to count blood cells

and, in some cases, identify their type. It is also possible to estimate the mean cell volume. As was discussed in the section *Cell Counters Based on Electrical Resistance Measurements,* it is necessary to know the volume of the blood that has been counted in order to record the cell count as number of cells per liter of blood.

THE HEMATRAK

Before automated instruments were used in the practice of hematology, blood smears were pre-

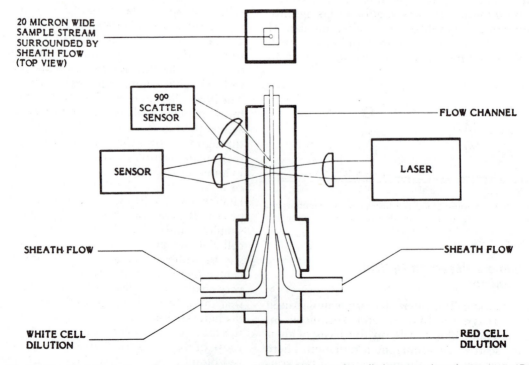

Figure 23–15. Simplified illustration of an ELT-15 and an ELT-1500 (low-angle radiation-scattering photometer). (Courtesy of Ortho Diagnostic Systems Inc., Westwood, Mass. Reprinted with permission.)

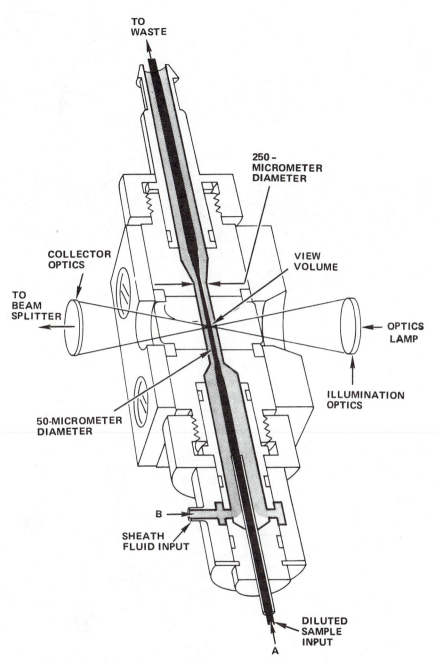

Figure 23–16. Flow cell on the Technicon H6000. (Courtesy of Technicon Instruments Corporation, Tarrytown, New York. Reprinted with permission.)

pared on slides, and the cells were then stained with Wright's stain. A technician would examine the smear with a microscope and count the number of white cells, estimate the number of platelets, study the morphology of the red blood cells, white blood cells, and platelets, and report the results.

The Hematrak is an instrument that uses a form of microscopy and several microprocessors to mimic the manual white cell count and the cell evaluations that are described in the previous paragraph. The sample of blood is spread to give a monolayer of cells. The smear is treated with Wright's stain and then examined by a special electronic scanning microscope, as illustrated in Figure 23–17.

The microscope itself is of classic design. The light source used is a cathode ray tube much like the one illustrated in Figure 1–18, p. 10. The computer controls the voltages applied to the tube in such a way as to move the small spot of white light across the slide. In the search mode the small light spot moves across the slide in linear sweeps 1 μm apart. The light transmitted through the blood smear passes through two dichroic mirrors that separate the transmitted radiant energy into its green, red, and blue components. The intensity of these

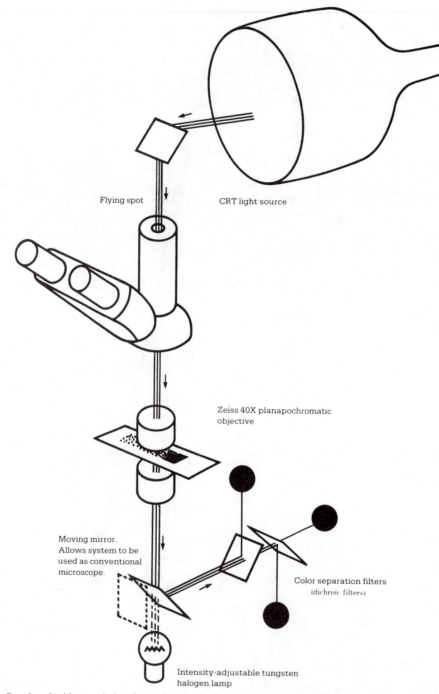

Flying spot

CRT light source

Zeiss 40X planapochromatic objective

Moving mirror. Allows system to be used as conventional microscope.

Color separation filters (dichroic filters)

Intensity-adjustable tungsten halogen lamp

Figure 23–17. Reprinted with permission from: Geometric Data, The Search for Clinical Confidence, Wayne, Penn., SmithKline Beckman.

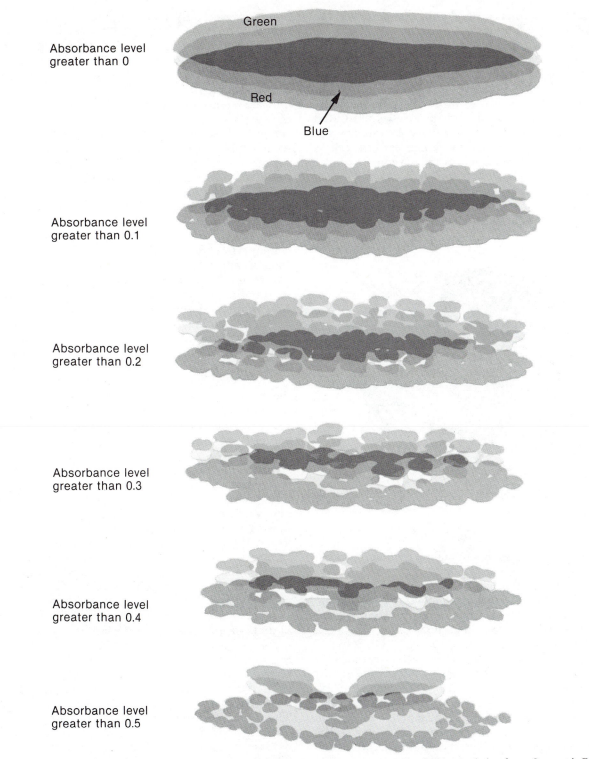

Figure 23–18. Note: Hypothetical values were selected to illustrate the principle. (Reprinted with permission from: Geometric Data, The Search for Clinical Confidence, Wayne, Penn., SmithKline Beckman.)

three components is used by the computer to help identify the cell under examination.

When a white blood cell is encountered, the computer scans the cell in linear sweeps 0.25 μm apart. The computer also classifies the absorbance of the green, red, and blue radiation into six different levels. In this way the computer has a three-color absorbance profile of the cell, as illustrated in Figure 23–18. Once this three-color, six-absorbance range, two-dimensional map of the cell is known, the computer makes measurements on the size of the cell's morphological features by electronically placing lines of different lengths and at different angles across the map. This is illustrated in Figure 23–19. On the basis of the 96 measure-

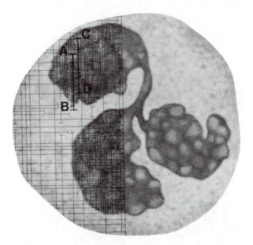

Figure 23–19. Reprinted with permission from: Geometric Data, The Search for Clinical Confidence, Wayne, Penn., SmithKline Beckman.

ments that the instrument makes, the microprocessor classifies the cell examined.

The Hematrak can be programmed to examine between 200 and 1000 white blood cells in a single blood smear. The normal cells are classified and are presented as the differential white blood cell count. The location of the unclassified cells is stored in the computer so that a technician can visually examine the cells by traditional optical microscopy. The instrument can also be programmed to examine the morphology of up to 5000 red cells. These data are used to produce a cell histograph (Price-Jones curve). The instrument estimates the number of platelets. The readout is on a color video monitor or printer.

CLINICAL FLOW CYTOMETRY: THE TECHNICON H6000

The Technicon H6000 is a clinical flow cytometer that operates at the rate of 90 samples per hour. The instrument aspirates 220 μl of ethylene diaminetetraacetic acid (EDTA)-anticoagulated whole blood. The sample is diluted, air-bubble segmented,* and routed to a proportioning pump with three manifolds. These manifolds treat the sample and route it to four optical systems. One of the optical systems identifies and counts various types of white blood cells. This is called the peroxidase channel. A second optical system identifies and counts basophiles. The third manifold splits the sample and sends it to two optical systems. One of these measures the hemoglobin. The other counts red blood cells and platelets. All of the optical systems use the same quartz halogen source.

The data generated allow for the following information to be reported: white blood cell count, lymphocytes, monocytes, eosinophils, large unstained cells, cells with high peroxidase activity, and basophiles. These cell types are counted as well as reported as to their percentage of the white blood cell count. Red blood cells and platelets are also counted. The hemoglobin concentration is determined by the cyanmethemoglobin method, the hematocrit value is measured, and the red cell indices are calculated.

Now, let's consider the three flow paths and the measurements that are made in each. The peroxidase channel counts and identifies the various types of white blood cells by measuring cell size and cell peroxidase activity. The different types of white blood cells have different levels of peroxidase activity. The diluted whole blood that flows into the peroxidase channel is treated with a surfactant that lyses the red cells. The blood is also treated with a mildly alkaline fixative that fixes the enzymes. The flowing blood cells are heated to 59°C. This inactivates the cellular catalyses. The sample is mixed with pH 7.3 buffer, hydrogen peroxide, and 4-chloro-1-naphthol. The peroxidase catalyzes the destruction of the hydrogen peroxide. The oxygen radicals released react with the 4-chloro-1-naphthol to generate naphthoquinone, a gray-black compound. In this system, the higher the peroxidase activity of a cell, the blacker the cell stains. The flowing stream is adjusted to pH 9.7 to clarify the stream. The stream is debubbled, and a small constant fraction of the stream is passed through a hydrodynamically focused flow cell (Fig. 23–16). This causes the white blood cells to pass through the active volume of the cell detectors. The peroxidase channel uses a dual detection system (Fig. 23–20) and consists of a photometer and a low-angle light-scattering photometer. The photometer measures the absorbance and size of the white blood cells, and the computer receives the absorbance and size information on each cell, ultimately plotting the data as shown in Figure 23–21.

_____

*See Chapter 25, page 271.

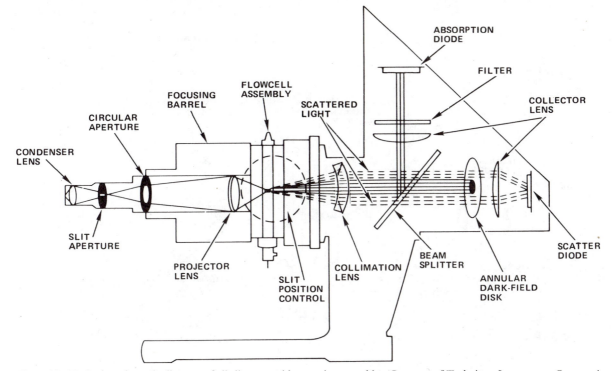

Figure 23–20. Optics schematic diagram of alkaline peroxidase optics assembly. (Courtesy of Technicon Instruments Corporation, Tarrytown, New York. Reprinted with permission.)

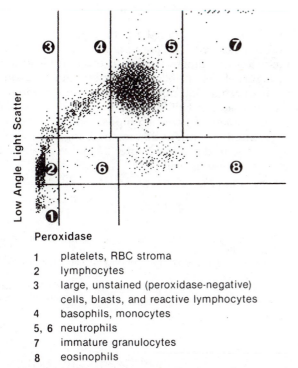

Peroxidase

1   platelets, RBC stroma
2   lymphocytes
3   large, unstained (peroxidase-negative)
    cells, blasts, and reactive lymphocytes
4   basophils, monocytes
5, 6   neutrophils
7   immature granulocytes
8   eosinophils

Figure 23–21. A normal WBC histogram, showing the three general areas used in cell classification: lymphocytes, mononuclear cells, and granulocytes. Differential (instrument): 62% granulocytes, 9% mononuclear, 29% lymphs; 100-cell conventional differential: 58% segs, 3% bands, 7% monos, 32% lymphs. (Reprinted with permission from: L.J. Williams, J. Med. Tech., *1* (3):194, Figure 8, March 1984.)

Monocytes and basophiles both have about the same peroxidase activity and are also about the same size. Because of these characteristics, the Technicon H6000 has a separate channel for the basophile count. It works in the following way: A portion of the prediluted whole blood sample is air segmented and is sent to the basophile channel. There, it is further diluted and then treated with Alcian blue, cetylpyridinium chloride, and lanthanum chloride. The cetylpyridinium chloride lyses the red cells and prevents the red cell membrane fragments from clumping. Lanthanum chloride selectively inhibits the staining of nucleic acids. The flowing suspension is next treated with malic acid. The resulting suspension contains Alcian blue–stained basophiles. All the other cells are unstained. The suspension is debubbled and passed through a hydrodynamically focused flow cell, which is irradiated with polychromatic radiant energy. The radiant energy is scattered off the white blood cells as they pass through the flow cell.

The optical system consists of two low-angle light-scattering photometers. One of the photometers has a red filter; the other has a yellow–blue-green filter. The channel that transmits yellow radiant energy receives a lower intensity when a blue-stained basophile passes than it does when an unstained cell passes through the flow cell. This fact allows the instrument to count the basophiles.

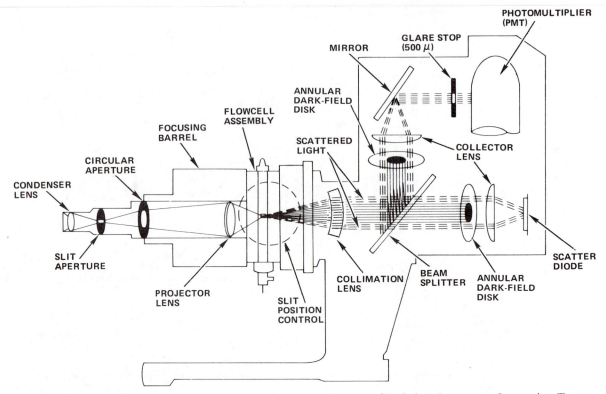

Figure 23–22. A dual-channel, low-angle light-scattering photometer. (Courtesy of Technicon Instruments Corporation, Tarrytown, New York. Reprinted with permission.)

The third prediluted stream of air-segmented whole blood is mixed with an isotonic diluent. A small but constant fraction of this suspension is diluted again with more isotonic diluent. The overall dilution is 1:1300. This suspension is debubbled and passed through a hydrodynamically focused flow cell. The scattered radiant energy is examined by a dual-channel, low-angle light-scattering photometer (Fig. 23–22). The platelets do not scatter enough radiant energy to activate an output from the diode-detected photometer. They do activate the photomultiplier-detected photometer. Red blood cells scatter enough radiant energy to activate an output from both detectors. The computer uses this data to identify and to count red blood cells and platelets. The pulse heights are also used to calculate the hematocrit.

Part of the diluted blood sample from the red blood cell–platelet channel is routed to a photo-metric detector where the hemoglobin concentration is measured. The photometric system is a double-beam instrument similar to the one illustrated in Figure 4–44, p. 52. Hemoglobin concentration is measured at 550 nm using the cyanmethemoglobin method. The H6000 is calibrated with stabilized human blood of known composition and uses air-segmented continuous flow methodology. This topic is covered in detail in Chapter 25 and is applicable to this section as well.

REFERENCES

1. Miale J.B.: Laboratory Medicine Hematology, 6th Ed. St. Louis, C.V. Mosby, 1982.
2. Gulati G.L., Hyun B.H., and Ashton J.K.: Lab. Med., *15*:395–401, 1984.
3. Williams L.J.: Cell histograms: new trends in data interpretation and cell classification. J. Med. Tech., *13*:189–197, 1984.

Instrumentation in Hematology: Coagulation Measurements

HEMOSTASIS

Hemostasis is the term used to describe the self-sealing property of the circulatory system. It is the result of the actions of a number of body systems. For example, when a blood vessel is damaged, the endothelial cells that line the inside of the blood vessel are disrupted, exposing the underlying subendothelial cells to the blood. The disruption also exposes collagen that is located in the blood vessel wall. At this point platelets respond, traveling from the circulating blood to adhere to the site of the damage and, under the influence of collagen and other materials, begin to stick to each other. This process is called platelet aggregation. Aggregated platelets secrete various biochemicals that help stimulate the formation of a blood clot.

Blood clots are formed by the interaction of about a dozen proteins, which are called coagulation factors. Coagulation factors work together to convert the protein prothrombin into a proteolytic enzyme called thrombin. Thrombin in turn acts on a circulating protein called fibrinogen. This reaction converts fibrinogen into a fibrin monomer. The fibrin monomers then polymerize to form macromolecules called fibrin. The fibrin strands in conjunction with the aggregated platelets and red blood cells form a clot that frequently can prevent excessive blood loss.

Hemostasis can be compromised by a number of diseases. Any disease that reduces the number or effectiveness of platelets can lead to bleeding. For this reason, hematologists are interested in platelet counters and platelet aggregometers—instruments that quantitatively study the effectiveness of platelets in their ability to aggregate.

As with platelets, any disease state that reduces the effectiveness of the coagulation factors can lead to bleeding. Instruments called coagulation analyzers can be used to study the functional state of the coagulation system. These instruments assess the competence of the coagulation system by measuring the amount of time necessary for the formation of fibrin strands to occur. These tests are done under carefully controlled conditions. The time test can be done using different reagents in order to evaluate various parts of the coagulation system.

The most common tests done are the prothrombin time and the activated partial thromboplastin time tests. In both cases the proper reagents are mixed with the sample; when the last reagent is added, the timer is started. The timer is stopped when the instrument detects the formation of fibrin strands.

In the past few years there has been increasing interest in the monitoring of the kinetics of fibrin formation. A number of these instruments are commercially available.

OPTICAL COAGULATION TIMERS

There are two common approaches to the determination of coagulation time. The first is an optical approach founded on turbidimetry and called the turbidimetric method. It is based on the fact that the fibrin strands are large enough to scatter or to block the transmittance of radiant energy. The optical system used in many analyzers is illustrated in Figure 24–1. The source is frequently a light-emitting diode (LED) that emits green radiant energy; tungsten lamps are also used on some instruments. The cell is often a disposable square or round plastic device or a disposable glass cell. The detector is frequently a cadmium sulfide photoresistor. This is almost always the case when an LED is used as the source. Other instruments use silicon photodiodes.

When a coagulation time is to be measured, the plasma sample is mixed with all of the reagents except one. When the last reagent is added, the timer is started. When the detector detects a sharp decrease in the transmitted light, P, the timer automatically stops. The resulting time is used as a measure of coagulation competence. Coagulation measurements are temperature dependent. Most instruments have provisions for maintaining the sample at $37° \pm 0.5°C$.

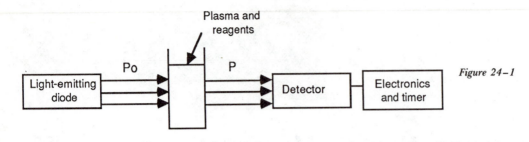

Figure 24–1

CONDUCTANCE COAGULATION TIMERS

The other way of measuring coagulation time is the electrical conductivity approach. In instruments using this technique, two electrodes are immersed in the reaction mixture (Fig. 24–2). As the last reagent is added to the test well, the timer is started and the moving electrode sweeps through the solution. It moves very close to the fixed electrode but does not touch it. The moving electrode then returns to its original position. It repeats this cycle about every half second, keeping the solution well mixed. When fibrin strands start forming, they connect the two electrodes, causing a change in electrical conductivity that in turn stops the timer.

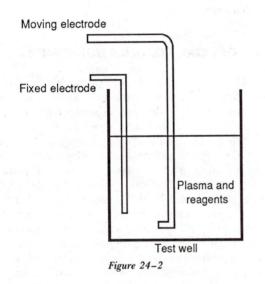

Figure 24–2

THE BIO/DATA COAGULATION PROFILER

Bio/Data Corporation manufactures a unique coagulation analyzer called a Coagulation Profiler. It is an optical instrument that is actually a turbidimeter. It uses a small tungsten lamp as a light source and a cadmium sulfide photoresistor as a detector. Plasma is the sample run. The instrument's ability to measure not only the coagulation time parameters but also changes in turbidance (apparent absorbance) per unit time and to present these data graphically sets it apart from other instruments. The resulting plot is called a Thrombokinetogram, or TKG. The shape of the Thrombokinetogram provides the hematologist with more information than the coagulation time parameters alone. A small insight into the use of the TKGs can be inferred from the next two figures. Figure 24–3 is an example of a TKG for normal plasma. Figure 24–4 shows a TKG for an abnormal sample.

OPTICAL PLATELET AGGREGOMETERS

There has been an increasing amount of interest in the quantitative measurement of platelet aggregation. Most of the instruments used are basically turbidimeters. There are a number of configurations used, but double-beam operation is common. Light-emitting diodes are frequently used as the source of radiant energy, although some instruments use tungsten lamps. The instruments with tungsten lamps generally use filters to provide for reasonable monochromatic radiant energy. The wavelength of operation is generally between 500 nm and 700 nm. Silicon photodiodes are popular detectors on instruments with tungsten sources. Cadmium sulfide photoresistors are popular on instruments using LED sources. The optical cells on these instruments have provisions for magnetic stirring. Stir rates between 500 and 2000 rpm can be used. Both single-beam and double-beam turbidimeters are used. The optical paths of the single-beam instruments are very similar to the path illustrated in Figure 24–1. The double-beam instruments are often used by placing a patient's platelet-rich plasma in one cell and his or her blood cell-free plasma in the other cell. This approach removes interferences due to lipemia or icterus.

The measurements are made by adding an aggregate to the sample of platelet-rich plasma. The sample is stirred at a constant speed and maintained at 37° ± 0.5°C. Several aggregates may be used, such as collagen, arachidonic acid, ADP, and adrenaline. The output of a platelet aggregometer is illustrated in Figure 24–5. Figure 24–6 shows the effect of different concentrations of aggregate.

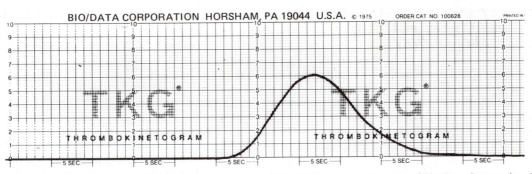

Figure 24-3. A normal activated partial thromboplastin time (APTT) of 30 seconds. (Courtesy of Bio/Data Corporation, Hatboro, Penn. Reprinted with permission.)

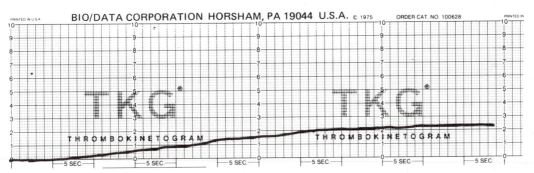

Figure 24-4. The APTT on a factor IX deficient plasma; clotting time is 102 seconds. (Courtesy of Bio/Data Corporation, Hatboro, Penn. Reprinted with permission.)

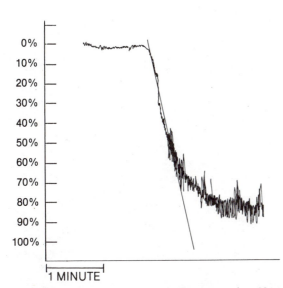

Figure 24-5. Collagen induced platelet aggregation. Normal control platelets: % aggregation = 86; slope = 66. (Courtesy of Bio/Data Corporation, Hatboro, Penn. Reprinted with permission.)

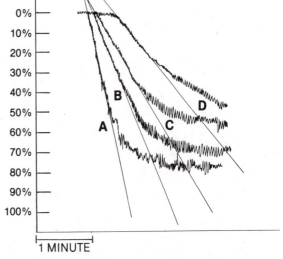

Figure 24-6. Arachidonic acid induced platelet aggregation. (A) conc. = 750 μg/ml; % aggregation = 79; slope = 54. (B) conc. = 500 μg/ml; % aggregation = 74; slope = 43. (C) conc. = 250 μg/ml; % aggregation = 54; slope = 38. (D) conc. = 100 μg/ml; % aggregation = 50; slope = 21. (Courtesy of Bio/Data Corporation, Hatboro, Penn. Reprinted with permission.)

ELECTRICAL IMPEDANCE OR RESISTANCE PLATELET AGGREGOMETERS

A unique approach to platelet aggregation measurement is possible with the electrical impedance channel on the Chrono-log instrument. The Chrono-log has optical capability as well. The instrument uses a pair of platinum electrodes in a stirred sample to detect platelet aggregation. The electrodes are powered with a 5-volt sigmoidal power supply, with a frequency of 15 kHz. As the platelets aggregate, they adhere to the electrodes. Because the platelets have a relatively high electrical resistance, the resistance between the platinum electrodes increases during aggregation.

The instrument provides a plot of resistance versus time. These plots are similar to the ones in Figures 24–5 and 24–6. The advantage of the impedance approach is that it can be used with whole blood.

LEUKOCYTE AGGREGOMETER

Currently there is some interest in the use of neutrophil aggregation measurements. According to the manufacturer, the Bio/Data PAD-4 Platelet Aggregometer can be used for these studies.

Automation in Clinical Chemistry

INTRODUCTION: A MANUAL METHOD

In order to better understand how automation has been developed for the clinical laboratory, it is helpful first to consider a common nonautomated analysis. I have chosen the determination of serum protein by the biuret method as the teaching example. The chemical reactions involved in the biuret method are discussed in Chapter 4. Let's start by considering the solutions that are needed to do the protein analysis. These are listed in Table 25–1.

TABLE 25–1. Solutions Needed for a Total Protein Analysis

Reagent	Purpose
Biuret	Chromogen
Distilled water	Diluent
7% protein	High standard
5% protein	Middle standard
3% protein	Low standard
High control	Quality control
Low control	Quality control

The analysis of serum protein consists of six steps, which we shall consider individually.

Step One: Dilute the Standards, Sample, and Controls. Carefully pipet 0.10 ml of the three standards, 0.10 ml of the reagent blank (distilled water), 0.10 ml of the sample, and 0.10 ml of the two control sera each into their own test tubes. This requires seven clean pipets and seven clean test tubes. Next, using an additional clean pipet, transfer 1.90 ml of distilled water into each of the seven tubes.

Step Two: Add the Chromogen. Using another clean pipet, transfer 2.00 ml of biuret reagent into each of the seven tubes.

Step Three: Mix the Tubes. Carefully mix the contents of each tube.

Step Four: Incubate the Tubes. Allow the tubes 5 minutes to react at constant temperature.

Step Five: Measure the Absorbance. Determine the absorbance of all of the tubes by using the reagent blank tube (distilled water) to calibrate the instrument to 0 absorbance. The remaining six test tubes can then be analyzed.

Step Six: Do the Paperwork. In this last phase of the analysis, plot out a standard curve such as the one illustrated in Figure 25–1. This curve is then used to find the protein content of the high control, the low control, and the sample. If the values for the high and low controls are within the

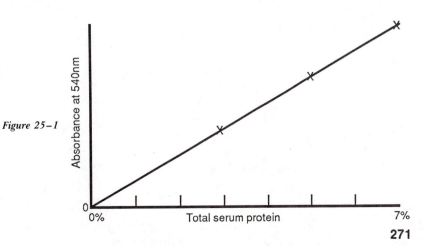

Figure 25–1

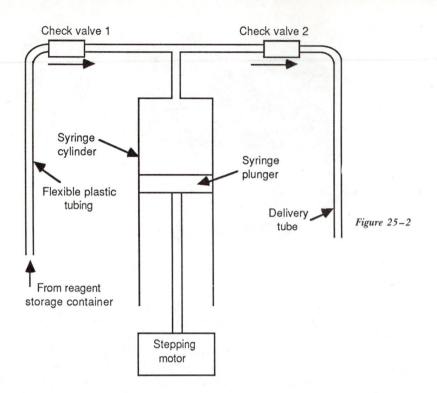

Check valve 1   Check valve 2

Syringe
cylinder

Syringe
plunger

Flexible plastic
tubing

Delivery
tube

Figure 25–2

From reagent
storage container

Stepping
motor

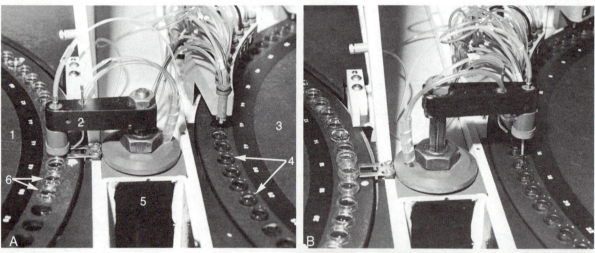

Figure 25–3. A, The robotic sampling arm in the sampling position. *Key:* 1, sample turntable; 2, robotic arm; 3, analytical turntable; 4, cells in the analytical turntable; 5, sampling pipet wash station; 6, sample cups containing samples. *B,* The robotic sampling arm delivering sample to a cell in the analytical turntable. *C,* The robotic sampling arm in the wash position.

accepted limits, the total serum protein for the patient can be reported.

In summary, the analysis of serum protein for one patient requires seven solutions, seven test tubes, nine clean pipets, and a visible photometer to carry out the steps listed in Table 25–2. In the

TABLE 25–2. Steps of Serum Protein Analysis

Dilution of the sample
Chromogen addition
Solution mixing
Incubation for the development of color
Measurement of absorbance
Paperwork

hands of a good analyst, the entire process can be completed in about 50 minutes.

THE AUTOMATION OF THE BIURET PROTEIN ANALYSIS: A HYPOTHETICAL INSTRUMENT

In this section I will discuss how a hypothetical, automated biuret protein analyzer could be made. Such an instrument would automate each of the six steps described in the manual method.

When one evaluates the steps of the manual method, it becomes apparent that they could be done mechanically. For example, all of the reagent pipetting could be done with precision syringes that are suitably valved and driven (Figure 25–2). The syringe plunger could be mechanically linked to a stepping motor — a motor designed to turn a small, fixed fraction of one shaft revolution every time a voltage pulse is applied to the motor. The pipetting machine could be designed to give 1 (μl) microliter of output per voltage pulse. Thus, a 1.00-ml liquid delivery could be caused by sending the stepping motor 1000 closely spaced voltage pulses. A microprocessor could be used to control the stepping motor and hence the pipetter.

For application in our hypothetical biuret protein analyzer, the pipetter would be fitted with two check valves. Check valve 1 would allow flow from the reagent storage container to the syringe but not from the syringe to the storage container. Check valve 2 would allow flow from but not to the syringe. (Check valves are illustrated in greater detail in Figure 17–6, page 175.) The delivery tube could be either fixed in position or movable. For the sake of simplicity, this discussion assumes that it is fixed in position. A system that would perform the same serum protein analysis as was done manually would need two of these precision pipetters—one for distilled water and one for biuret reagent.

The hypothetical biuret protein analyzer would also need a system for handling the standards, sam-

ple, and controls. This might be achieved through the use of a turntable containing the samples in cups. Such an arrangement is shown in Figure 25–3. A robotic sampling arm in conjunction with a sampling pipet could be used to pick up a measured volume of the sample from the cup. The robotic arm would then pivot about 180° and deposit the sample into a cell. Before we consider this in more detail, let's examine two of the ways in which the sampling pipetter might be built: an air displacement pipetter and a positive displacement pipetter.

The syringe in an air displacement pipetter does not have direct contact with the liquid it is pipetting. Instead, the syringe is connected to the sampling pipet by a plastic tube (Figure 25–4).

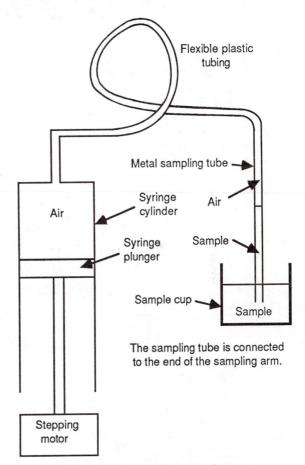

Figure 25–4. Air displacement pipetter.

When a sample is measured into the pipet, the microprocessor, via the stepping motor, withdraws the syringe plunger by the appropriate amount, causing the sample to be sucked into the sampling pipet. When the sample is delivered, the microprocessor, again via the stepping motor, moves the plunger into the cylinder. The increased air pressure drives the sample out of the sampling pipet. Pipetters of this type (Figure 25–5) are very accurate and trouble free.

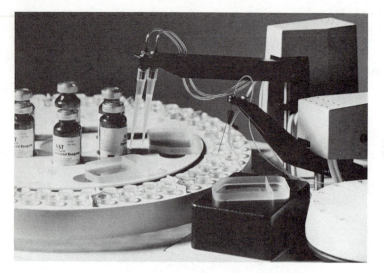

Figure 25–5. Courtesy of Baker Instruments Corporation, Allentown, Penn. Reprinted with permission.

A second approach to the measurement and transfer of the sample is the so-called positive displacement pipetter, illustrated in Figure 25–6. This type of pipetter does not allow for the sample to be in contact with the precision syringe. Instead, the syringe is filled with distilled water or some other liquid diluent that is supplied from a reservoir. The liquid enters the syringe through a microprocessor-controlled valve. Before a sample is picked up, the liquid fills the sampling pipet to the tip. When a sample is picked up, the sampling pipet is dipped into the sample by the robotic sampling arm. The stepping motor then withdraws the plunger, thus drawing the sample into the sampling pipet. Then the robotic arm withdraws from the sample and moves the sampling pipet to the reaction tube

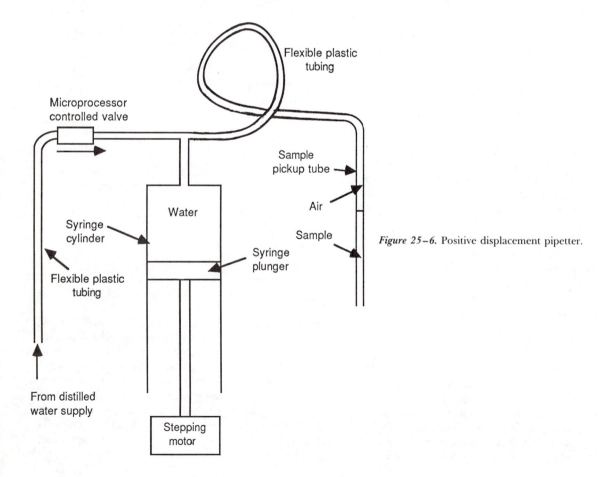

Figure 25–6. Positive displacement pipetter.

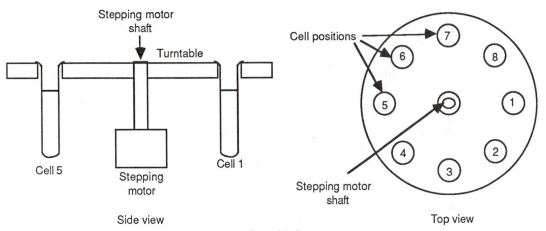

Figure 25–7

or cell. Next the stepping motor drives the sample out of the pipet into the cell. The necessary diluent is also dispensed at this time, rinsing the sampling pipet into the reaction tube or cell. The sampling pipet is then withdrawn from the sample tube and either is dipped into flowing distilled water in a wash basin or is sprayed with distilled water as it is returned to the sample turntable. The process removes any sample from the outside of the sampling pipet. On some pipetters, an air bubble is drawn into the sampling pipet just before the sample is drawn. This helps to prevent the sample from mixing with the liquid in the pipet.

In order to automate the manual method of serum protein analysis, one would replace the seven test tubes with seven cells arranged in a turntable as illustrated in Figure 25–7. The system would need two reagent pipetters, one sample transfer

pipet with robotics, and a sample turntable as diagrammed in Figure 25–8. The discussion that follows describes how the biuret protein analyzer would automate the six steps of manual serum protein analysis.

Step One: Dilute the Standards, Sample, and Controls. As depicted in Figure 25–8, the sampling arm would move to the sample turntable and position itself over the sample cup. The sampling pipet would be lowered into the sample, and the stepping motor on the air displacement pipetter would draw in 0.1 ml of sample. The sampler would then withdraw from the sample cup and pivot about 90°, placing the sampling pipet over the analytical turntable. The microprocessor, via the turntable stepping motor, would then position cell 1 under the sampling pipet, the sampling arm would drop, and the sample would be forced out of the sampling

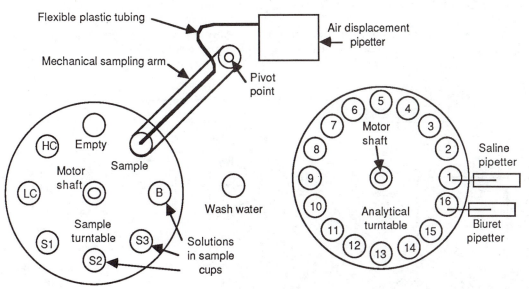

Figure 25–8

pipet by the syringe. A command from the microprocessor would cause the sample probe to wash itself in the wash basin. This would be done by the stepping motor's drawing water into the sampling pipet, then expelling it several times. The microprocessor would then command the sampling arm to move to the sample turntable. It would also advance the sample turntable one position. While the sampling pipet was being washed, the microprocessor would advance the analytical turntable one position. The sampling arm could then pick up the high control and deposit 0.1 ml of it into cell 2. This cycle of sample, deliver, and wash would continue until the seventh cell had received 0.1 ml of the blank. The analytical turntable would then be rotated so that 1.9 ml of distilled water could be added to the cells. This would complete step 1 in Table 25–2.

Steps Two and Three: Add the Chromogen and Mix the Tubes. The microprocessor would direct the biuret reagent pipetter and the analytical turntable to work together to deliver 2 ml of biuret reagent to each of the seven cells. This would be done with enough force to mix the contents of each cell. This would complete steps 2 and 3 in Table 25–2.

Step Four: Incubate the Solution. The microprocessor would then give the samples, standards, and controls 5 minutes to incubate. During this time the chromogen would react with the analyte and develop the colored product. This would complete step 4 in Table 25–2.

Step Five: Measure the Absorbance. The hypothetical biuret protein analyzer would need a photometer that could measure the absorbance of the cells at 540 nm. The photometer, located under the analytical turntable, would generally be a single-beam photometer that had been designed for great stability. An example of this instrument is illustrated in Figure 25–9.

The microprocessor would position the turntable so that the blank in cell 7 could establish the

value of Po. The microprocessor would store the value. Next it would advance the turntable one position counter-clockwise. In this position, P for the 3 per cent standard would be measured and stored. The turntable would then be advanced one more position and the value for P for the 5 per cent standard could be measured and stored. This process would continue until the value of P for all cells is known, thereby completing step 5 in Table 25–2.

Step Six: Do the Paperwork. At this point our hypothetical automated biuret protein analyzer would have recorded the absorbance of all the cells as well as the concentration of all the standards. The microprocessor would use this information to calculate the concentration of the sample and the two controls and to compare the control values with the accepted values. If the controls were within an acceptable tolerance range, the concentration of the sample would be printed out, completing the analysis.

Carefully consider the six steps in the manual method and correlate them with the six steps in the automated method because this introduction will be frequently cited in the rest of this chapter.

AUTOMATED BATCH ANALYZERS (NONCENTRIFUGAL)

The hypothetical biuret protein analyzer, sometimes called a test tube analyzer, was designed to automate the biuret protein method. Because it could only analyze one sample, such an instrument would be impractical in an actual laboratory. On the other hand, if the analytical turntable had been designed to hold 60 cells and the sample turntable 60 samples, standards, and controls, the instrument would have been more practical. Such an instrument could easily analyze 50 samples, 4 standards, and 6 controls in 1 hour.

Automated instruments like our hypothetical biuret protein analyzer are called batch analyzers

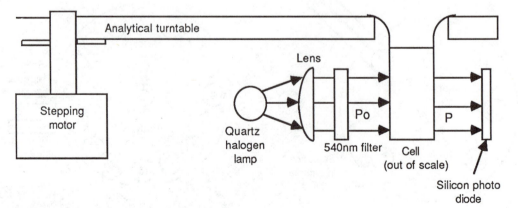

Figure 25–9. Side view of the analytical turntable and single-beam photometer. Only one cell is illustrated.

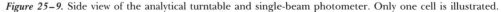

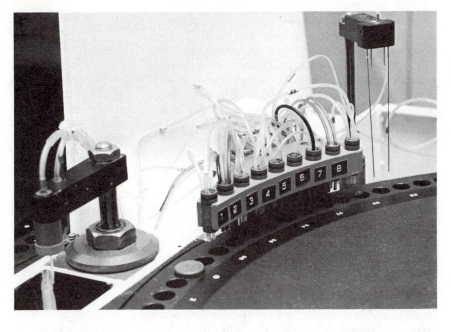

Figure 25–10. A bank of delivery tubes.

because they can analyze a batch of samples for one analyte at a time. It is very common to build batch analyzers that can analyze for a number of different analytes. These instruments are controlled by one or more microprocessors. The computer program that runs the biuret protein method has control of the instrument while the protein samples are being run. The computer then loads the program that runs the next determination. These programs are generally stored on a magnetic disc. It is not uncommon for a batch analyzer to be capable of measuring 30 or more analytes, one analyte at a time. Instruments of this type generally have a reagent delivery system for each reagent the instrument will need. A bank of delivery tubes is shown in Figure 25–10.

The photometer on these instruments uses a filter wheel, like the one illustrated in Figure 25–11, with a stepping motor connected to it. The filter wheel allows for the measurement of absorbance at any wavelength for which there is a filter. Since the microprocessor knows the location of each filter, it can position the proper filter in the optical path. After the first analyte has been determined on the scheduled samples, the microprocessor positions the filter for the next analyte in the optical path.

The presence of a filter wheel makes it possible to measure the absorbance of a sample at more than one wavelength. For example, some clinical analyzers determine protein by the biuret method at 540 nm (the analytical wavelength). The instrument also measures the absorbance of each protein sample at 510 nm (a nonanalytical wavelength). The biuret product does not absorb radiant energy at 510 nm. The assumption is that any interference

that would contribute to a high absorbance at 540 nm would also have the same absorbance at 510 nm. The microprocessor subtracts the absorbance at 510 nm from the absorbance at 540 nm. In practice, this difference in absorbances is often a better value for the determination of protein than is the value measured at 540 nm alone. This type of method is called a bichromatic method. In these methods, it is hoped that the absorbance at the nonanalytical wavelength can be used to correct for any nonanalyte absorbance at the analytical wavelength. This is the equivalent of measuring a serum blank in a manual method.

Some instruments use other approaches for the pipetting, washing of sampling pipets, and mixing

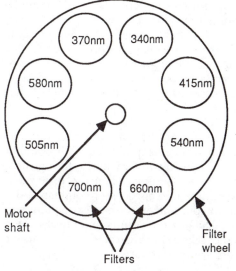

Figure 25–11. A filter wheel.

of samples. There are a number of ways to arrange an analyzer. Some instruments use test tubes instead of photometer cells for the actual reactions. The solutions are then pumped through a flow cell for the photometric measurement of the absorbance. The flow cells used are a little larger than those used in liquid chromatography. They are, however, much the same in appearance. The used test tubes are frequently thrown away.

It is not necessary to do the analysis of every analyte on every sample. If the sample in position 6 in the sample turntable does not need to be analyzed for glucose, the instrument skips tube 6 when the glucose method program is running.

RANDOM ACCESS ANALYZERS OF CIRCULAR CONFIGURATION

In contrast with the batch analyzer discussed in the previous section, there are also random access analyzers, which will do all the scheduled determinations on a patient's serum before moving on to the next patient's serum sample. The microprocessor may have a menu of up to 30 determinations that it can perform. Once the physician orders the desired tests, the instrument operator enters this information into the microprocessor by keyboard or a bar code that is placed on the sample tube. The ordered tests are then done. The unordered tests are not done during the analysis of that particular patient's serum.

Random access analyzers can be built in a turntable configuration such as the one in Figure 25–8. The analytical turntable still starts out with empty cells; the sample turntable, with samples. The sampling arm, under microprocessor control, pipets portions of a given patient's serum into the required number of different cells. The amount of sample placed in each cell depends on which analysis is to be done. Once the cells have received their sample of a given patient's serum, the microprocessor proceeds to ensure that the proper diluents and reagents are added to each tube. After the reactions are ready to be used, the microprocessor begins making spectroscopic measurements of the various cells.

A photometer with a multiple filter wheel is used to make the photometric measurements. (Some instruments use multiple photometers.) The filters are changed many times during the analyses of a patient's serum. When the ordered tests are finished, the microprocessor commands the instrument to continue on to the next patient's serum. It also calculates the results and checks the controls for the sake of analytical quality assurance. The results are then reported.

RANDOM ACCESS ANALYZERS OF PARALLEL CONFIGURATION

Although there is much to be said for the circular configuration used on many random access analyzers, it is also possible to build fine instruments in the so-called parallel configuration. In order to demonstrate the parallel approach, I will once again develop a hypothetical instrument. This time it will be a test tube analyzer designed to do a biuret protein analysis on up to 10 serum samples and will have a sample tube conveyer system and a reaction tube conveyer system. (Figure 25–12). The arrows indicate the direction of movement of the tube conveyer.

The analytical conveyer system would move a tube one position down the analytical path every minute. For example, tube 23 would move to the vacant position. Tubes 1 through 24 would shift one position toward the rear of the instrument. Tubes 8 to 2 would shift one position toward the right side of the instrument. This would leave the front corner on the left side vacant. Tubes 9 through 17 could then shift one position toward the front of the instrument. This would leave a vacant position on the left rear of the conveyer. Tubes 18 through 23 could then shift one position toward the left. This would recreate the vacant position as shown in Figure 25–12. Each area on the analytical conveyer system would have a function. The tubes on the right side would be involved in the actual analysis. The tubes across the rear of the instrument would be washed for reuse by a washing system. The tubes on the left side of the instrument would pass through a dryer. Tubes on the front would be cooled to 37°C for further use. This approach would make using disposable test tubes unnecessary.

The sample tube conveyer would advance one position every minute, at which time a sampler would pipet a measured quantity of serum, standard, or control into the tube on the right front corner of the analytical conveyer. The location of the sampler and a saline pipetter and a biuret reagent pipetter are illustrated in Figure 25–13. There is also a mixer and a sample aspirator for the photometric detector.

With this basic description of how our hypothetical instrument works, let's refer to the first section of the chapter and determine how we could automate the manual biuret method. The test tubes would have the following configuration: distilled water blank, sample tube 1; the 7 per cent protein standard, sample tube 2; the 5 per cent protein standard, sample tube 3; the 3 per cent protein standard, sample tube 4; the sample, tube 5; the high control, tube 6; and the low control, tube 7. The microprocessor would start the analysis by

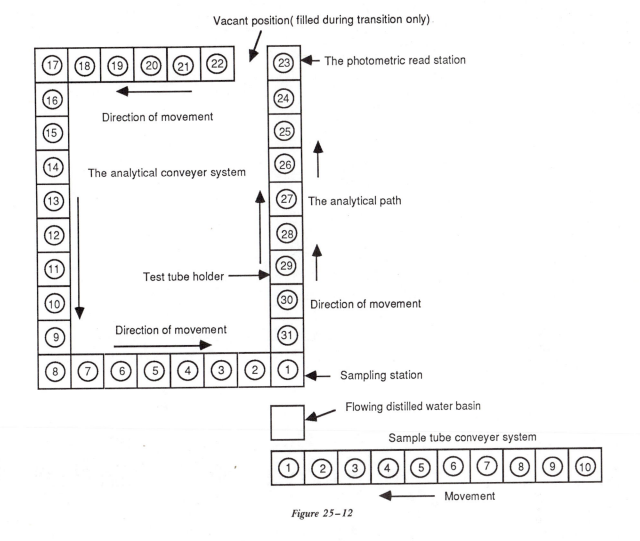

Figure 25–12

causing 0.1 ml of the blank to be pipetted into tube 1 on the analytical conveyer. The analytical conveyer would shift all tubes one position. The microprocessor would then cause 1.9 ml of distilled water to be pipetted into tube 1 and 0.1 ml of 7 per cent protein standard into tube 2, which would now be in the front right corner. The conveyer would then shift all the tubes on the conveyer one position. Tube 3 would now be in the sampling station and would receive 0.1 ml of 5 per cent protein standard. Tube 2 would receive 1.9 ml of distilled water, and tube 1 would receive 2 ml of biuret reagent. The microprocessor would then cause the tubes to move one more position on the analytical conveyer. Tube 4 would now be in the sampling station. It would receive 0.1 ml of 3 per cent protein solution, tube 3 would receive 1.9 ml of saline, tube 2 would receive 2 ml of biuret reagent, and tube 1 would be mixed by the mixer. The microprocessor would signal the tubes to move one more position

on the analytical conveyer. This process would continue until all seven of the tubes on the sample conveyer had been sampled and were progressing down the analytical conveyer. When tube 1 reached the photometric read station, the microprocessor would direct the sample aspirator to enter tube 1 and aspirate the reagent blank through the flow cell in the photometer. It would determine the absorbance of the contents of the tube using the bichromatic approach. The value of Po would be determined at 510 nm and the value of P, at 540 nm. Once the difference in absorbance was calculated and stored, the microprocessor would cause both conveyers to advance one position. By this time, tube 2, containing the 7 per cent standard, would have reached the photometric read station and would be aspirated through the photometer.

This process would continue until the low control from sample tube 7 had been aspirated through the photometer. Please note that more samples and

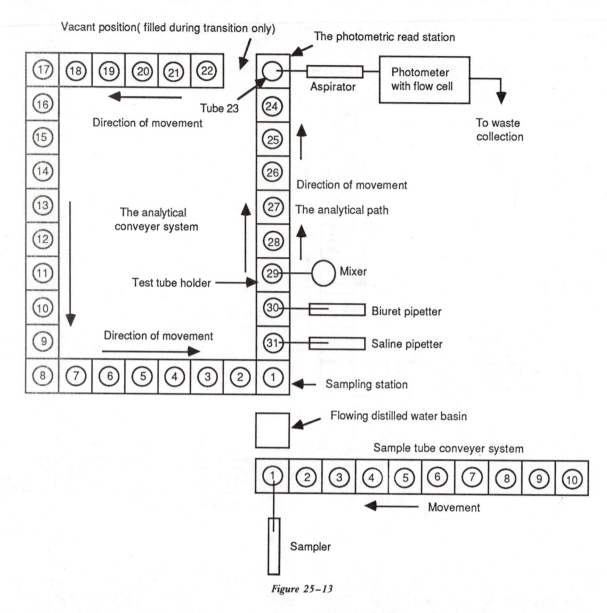

Figure 25–13

controls could be loaded onto the sample tube conveyer. The biuret analyzer in parallel configuration would continue analyzing samples as long as they were supplied. The instrument's computer would print out the concentrations of the samples and controls. I recommend that you review the components of this system because an understanding of it is needed in order to comprehend the material presented in the next paragraph.

With the previously described hypothetical biuret analyzer in mind, consider a second hypothetical analyzer. This instrument, illustrated in Figure 25–14, would analyze protein and uric acid. The chemistry of the uric acid method is presented in Chapter 3, p. 22. As this instrument begins to operate, the sample in tube 1 would be pipetted into the uric acid tube in rack 1. The protein tube in

rack 1 would not receive any sample. The microprocessor would advance rack 1 one position toward the rear of the instrument. Rack 2 would then move as a unit into the sampling station. The samplers would place serum from the first patient into the protein analysis tube, and serum from the second patient would move into the uric acid tube. The uric acid sample for the first patient would receive phosphotungstic acid reagent at this time. The microprocessor would move the first and second racks back one position. Rack 3 would move into the sampling station. The sample tube conveyer would move to the left one position. A serum sample from the second patient would be pipetted into the protein tube of rack 3. A serum sample from the third patient would be pipetted into the uric acid tube in rack 3. The uric acid sample in rack 1 would now

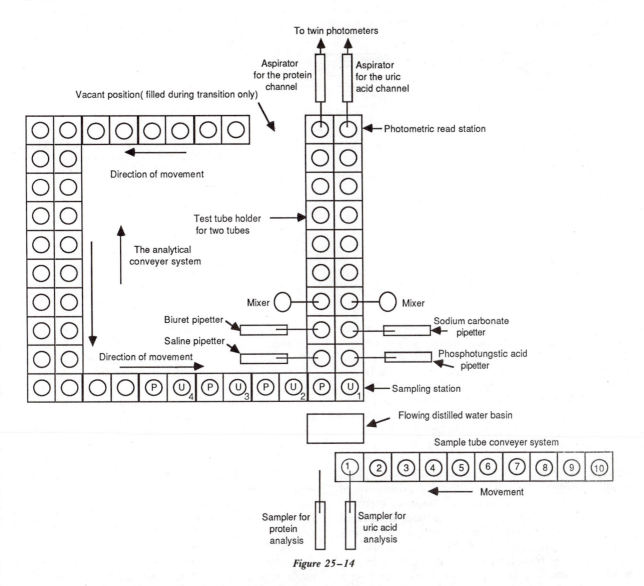

Figure 25-14

receive carbonate. The protein tube in rack 1 would be empty. The microprocessor would not add biuret reagent to this tube. This process would continue until all the samples had been analyzed. The microprocessor would keep track of which tube contained each patient's serum. If a uric acid sample on the sixth patient was not scheduled, a sample would not be pipetted into the uric acid tube for rack 6. Thus, instruments of this type can be built as random access analyzers.

The greatest extent to which this approach has been carried can be seen in the American Monitor PARALLEL, which functions on a 15-second cycle and has 30 analytical channels—one channel reserved for a serum blank, one reserved for a sodium/potassium flame photometer, and the remaining 28 channels for photometric methods. It can analyze 30 analytes on 240 samples per hour. The instrument performs all reagent addition from above the tubes instead of from beside them (Figure 25-14). There are 14 double photometers on the instrument. These are two-channel photometers that share a source. In some cases, rate methods are used for the chemical analysis. In other cases, they are end-point methods.

THE DU PONT ACA V

Several years ago Du Pont introduced a very unique clinical analyzer called the aca I (aca stands for automated clinical analyzer). The instrument was designed to mechanically perform in a plastic bag all the functions an analyst normally did manually. The most recent instrument in this venerable series is the aca V, which has the capability of about 70 bioanalytes. The instrument is discrete and requires a reagent pack for each test it performs.

Recalling the manual biuret method that was

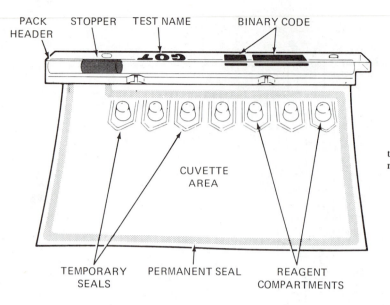
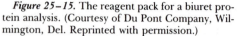

Figure 25–15. The reagent pack for a biuret protein analysis. (Courtesy of Du Pont Company, Wilmington, Del. Reprinted with permission.)

outlined at the beginning of the chapter, consider the biuret reagent pack and the aca V instrument. The instrument has the ability to automate this classic protein analysis. The reagent pack for a biuret protein analysis is illustrated in Figure 25–15.

The pack has a heavy permanent seal except in the sample introduction area. It also has seven reagent compartments with temporary seals. A plastic header is used to hang the pack on a conveyer. The test name and binary code are written on the header. For the determination of total protein (TP), reagent compartments 1 and 2 contain solutions of sodium hydroxide and sodium potassium tartrate. These compounds are two of the constituents of biuret reagent. The third and fourth reagent compartments contain a copper sulfate solution, the last constituent of biuret reagent. The pack is placed behind a sample cup (Fig. 25–16) that contains a macro and a micro sample cup. The micro cup is

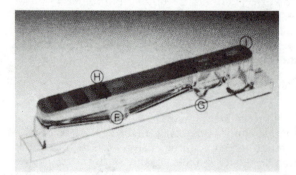

Figure 25–16. The sample cup. The sample cup contains separate compartments for either a macro sample (maximum fill 3 mL) (F), or a micro sample (maximum fill 0.5 mL) (G). Patient identification is provided by a pre-printed optically-scanned bar code (H) and corresponding numerals (I) on the sample cup lid. (Courtesy of Du Pont Company, Wilmington, Del. Reprinted with permission.)

used for pediatric analysis. The sample cup and pack are loaded into the sample feed tray as shown in Figure 25–17. The sample cup is moved to the left of the fill station, and its bar-coded lid is read. The needle on the right of the fill station (1) picks up the serum sample and delivers it into the pack. The amount of serum depends on the analysis. For a total protein analysis, the amount of sample added is 0.16 ml; this is followed by 4.84 ml of distilled water, which passes through the sampling needle. This effectively washes the sample out of the needle and into the bag. The cup is then moved to the sample cup exit tray (3), and the pack is shuttled to the transport belt (4), where it is moved to the seal breaker/mixer (5). The first four temporary seals are broken and the reagents are mixed with the sample. The reaction mixture is given time to react in position 6. Breaker/mixer II is not needed in the biuret method. The pack then moves to the photometer, where the plastic bag part of the pack is pressed between two machined metal plates. There is a hole in each of these plates that allows for the transmission of radiant energy and for the formation of an optical cell. There is also a provision for excess fluid to be channeled from the cell into an overflow chamber. The formed cell, the overflow chamber, and the channel are shown in Figure 25–18 and the photometer in Figure 25–19.

The photometer is a unique double-beam instrument. The reference path receives 10 per cent of the radiant energy. It uses this radiant energy to correct for instrument drift and, in some cases, to measure P_o. The filter wheel is connected to a stepping motor, which is under microprocessor control. When the photometer measures the amount of protein in a sample, it measures P at 540 nm and P_o at 510 nm. From this information the microprocessor can then calculate the absorbance. This

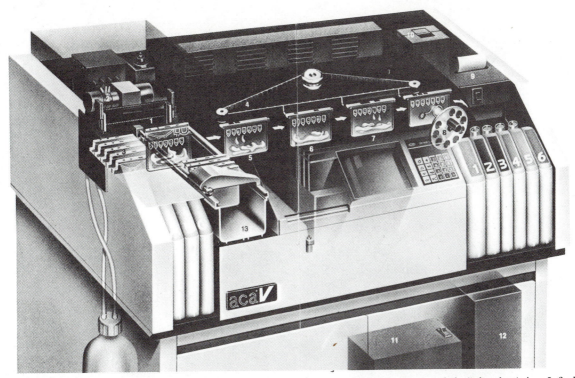

Figure 25–17. Key: 1, fill station; 2, ion-selective electrodes; 3, sample cup exit tray; 4, transport belt; 5, breaker/mixer I; 6, delay station; 7, breaker/mixer II; 8, photometer; 9, printer; 10, disk drive; 11, power supply; 12, used pack receptacle; 13, sample feed tray. (Courtesy of Du Pont Company, Wilmington, Del. Reprinted with permission.)

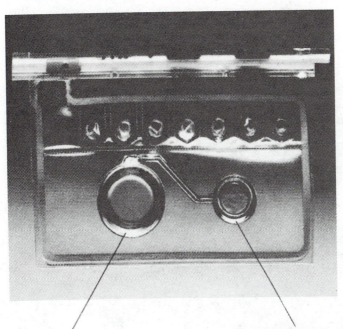

Photometric cell   Overflow chamber and channel

Figure 25–18. Courtesy of Du Pont Company, Wilmington, Del. Reprinted with permission.

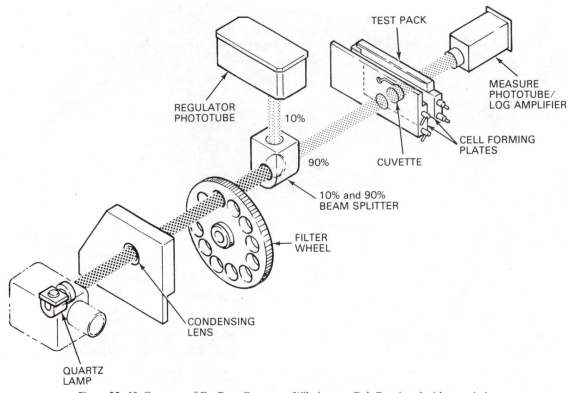

Figure 25–19. Courtesy of Du Pont Company, Wilmington, Del. Reprinted with permission.

technique is called a bichromatic method. The instrument's computer then calculates the patient's protein concentration on the basis of a previous calibration that uses three standards and a blank. The used packs are ejected into the used pack receptacle (12). The instrument measures the absorbance of some analytes at only one wavelength. The rate methods that are performed using the NAD^+–

NADH system are an example of this. The absorbance at 340 nm is measured during the 17.07-second read cycle. The change in absorbance is used to calculate the results.

Some of the reagent packs use pilled solid reagents such as those illustrated in Figure 25–20. The creatine kinase pack illustrated contains liquid reagent in compartments 1 and 5 and solid pills in

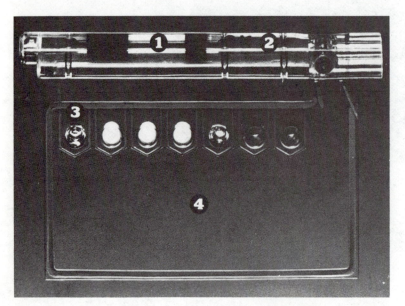

Figure 25–20. Courtesy of Du Pont Company, Wilmington, Del. Reprinted with permission.

positions 3, 4 and 5. The breaker/mixers can crush the pills and expel them into the pack. The availability of seven reagent compartments and two breaker mixer stations makes possible some fairly complex analytical methods.

In addition to approximately 68 photometric methods that the instrument utilizes, it can also measure sodium and potassium by ion-selective electrodes (ISE) (see Chapter 13). The sample is picked up by the sample needle near the left side of the instrument (see Fig. 25–17). The sample is passed through the flow through ISEs. The instrument uses the bubble-segmented continuous flow method (discussed in a later section of this chapter) for sample transport in this potentiometric analyzer.

The aca family of instruments has been very popular over the past decade.

AN AUTOMATED ANALYZER BASED ON REFLECTOMETRY—THE KODAK EKTACHEM INSTRUMENTS

In this section I will use the biuret protein method to show how the KODAK EKTACHEM Analyzers automate the manual clinical methods described in the beginning of the chapter.

The KODAK EKTACHEM 100, 400, and 700 Analyzers are random-access, highly automated clinical analyzers. The EKTACHEM 100 Analyzer,

now out of production, was designed to perform about 16 different analysis methods, all of which were reflectance photometry methods. The EK-TACHEM 400 Analyzer is designed to perform 20 different analysis methods and the 700 analyzer, 26 methods. Both the 400 and 700 models use ion-specific electrodes for the measurement of sodium, potassium, chloride, and carbon dioxide. These methods are discussed in Chapter 13, **Potentiometric Methods of Analysis**.

Both the 400 and 700 analyzers use reflectance photometry as the method for the determination of non-electrolytes. The reflectance methods are based on chemical reactions that occur in a thin layer of gelatin mounted on plastic slides. Hence, the devices are called "slides." (See Chapter 12, Reflectance Photometry; Reflectance Densitometry.)

In contrast to the 400 model, which has a single reflectometer, the 700 model has two reflectometers. The second reflectometer, a rate reflectometer, is used for the determination of serum enzymes by rate methods. It is configured to measure the reflection density of NADH at 340 nm. This is done 54 times during 5-minute incubation period. The computer uses the change in reflection density per 5.55 seconds in order to calculate the enzyme concentrations. The physical arrangement of the EKTACHEM 700 Analyzer is shown in Figure 25–21.

The EKTACHEM instruments are provided with a number of fail-safe devices: Optical sensors

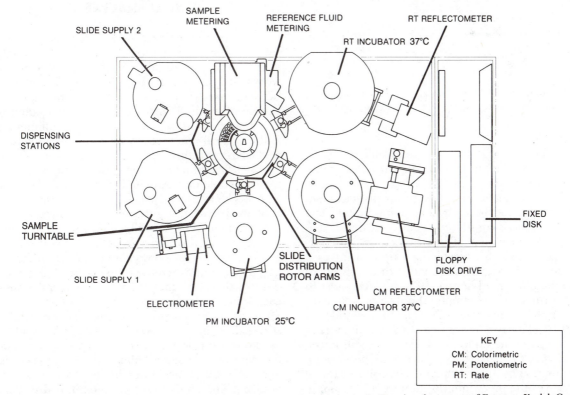

Figure 25–21. A schematic diagram of the EKTACHEM 700 Analyzer (top view). (Reprinted courtesy of Eastman Kodak Company. Copyright © by Eastman Kodak Company.)

ensure that slides are dispensing; additional optical sensors ensure that a slide has reached the sample metering station; pressure sensors detect clogged sample pipet tips; and 18 microprocessors are used in various ways to ensure that the instrument works properly.

Let's consider the analysis of total serum protein on the EKTACHEM Analyzer with the assumption that the instrument has already been calibrated using the three standard solutions purchased from Kodak. In addition, let's examine the progression of a high and low control through the instrument. The sample and two control sera are placed in sample cups that are placed on the sample turntable. The turntable is shown in Figure 25–22. The sample cups are covered with an easily

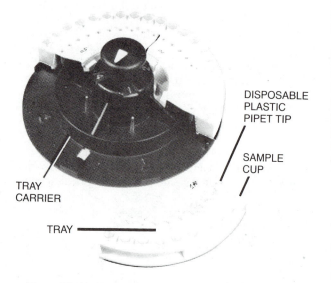

Figure 25–22. Tray and tray carrier. (Reprinted courtesy of Eastman Kodak Company. Copyright © by Eastman Kodak Company.)

penetrable top that reduces evaporation of the sample, and each sample cup has its own disposable plastic pipet tip (Fig. 25–22).

The metering station has an air-displacement precision pipetter driven by a stepping motor. The pipet proboscis is connected to a robotic device that moves it. In the course of an analysis, the proboscis picks up a disposable plastic pipet tip and drives it into the sample cup. The pipetter draws in 30 μl of sample plus an additional 10 μl for every test to be performed on the sample. If only a protein content were being determined, the pipetter would pick up 40 μl. The robotic device then withdraws the pipet tip from the liquid, and while it is still in the sample cup it expels 10 μl of sample. This is done to ensure that no air is in the tip. The robotic device next moves the pipet into position to apply sample to the slide. The slide supply module places a protein slide on the slide distribution rotor which

brings the slide to the sample metering station. The pipetter dispenses 10 μl of sample onto the slide and the pipetter ejects the plastic tip into the instrument's used tip receptacle. The instrument then ejects a second protein slide from the slide storage module, which is received by the distribution rotor. The rotor moves the slide treated with the sample to the incubator that serves the optical system. In doing this, the second (new) protein slide is moved to the sample metering station, and the sample turntable advances one test tube. The high control is sampled and applied to the new slide. This process continues until the low control has been sampled and its slide taken to the incubator.

The incubators are maintained at 37° ± 0.4°C so that enzymatic reactions can be performed. The slides are stored there for five minutes then transported to the reflectometer read station located at the back of the incubator.

Reflected radiant energy is measured and relayed to a microprocessor for data reduction. The biuret protein slide is read at 540 nm, just as it is when the manual method is used. The results of the protein analysis of the sample and controls are printed out. If the controls are within limits, the protein analysis of the patient is reported. This sequence of events is typical for the end-point methods on the EKTACHEM instruments.

IN SITU ANALYZERS

An in situ analyzer is an instrument that automates a manual method developed by an analyst with only one test tube. Such a manual method works well as long as the test tube is washed well between uses. Instruments of this kind can use a variety of methods to detect the analyte. Photometry, potentiometry, amperometry, voltammetry, and electrical conductivity have all been exploited in analyzers of this type.

Let's examine how the manual protein analysis method could be automated with an in situ analyzer. The Beckman ASTRA family of instruments does protein analysis using this method and will serve as our example (Fig. 25–23). The fill pump pumps biuret reagent through a preheater into the reaction cell (cup). The fill pump slightly overfills the cell, and the sip pump removes the excess. This process ensures that the same amount of reagent is added each time. A sampling pipet (not illustrated) pipets 13 μl of serum into the open top of the cell. The stirrer then mixes the solution. The cell has a built-in photometer, the light source is a tungsten lamp, the detector is a silicon photodiode, and there is a 545-nm interference filter built into the detector housing. The photometer monitors the absorbance from the moment the sample is added until 11 seconds have passed. The rate of the absorbance

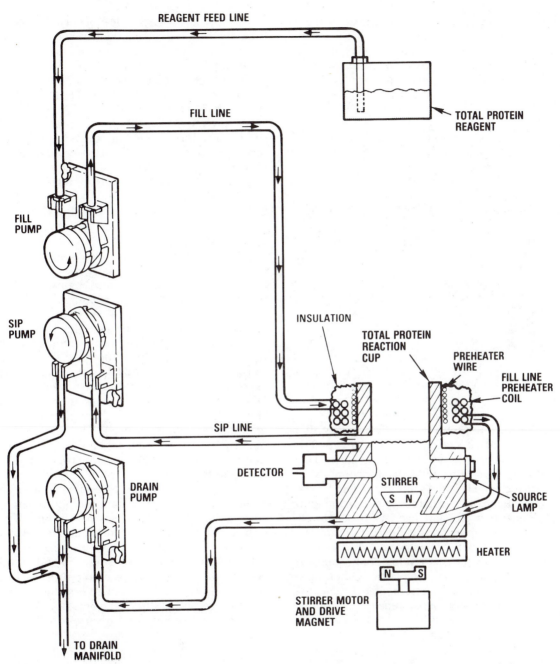

Figure 25–23. ASTRA-8 protein channel. (Copyright 1983 Beckman Instruments, Inc. Used by permission.)

increase ($\Delta A/\Delta$time) has been found to be proportional to the protein concentration. Once the absorbance measurements have been made, the biuret reagent is drained out of the cell, which is then cleaned with fresh reagent solution. The sampling pipetter is flushed with wash solution at a separate wash module to clean the pipet tip. The ASTRA protein analysis system can analyze 70 samples per hour.

The in situ method of automation is also common for blood cell counters.

CENTRIFUGAL ANALYZERS

Centrifugal analyzers are instruments that automate manual analysis methods by managing fluid transfer with centrifugal force. They are batch analyzers that use optical methods for the detection of the analyte concentration. They have been used for a large variety of analytical methods, including endpoint, rate, and EMIT methods. The Multistat III F/LS can also do fluorescent and nephelometric measurements.

The centrifugal analyzers are made up of three major modules: the centrifugal rotor, the photometer, and the microprocessor-controlled rotor loading system. Let's first consider the rotor; it is the heart of the centrifugal analyzer. Figure 25–24 il-

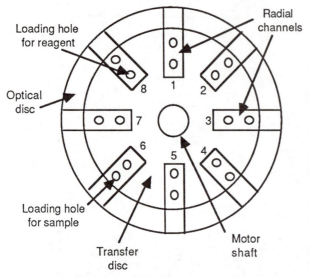

Figure 25–24. An eight-channel rotor. Top view.

lustrates the top view of a hypothetical eight-radial channel rotor. The rotor generally consists of a transfer disc and an optical disc. These two discs may or may not be physically separate parts. They do, however, have quite different functions. Each of the radial channels is the same shape and length. A side view of one of the channels is shown in Figure 25–25. The two wells and barriers make up the transfer disc. The optical disc is the part of the rotor that is beyond barrier 2. The optical disc must be made with optically transparent materials; some optical discs are made totally of plastic. Others have glass or quartz top and bottom plates. The two transparent plates are separated by a window spacer.

Before an analysis, well 1 of each radial channel is loaded with the chromogenic reagents needed for the analysis. For example, if a total protein analysis were done, well 1 would contain biuret reagent. Well 2 of each radial channel would be loaded with a serum sample, control serum, or standard. When the analysis is done, the loaded rotor is quickly accelerated to about 300 rpm. The centrifugal force sends the chromogen into the sample and then forces them both into the cuvette and holds the solution against the outside wall of the optical rotor (Figure 25–26). The solutions are mixed by air bubbles or by changing the rotor speed abruptly. If air bubbles are used, they are introduced through siphon tubes (Fig. 25–27) by a vacuum at the center of the rotor.

Once the reagents are mixed and the rotor is operating at a constant speed, the process of obtaining the photometric data begins. The photometric data are obtained by passing the optical disc through a photometer or a spectrophotometer. A typical photometer is illustrated in Figure 25–28. The photometric measurements are made while the rotor is spinning at about 300 rpm. The instrument has a built-in system that lets the microprocessor know what cell the photometer is viewing. Generally, this system is an optical encoder that monitors slots or holes in the rotor, although many different approaches have been used. The photometer frequently has a thermistor and a heater located near the rotor, which maintains the rotor at constant temperature.

If, for example, the rotor is a 30-chambered device and spins at 300 rpm, the photometer measures P for each cell 5 times per second. The microprocessor generally averages about 10 measurements on each cell. The blank cell provides the P_o value while the value of P for the other cells is measured. An oscilloscopic readout of the P values is shown in Figure 25–29. The first and last signals are voltage pulses directly proportional to P_o; the other pulses are P values for various solutions. The

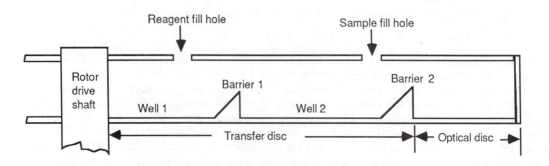

Side view of a rotor channel

Figure 25–25. Side view of a rotor channel.

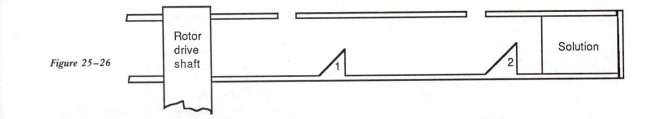

Figure 25-26

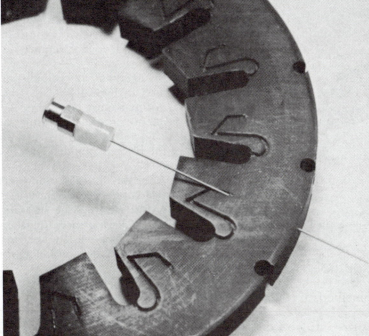

Figure 25-27. Cuvette window spacer with siphons to show path of fluid evacuation. Spacer for 15-cuvette ORNL centrifugal analyzer rotor. (Reprinted with permission from: C.P. Price and K. Spencer (eds.), Centrifugal Analysers in Clinical Chemistry, New York, Praeger Publishers, 1980.)

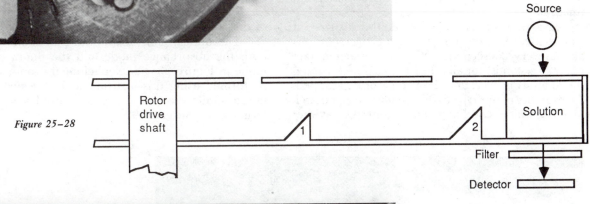

Figure 25-28

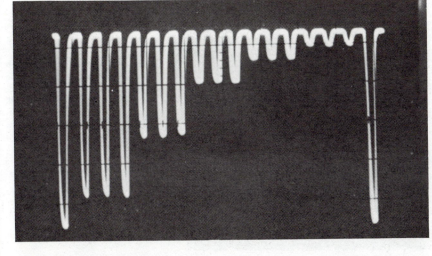

Figure 25-29. Transmission signals from a 17-cuvette miniature centrifugal analyzer. (Reprinted with permission from: C.P. Price and K. Spencer (eds.), Centrifugal Analysers in Clinical Chemistry, New York, Praeger Publishers, 1980.)

Figure 25–30. Courtesy of Allied Instrumentation Laboratory, Lexington, Mass. Reprinted with permission.

microprocessor can use these data to calculate the absorbance for each of the cells.

If end-point chemistry methods are used, the rotor is spun until all the samples have reached their terminal absorbances. This would be the case for the analysis of total protein by the biuret method, for example.

If a rate method were used, the instrument would measure the absorbance of the solutions about every 20 seconds. When the reaction rate ($\Delta A / 20$ seconds) for all cells reaches a steady state, the analysis is over, and the rotor can be stopped. An example of this type of determination is the AST method, which is outlined in Chapter 6, **Ultraviolet**

Absorption Spectroscopy. The microprocessor uses the steady state rate to calculate the enzyme concentration.

The other major component of a centrifugal fast analyzer is the automated rotor filling system. Such a system generally has a turntable for samples and a turntable for chromogenic reagents. The rotor fill system has pipetting machines that pipet samples and chromogens, as shown in Figure 25–30. The microprocessors used with these instruments are very basic and do not require any special consideration.

Some of these instruments use plastic disposable rotors as illustrated in Figure 25–31; other instruments have rotors that must be washed between uses. They frequently are washed while the instrument is running. To perform this process, the hollow drive shaft of the rotor is supplied with distilled water. The optical disc has siphons (Fig. 25–27) on each cell. After the optical readings are taken, the rotational speed of the rotor is increased. This causes the liquid in the cells to be siphoned out of the rotor into a catch ring that drains the waste from the instrument. The wash water then washes the transfer and optical discs and exits the rotor through the siphons.

The Multistat III F/LS Analyzer can also make fluorescent measurements. This is possible by having the window for excitation in the outside walls of the rotor, such as the rotor in Figure 25–31.

The optical path of the Multistat is shown in both the absorbance mode and the fluorescence mode in Figure 25–32. Nephelometry can also be performed using this instrument. In the nephelometric mode the xenon source is used, and scattering is measured at 90°.

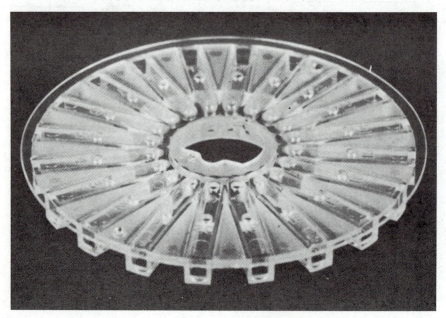

Figure 25–31. Instrumentation Laboratory 20-cuvette disposable U.V. transmitting plastic rotor. (Reprinted with permission from: C.P. Price and K. Spencer (eds.), Centrifugal Analysers in Clinical Chemistry, New York, Praeger Publishers, 1980.)

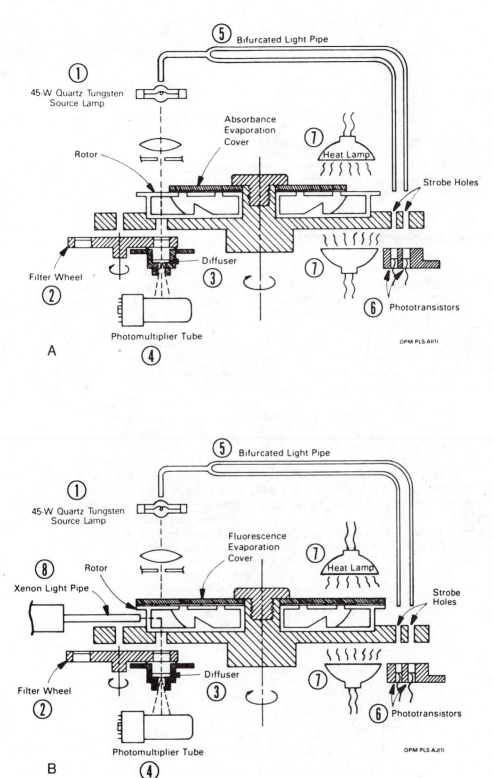

Figure 25–32. Optical path of the Multistat. *A,* absorbance mode. *B,* fluorescence mode. (Courtesy of Allied Instrumentation Laboratory, Lexington, Mass. Reprinted with permission).

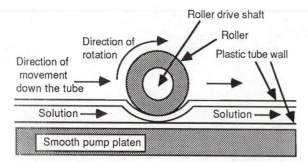

Figure 25–33. Redrawn from: Allied Instrumentation Laboratory, Lexington, Mass.

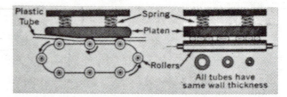

Figure 25–34. Courtesy of Technicon Instruments Corporation, Tarrytown, New York. Reprinted with permission.

SEGMENTED CONTINUOUS FLOW ANALYZERS

In the early 1950s, a visionary inventor by the name of L.T. Skeggs, Jr. built a system that could automate many clinical analytical procedures. The system consisted of a number of components, each designed to replace one or two steps of a manual method. The results of his work were commercialized by the Technicon Instruments Corporation, and the instrument became known as the Auto-Analyzer.

Since Skeggs's instrument automated the manual biuret protein method discussed at the beginning of the chapter, it will conveniently serve as our teaching example. Let's start by discussing how to develop a continuous flow system that would perform the biuret method for *one* sample. We will then upgrade the system to a form suitable for multisample analysis.

The biuret method involves adding a sample to biuret reagent. The relative amounts of the two solutions are selected so that the resulting solution has a reasonable absorbance. In the manual method this is done with pipets. In a continuous flow analyzer it is done with a peristaltic pump. A pump presses a flexible tube between a roller and a platen as shown in Figure 25–33. When the roller is moved down the platen by the roller drive mechanism, the solution in front of the roller is pushed on down the tubing. Solution behind the roller is pulled down the tube. The analyzer built by Skeggs used a multichannel peristaltic pump that had a number of rollers that were connected on each end to an endless bicycle chain. This is illustrated in Figure 25–34. The pump in this illustration has eight rollers, with at least two of them pumping fluid down the tube at all times. Skeggs's pump had provisions for a number of plastic tubes (Fig. 25–35). The multiple channel peristaltic pump could be fitted with tubings that had identical wall thicknesses but different internal diameters. This allowed the pump to combine different volumes of solutions. The tubing used to pump the sample had an 0.02-inch internal diameter. The biuret reagent was pumped by a pair of 0.1-inch internal diameter tubes. The pump, tubing, and connections are illustrated in Figure 25–36. Together, this array of parts is called a pump manifold.

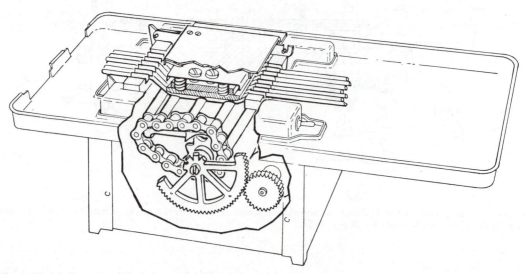

Figure 25–35. Courtesy of Technicon Instruments Corporation, Tarrytown, New York. Reprinted with permission.

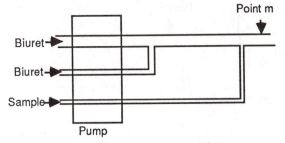

Figure 25–36. A pump manifold.

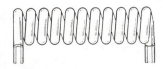

Figure 25–37. A mixing coil.

The ratio of sample to total solution passing through the tube at point M is almost identical to the ratio found in the manual method. The solution exiting the manifold would be routed to a mixing coil made of glass such as the one illustrated in Figure 25–37. The mixing coil provides the time and the conditions necessary for allowing the biuret reagent to mix well with the much more viscous and dense sample. After the solution passes through the mixing coil, it moves on to a photometer where its transmittance is measured.

This system allows one to start with biuret reagent and a sample and ultimately to obtain an absorbance that is proportional to the protein in the original sample. In the next few paragraphs I will illustrate how this system can be modified to do a protein analysis on about 60 samples per hour.

In order to truly automate the biuret protein method, one needs a sample turntable and sampling pipetter. The system originally used by Technicon is shown in Figure 25–38. In this instrument the turntable holds 40 sample tubes. The sampler has a built-in water basin that is supplied with a continuous flow of clean distilled water. In Figure 25–38, the distilled water supply line is marked with

a C, the basin overflow with a D, the sampling probe with an A, and the 0.02-inch sample tubing connecting with the pump, with an A-1. Some models use mechanical mixers to mix the serum before sampling. The mixer is the white device with two probes. The sampling probe is carried by a robotic device (marked with a B in the illustration). The sample probe is connected to the proportioning pump, and consequently, it is always drawing in fluids or air. For most work the sampler is mechanically programmed for the probe to sample the serum sample for 45 seconds after which it withdraws and moves to the wash basin. While in transit the probe picks up an air bubble. It then draws in distilled water for 15 seconds. During this time, the sampler advances one position. The system is then ready for the next sampling cycle. In this way, the biuret reagent is "injected" with samples and wash water. The wash water separates the samples from each other as they pass through the instrument.

The manifold in Figure 25–36 would also need to be modified. It has been found experimentally that the operation of a continuous flow analyzer can be greatly improved by segmenting the flowing stream with air bubbles. This is done by pumping air into the manifold through a small tubing (Fig. 25–39). The air bubbles tend to even out the flow rate as the liquid moves down the tube. Without the bubbles, the flow rate of the liquid at the center of a tube would be much faster than the flow rate

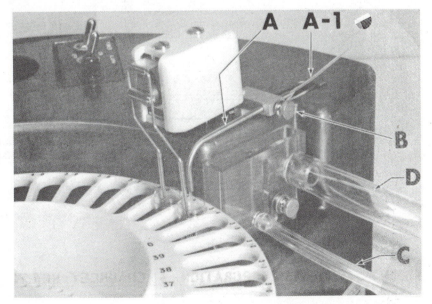

Figure 25–38. Courtesy of Technicon Instruments Corporation, Tarrytown, New York. Reprinted with permission.

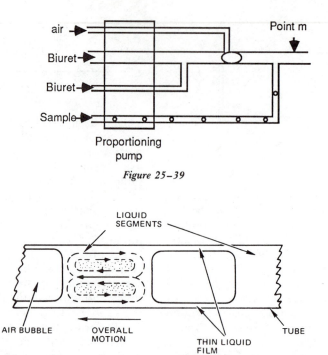

Figure 25-39

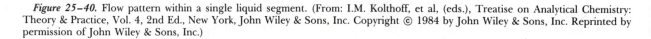

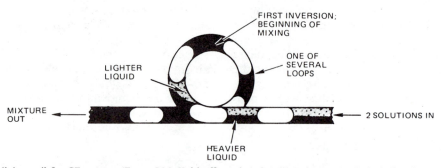

Figure 25-40. Flow pattern within a single liquid segment. (From: I.M. Kolthoff, et al, (eds.), Treatise on Analytical Chemistry: Theory & Practice, Vol. 4, 2nd Ed., New York, John Wiley & Sons, Inc. Copyright © 1984 by John Wiley & Sons, Inc. Reprinted by permission of John Wiley & Sons, Inc.)

Figure 25-41. Mixing coil for CF system. (From: I.M. Kolthoff, et al, (eds.), Treatise on Analytical Chemistry: Theory & Practice, Vol. 4, 2nd Ed., New York, John Wiley & Sons, Inc. Copyright © 1984 by John Wiley & Sons, Inc. Reprinted by permission of John Wiley & Sons, Inc.)

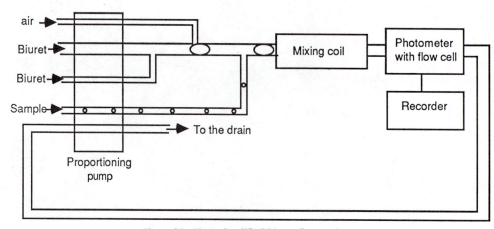

Figure 25-42. A simplified biuret flow path.

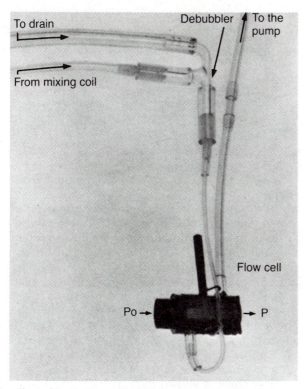

Figure 25–43. Courtesy of Technicon Instruments Corporation, Tarrytown, New York. Reprinted with permission.

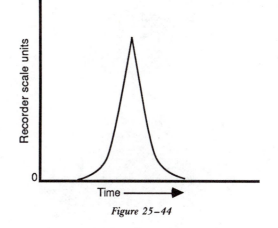

Figure 25–44

of the liquid next to the tube wall. The presence of the bubbles also enhances mixing (Fig. 25–40). The mixing coil also performs better in the presence of air bubbles (Fig. 25–41). The whole system except for the sampler is shown in Figure 25–42. The output line from the photometer is shown routed back through the pump. This arrangement makes the flow system pump smoothly.

The photometric system in an AutoAnalyzer is a double-beam photometer much like the one illustrated in Figure 4–44, p. 52. The photometer does not contain a reference cell, however. The function of the reference path is to allow the instrument to compensate for drift in Po. The liquid from the mixing coil is routed through a flow cell and debubbler (Fig. 25–43).

If a single-protein solution were passed through this system, the recorder would draw a single peak such as the peak illustrated in Figure 25–44. When many samples, standards, and controls are run, the output resembles the graph in

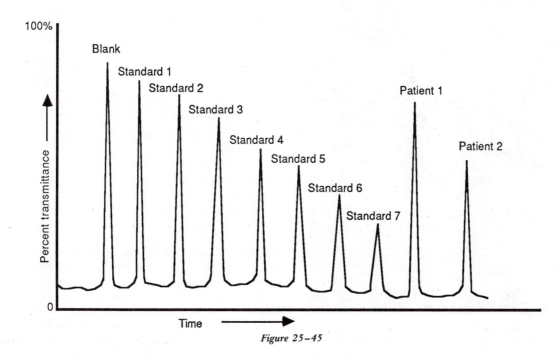

Figure 25–45

Figure 25–46. Courtesy of Technicon Instruments Corporation, Tarrytown, New York. Reprinted with permission.

Figure 25–45. The peak height for the blank is assumed to be proportional to Po. The height of the various other peaks is assumed to be proportional to P for that sample. These data allow the absorbance to be calculated for each peak. A standard curve can be prepared, and the protein concentration in the patient's sera can be calculated.

The AutoAnalyzer had the capacity to perform many other kinds of reactions. If a chromogen needed heat for its reaction, a high temperature bath was provided such as the one illustrated in Figure 25–46, which shows a bath for the glucose method.

If the analyte is a small molecule and if serum proteins interfere with the analysis, a dialyzer is used to separate the small molecules from the protein. The dialyzer is made from two plastic plates that have carefully matched spiral flow channels machined into their faces. The face of a dialyzer plate is shown in Figure 25–47. The two plates are separated by a cellophane semipermeable membrane. The sample is pumped with air-bubble segmentation through the flow channel in one of the plates. Air bubble–segmented chromogen solution is pumped through the other. An end view of one of the flow paths is shown in Figure 25–48. The di-

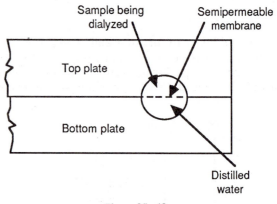

Figure 25–48

alyzer is very effective in separating large and small molecules.

AutoAnalyzers have been built that use a flame photometer for the determination of sodium and potassium. They have also been configured with filter fluorometers in place of the absorption photometer in Figure 25–42.

Once the first single-analyte instrument was produced and proved to be effective, it was only a matter of time before multianalyte instruments were produced. The first so-called two-channel Auto-Analyzer shared the same pump and sampler but had two separate manifolds, two separate reaction systems, and two photometric systems. The sample was split and entered into each of the two manifolds.

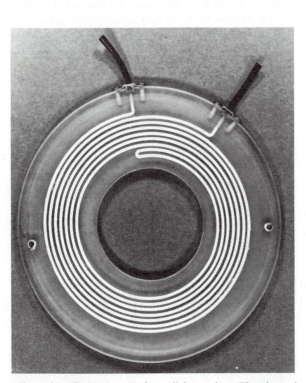

Figure 25–47. An AutoAnalyzer dialyzer plate. The channel has been filled with a white powder in order to provide contrast.

The AutoAnalyzer II, as it was called, was basically a combination of two of the original instruments.

As time passed, electronics improved and components were miniaturized. Improvements were also made in reagents and reactions. This led to the production of the SMA family of AutoAnalyzers (SMA stood for sequential multiple analyzer.) There was the SMA 12/30, the SMA 12/60, the SMA 6/60, and the SMA 18/60. The first number in the designation was the number of analytes determined. The second number indicated the quantity of samples that could be analyzed per hour. Thus, an SMA 18/60 could do 1080 separate chemical analyses per hour. These instruments used a single sampler, one proportioning pump per three analytes, one manifold per analyte, and a multiple photometer. Some of the SMA 12/60 instruments had either 10 or 12 flow cells radially located around a single source. There were either 11 or 13 detectors respectively; one was used to monitor Po, while the others were used to monitor the flow cells.

The latest air-segmented continuous flow analyzer is the Technicon SMAC. (SMAC stands for simultaneous multiple analyzer with computer.) The SMAC has been built from even more miniaturized parts than was the SMA generation of instruments. It uses a single sampler and can process 150 samples per hour. Because of the high processing rate of samples, the sampler is designed differently. It can hold 152 samples in racks of eight sample tubes; there is space for 19 racks. The sampler moves the racks as was discussed in the section *Random Access Analyzers of Parallel Configuration* (page 278). Air bubbles are still used to optimize the fluid dynamics. Proportioning pumps are used as they were in previous AutoAnalyzers. Dialyzers have been miniaturized so that paths are as short as 1 inch. Ion-selective electrodes perform the sodium and potassium determination. The photometric equipment on this generation of instruments is much different than that on previous Autoanalyzers. The visible photometers have a single source of radiant energy. The polychromatic radiant energy is transmitted by fiber optics to the various flow cells. The flow cells are about 1 mm in diameter and 1 cm in length. The radiant energy transmitted through the flow cells is taken to a channel selector (Fig. 25–49) by fiber optics. The U-shaped notch

in the channel selector signals the computer that the flow cells are about to be viewed by the photomultiplier tube. When the hole in the rotating channel selector aligns with the input fibers from flow cell 1, the input radiant energy passes through the hole and falls on the interference filter for analyte channel 1. The radiation transmitted by the filter is taken by an optical fiber to the photomultiplier tube. These data provide P for analyte channel 1. Immediately following this, the optical fibers originating from analyte channel 2 are viewed by the photomultiplier tube until all the visible wavelength channels have been interrogated for data. This all happens in less than a second. The data are stored, and the results are calculated by the computer. Although a bubble in the flow cell would produce a strange reading, there is no need to remove bubbles before a reading is taken. The computer is programmed to ignore the strange values produced by bubbles.

The instrument does all of its ultraviolet measurements at 340 nm. The same type of channel selector and fiber optical system make the ultraviolet measurements. The ultraviolet measurements are made with a second light source and detector.

Continuous flow systems have always had to correct for the carry-over of an analyte from one sample to the next. The percentage of the analyte from the first sample that carries over into the second sample can be experimentally calculated. The resulting correction factor can be applied to all of the measurements made on the channel. The SMAC instrument uses its computer to apply the correction factor automatically.

The latest instrument in this venerable family of instruments is the SMAC II. The instrument has improved computer capability, has fail-safe devices to shut down the instrument when a channel fails, and can do 24 analytes on 150 samples per hour. A block diagram of the instrument is shown in Figure 25–50.

SUMMARY

The question might be asked: Which of these automation approaches is best? The fact is that all of them are the best for some laboratories. The choice depends on lab volume, type of data profile to be generated, level of acceptable lab staffing, initial cost, cost of operation, and maximum time between sample collection and posting of the test results. Many laboratories have one instrument designed for high volume and another highly versatile low-volume instrument. The four hospitals with which I work have the following combinations: (1) a Technicon SMAC and a Du Pont aca; (2) two data-linked CFA 2000 centrifugal fast analyzers and an EKTACHEM 100; (3) an American Monitor Micro KDA circularly configured test tube analyzer and a

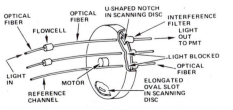

Figure 25–49. The channel selector. (Courtesy of Technicon Instruments Corporation, Tarrytown, New York. Reprinted with permission.)

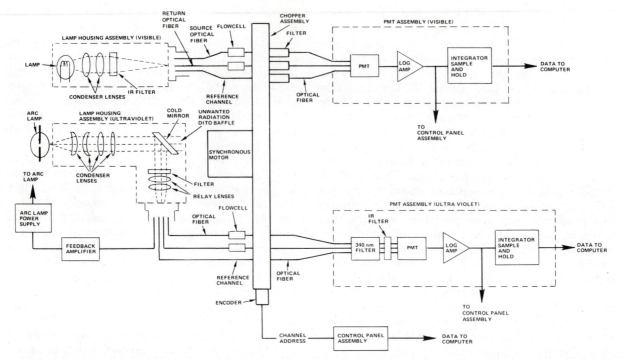

Figure 25–50. Block diagram of the optics and electronics of the SMAC. (Courtesy of Technicon Instruments Corporation, Tarrytown, New York. Reprinted with permission).

Du Pont aca; and (4) a Beckman ASTRA 8 and an Olympus DEMAND. This combination of a high-volume main instrument and a lower volume, highly versatile STAT analyzer is very common in medium-sized and large institutions.

I do not endorse any of these instruments for use in your laboratory. There are many fine instruments on the market, and only the leaders in your laboratory can make the decision as to which is best for your needs.

IMMUNOCHEMISTRY AND ITS APPLICATION TO QUANTITATIVE ANALYSIS

It has been known for a number of years that the bodies of animals may respond to the presence of disease-producing microorganisms by elaborating complex proteins called antibodies. These antibodies are one of the components of the serum proteins. Antibodies have the ability to bind tightly to the microorganisms that are of the same type as those responsible for their production. This binding phenomenon is the first step in the mechanism of the body's immune response.

The microorganism that is responsible for stimulating antibody production is called an antigen. Not only are microorganisms antigenically active, so are a number of complex molecules. For example, the injection of a foreign protein into an animal results in the production of antibody molecules. Even smaller molecules, such as drugs, if bound to a foreign protein and injected into an animal cause antibody formation. The antibody so produced binds to the drug molecules and holds on to them tightly. The product of an antigen-antibody reaction is an antigen-antibody complex. The reactions are frequently depicted in equation form.

Antibody + Antigen → Antigen-antibody complex

or

Ab + Ag → AgAb

Over the years scientists have developed some specialized nomenclature to describe antibody molecules. When a rabbit is injected with a human protein (antigen), the rabbit responds to the antigen (human protein) and produces antibody molecules. The antibody molecules that are produced are said to be directed against the human protein. The antibody molecules are also said to be rabbit-antihuman antibody. The rabbit serum proteins can be purified to produce a solution that contains mainly antibody molecules. Such a solution is called an antiserum.

Antibody molecules tend to be very specific in their ability to react with antigen. For example, a rabbit–antihuman protein antibody will not react with a goat protein. The specificity of antigen-antibody reactions has led to many analytical applications. This has been particularly true in recent years because chemists have been able to develop reactions that modify antigens and antibodies. This has allowed for the introduction of *tags*, which make it easy to follow the course of antigen-antibody reactions.

A number of these analytical procedures are discussed in the rest of this appendix.

NEPHELOMETRIC IMMUNOASSAY AND TURBIDIMETRIC IMMUNOASSAY

When the analyte to be determined is a large protein, it is possible to determine the analyte concentration by making use of an antibody that is specific for the analyte protein. In this approach the analyte protein is an antigen. The antigen-antibody complex that is produced by the reaction is large enough to scatter ultraviolet or visible radiation. In such a case, nephelometry or turbidimetry can be used to measure the amount of complex formed. For a review of nephelometry and turbidimetry see Chapter 10, **Radiation-Scattering Photometry: Turbidimetry and Nephelometry**. Familiarity with the material presented in Chapter 10 is necessary for understanding the following examples.

Consider, for example, the analysis of the human serum protein called immunoglobulin G (IgG). IgG is a large globular protein that is important in the immune system of the body. Its concentration has significance in diagnostic medicine. If serum is diluted with a buffer containing polyethylene glycol, and if the resulting solution is treated with an excess of antibody that is directed against IgG, the following reaction takes place:

IgG + Antibody directed → IgG-antibody complex
against human IgG

The polyethylene glycol enhances the formation of large particles of antigen-antibody complex. The particles that are formed scatter radiation and make it possible to measure the concentration of the IgG in the sample.

The actual determination can be done in two

ways. First, the amount of complex can be measured after the reaction has reached completion. This approach is called an end-point measurement. End-point procedures must be calibrated by determining the radiation scattering for several standards. The results are not strictly linear; a typical curve is illustrated in Figure A–1.

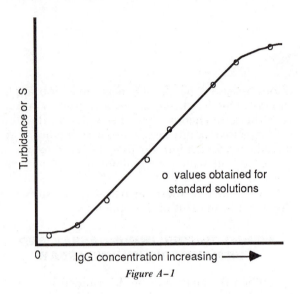

Figure A–1

The actual determination can also be done in a second way. When an antigen and antibody are placed together in solution, the molecules start to react and to produce antigen-antibody complex. The rate of this reaction quickly comes to a maximum value. The value of the maximum rate is directly proportional to the antigen concentration. Procedures that measure the maximum rate are called rate methods. Rate methods must also be calibrated with standards. The rate is measured by nephelometry or turbidimetry, and the results are plotted as shown in Figure A–2.

These two examples of an IgG determination illustrate the direct measurement of an antigen-antibody complex. The higher the concentration of IgG, the more radiation scattering detected.

By contrast, consider the measurement of gentamicin. Gentamicin is a drug that is measured by

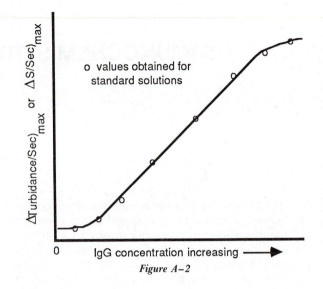

Figure A–2

an indirect procedure—a technique called inhibition immunoassay (IIA). In this technique, the patient's serum is mixed with a solution containing a protein that has several gentamicin molecules covalently bonded to its structure. We could depict such a protein as follows:

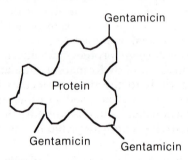

This material is placed in solution in limited amount. A divalent antibody that is directed against gentamicin is also added in a fixed, limited amount. When these three materials are incubated together, the following reactions take place. Consider Reaction A–1 (see below). Now consider the competing reaction, Reaction A–2 (see opposite page, top). If the patient's serum contains no gentamicin, the an-

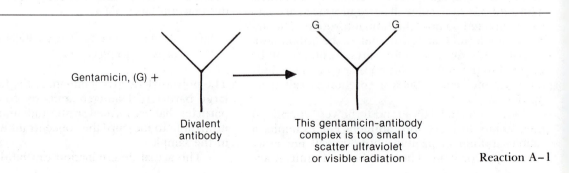

Reaction A–1

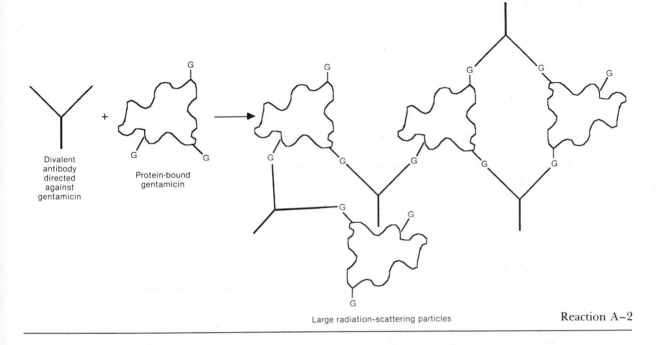

Large radiation-scattering particles

Reaction A–2

tibody, Y, reacts with protein-bound gentamicin and makes many large radiation-scattering particles, as shown in Reaction A–2. On the other hand, if the patient's serum contains an abundance of gentamicin, much of the antibody, Y, is used up by the gentamicin via Reaction A–1. When this is the case, there is not as much antibody available for Reaction A–2. This leads to the formation of fewer light-scattering particles. In this way the nephelometric or turbidimetric readout is lower for patients with high gentamicin concentrations.

Again, this type of measurement can be done by employing either an end-point method or a rate method. The method employed depends on how the instrument and its electronics were designed. In either case one would have to calibrate the instrument, and the curve would not be linear. One would expect the curve shown in Figure A–3.

The inhibitory immunoassay (IIA) approach is popular; it is also called a competitive binding immunoassay.

PARTICLE-ENHANCED TURBIDIMETRIC INHIBITION IMMUNOASSAY (PETINIA)

It is possible to use immunochemistry to determine the concentration of other drugs by using turbidimetry. Unlike the antigen-antibody complex that is produced in an IgG determination, most drugs, when reacting with antibodies, do not make large enough complexes to scatter light. One of the ways out of this dilemma is the PETINIA methodology. It is a competitive-binding immunoassay that uses an antigen (theophylline) that is covalently bonded to latex microspheres (Fig. A–4).

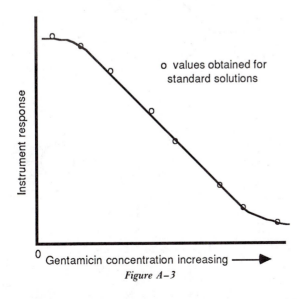

o values obtained for standard solutions

Instrument response

0 Gentamicin concentration increasing ⟶

Figure A–3

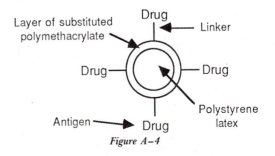

Layer of substituted polymethacrylate

Drug — Linker

Drug — Drug

Antigen ⟶ Drug

Polystyrene latex

Figure A–4

The purpose of the latex spheres is to increase the size of the antigen artificially. The increased size makes the particles of the antigen-antibody complex large enough to scatter ultraviolet radiation. Here is how the method works. The antigen-covered microspheres are incubated with divalent antibody that is directed against the drug, D. The reaction is shown at the bottom of the page. The reaction leads to the binding of many smaller particles. Such reactions are called agglutination reactions.

If no drug is present in the patient's serum, the added antibody agglutinates the coated microspheres into ultraviolet radiation–scattering particles. These particles are measured by turbidimetry. If, on the other hand, the patient's serum contains the drug, some of the antibody reacts with the patient's drug and less antibody is available for agglutination of microspheres. The reaction is

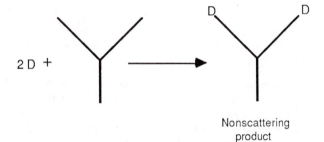

The drug-antibody complex is small and does not scatter radiation.

Thus, the higher the concentration of drug in the patient's serum, the lower the incidence of ra-

diation scattering. Methods of this type must be calibrated; one would expect a bit of nonlinearity in the calibration (Fig A–5).

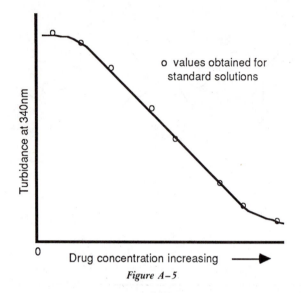

Figure A–5

This type of technique is used on the DuPont ACA for the theophylline procedure. The DuPont method makes use of a monoclonal antibody of great specificity and purity.

IMMUNOASSAY USING TAGGED MOLECULES

In contrast to the preceding methods of analysis, which rely on the size of antigen-antibody com-

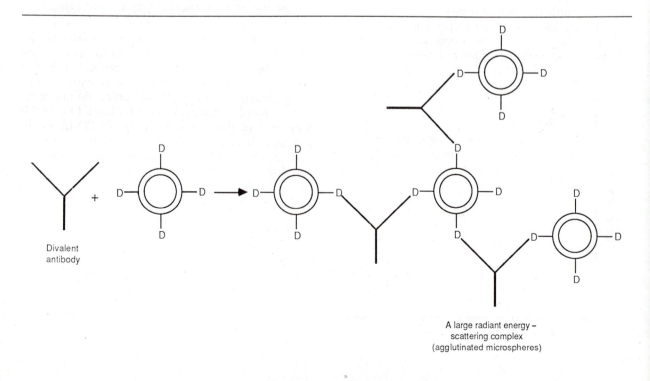

A large radiant energy –
scattering complex
(agglutinated microspheres)

plexes for their detection, several techniques rely on marked, or tagged, molecules. This tagging has been done by chemically bonding "tagging" moieties onto the antibody or antigen structures. The tags that have been introduced are radionuclides, enzymes, and fluorescent molecules. All of these tags will be discussed in detail in later sections of this appendix.

Radioimmunoassay (RIA)

Radioimmunoassay is one of the most important techniques used in analytical immunochemistry. These procedures frequently use what is called the competitive-binding principle of radioimmunoassay. In most current procedures, the "tag" that is used to follow the reaction is the radionuclide iodine-125, a reactor-produced gamma emitter. The principle involved in competitive-binding RIA can be best taught by example.

Human chorionic gonadotropin (hCG) is secreted by the placenta throughout normal pregnancy. The amount present in the serum is considerably increased during pregnancy, so a high level in the serum can be used to confirm the pregnant state. Human chorionic gonadotropin is a glycoprotein hormone with a molecular weight of about 40,000 and as such is a good antigen. It has been possible to raise a very specific rabbit anti–human chorionic gonadotropin antibody that makes possible this assay.

In the procedure* patient serum is allowed to incubate for 30 minutes at room temperature with a fixed, limited amount of antibody. The reaction proceeds to equilibrium during that time.

$$\text{Patient hCG} + \text{Rabbit-antihCG} \rightleftharpoons \text{patient hCG-rabbit}$$
$$\text{antibody} \qquad\qquad \text{antibody}$$

The hCG-antibody complex stays in solution. Next, human chorionic gonadotropin that has been tagged with ^{125}I, ^{125}IhCG, is added, and the tube is incubated for 90 minutes. During this incubation the ^{125}IhCG competes with the patient's hCG for the limited amount of rabbit antibody. The ensuing equilibrium reaction is the first reaction depicted below. After the 90-minute incubation, goat–anti-

rabbit protein antibody is added. This results in all the rabbit proteins in the tube precipitating out of solution. This includes the ^{125}IhCG–rabbit antibody complex as well as hCG–rabbit antibody complex. The precipitate is centrifuged to the bottom of the test tube and washed free of soluble materials. The radioactivity that is emitted by the ^{125}I in the precipitate is then counted with a scintillation counter. The higher the patient's hCG concentration, the smaller the amount of ^{125}IhCG that will be incorporated into the final precipitate. The less ^{125}IhCG, the lower the counts per minute.

If six standards are run through this same procedure, and the corrected count rate* is plotted against the concentration of hCG in each of the standards, the calibration curve shown in Figure A–6 will result. The standard curve can be used to find the concentration of hCG in the patient's serum. Note that the curve is nonlinear.

A second example might be helpful. The thyroid hormone 3,5,3′,5′-L-tetraiodothyronine (l-thyroxine, T_4) is synthesized by the epithelial cell monolayer of the thyroid gland follicles. The serum concentration of this hormone is measured as part of the assessment of thyroid gland function. The T_4 hormone is antigenically very active, so it is possible to raise antibody that is directed against the hormone.

In this procedure the patient serum is incubated in a test tube that is coated with T_4-specific immobilized antibody.† This antibody is actually chemically bonded to the walls of the test tube. After incubation the tube is washed to remove any unbound T_4 and serum. Next the tube is incubated with human T_4 that is tagged with ^{125}I. The reaction (bottom of page) begins and after a time comes to equilibrium. The tube is washed free of all unbound substances. The ^{125}IT$_4$ that is bound to the tube walls is then counted with a scintillation counter. If the patient has a high level of T_4 in the serum, less ^{125}IT$_4$ is taken up by the immobilized antibody. If, on the other hand, the patient has little T_4 in the serum, then the walls of the tube hold more ^{125}IT$_4$.

*Simplified for instructional use from the Becton and Dickinson booklet β-hCG[^{125}I] Radioimmunoassay Kit.
†From the booklet Gamma Coat [^{125}I] Free/Total T_4 Radioimmunoassay Kit. Clinical Assays—a division of Travenol Laboratories, Inc.

*Simplified for instructional use from the Becton and Dickinson booklet β-hCG[^{125}I] Radioimmunoassay Kit.

$$\text{Patient hCG-rabbit antibody complex} + {}^{125}\text{IhCG} \rightleftharpoons {}^{125}\text{IhCG-rabbit antibody} + \text{Patient hCG complex}$$

$$T_4\text{-antibody complex} + {}^{125}\text{IT}_4 \rightleftharpoons {}^{125}\text{IT}_4\text{-antibody} + T_4 \text{ complex}$$

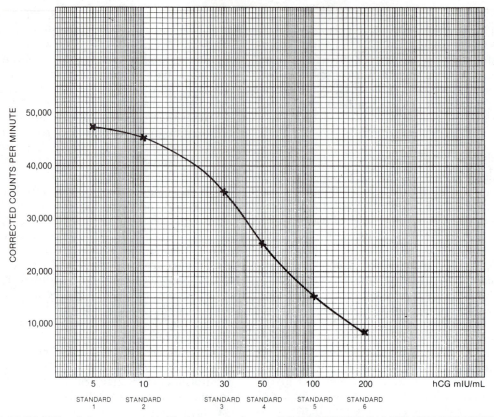

Figure A–6. β-hCG [^{125}I] radioimmunoassay kit. Typical standard curve. DO NOT USE TO CALCULATE UNKNOWNS. Corrected CPM versus concentration. (Courtesy of Becton Dickinson and Company, Orangeburg, New York. Reprinted with permission.)

If the same procedure is performed on a number of standards, and their count rates are plotted against the concentration of the standards, the standard curve in Figure A–7 results. The results for a patient X and a patient Y are also plotted on the curve.

These two examples of RIA are competitive-binding assays. They illustrate the two most common ways of separating the unbound radiotracer.

Enzyme Immunoassay (EIA)

A number of immunochemical analysis methods use an enzyme as a "tag" on an antibody. We shall examine two of these methods.

ENZME-MULTIPLIED IMMUNOASSAY TECHNIQUE (EMIT)

One approach is the enzyme-multiplied immunoassay technique of the Syva Company. Let's consider their quinidine method in some detail. In their procedure the patient's serum is mixed with a solution that contains a limited amount of antibody for quinidine. It also contains nicotinamide

adenine dinucleotide (NAD) and glucose-6-phosphate. The quinidine binds to some of the antibody; this is shown in Reaction A–3.

$$\text{Quinidine} + \begin{array}{c}\text{Antibody directed}\\\text{against}\\\text{quinidine}\end{array} \rightarrow \begin{array}{c}\text{Quinidine:antibody}\\\text{complex in}\\\text{solution}\end{array}$$

Reaction A–3

Next a solution of quinidine that is covalently bonded to the enzyme glucose-6-phosphate dehydrogenase (G6PDH) is added to the tube. The antibody that was left over from Reaction A–3 is now bound to the quinidine-enzyme complex; this is shown in Reaction A–4.

$$\text{Quinidine-G6PDH} + \text{Antibody} \rightarrow \begin{array}{c}\text{Quinidine}\\\text{G6PDH:}\\\text{antibody}\\\text{complex}\end{array}$$

Reaction A–4

This reaction goes until the supply of antibody is exhausted. Reaction A–4 leads to a product in which the G6PDH is deactivated. The remaining

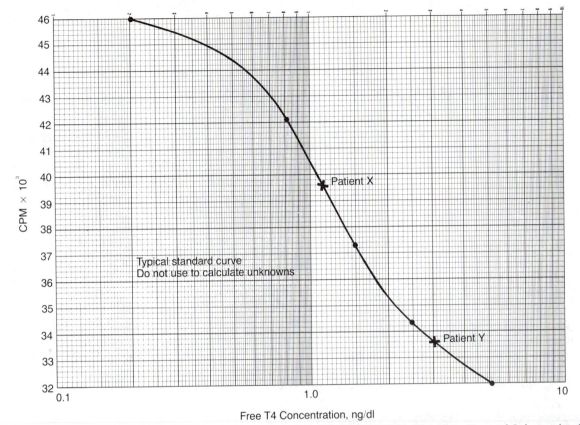

Figure A–7. Typical standard curve. DO NOT USE TO CALCULATE UNKNOWNS. (Data from Travenol Laboratories, Inc., Cambridge, Mass.)

quinidine-conjugated G6PDH remains active and can catalyze Reaction A–5. (see below).

The NADH$^+$ absorbs intensely at 340 nm, so the process can be followed by ultraviolet spectroscopy. The more quinidine there is in the patient's serum, the less quinidine-glucose-6-phosphatase will be bound and deactivated. The less quinidine-glucose-6-phosphate dehydrogenase is bound, the more there will be in solution. The more quinidine-glucose-6-phosphate dehydrogenase there is in solution, the faster the rate of Reaction A–5. Conversely, the less quinidine there is in the patient's serum, the more quinidine-glucose-6-phosphate dehydrogenase will be bound and deactivated. The more quinidine-glucose-6-phosphate dehydrogenase is bound, the less there will be in the solution. The less there is in the solution, the slower the rate of Reaction A–5. Thus, the *reaction rate* can be related to the original quinidine con-

centration in the patient's serum. The procedure that is frequently used measures the change in absorbance at 340 nm/30 sec and relates this value to the original quinidine concentration in the serum. The method must be standardized by running several sera of known quinidine concentration. If this is done, a plot like the one in Figure A–8 will result.

ENZYME-LINKED IMMUNOSORBENT ASSAY (ELISA)

Another enzyme-tag approach is the carcinoembryonic antigen method of Abbott Laboratories. This technique is an example of an enzyme-linked immunosorbent assay (ELISA). Carcinoembryonic antigen (CEA) is a tumor-associated antigen, and its serum concentration is monitored in the therapy of certain cancers. CEA is a glycoprotein

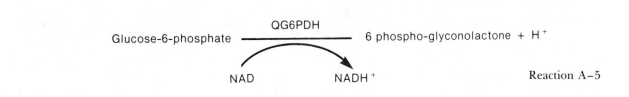

Reaction A–5

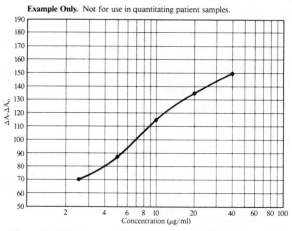

Example Only. Not for use in quantitating patient samples.

Figure A–8. Courtesy of Syva, Palo Alto, Calif. Reprinted with permission.

of about 200,000 molecular weight; hence good-quality antibody can be raised against it in a guinea pig. The procedure makes use of an immobilized antibody and is conducted as follows.

A plastic porous bead that is coated with guinea pig–antihuman carcinoembryonic antigen is placed in a well, and patient serum is added (Fig. A–9).

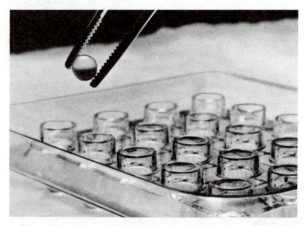

Figure A–9. From solid phase immunoassays for CEA clinical and laboratory experience. (Courtesy of Abbott Laboratories, Abbott Park, Ill. Reprinted with permission.)

The immobilized antibody and the serum are incubated for 2 hours. During that period the following reaction takes place:

Immobilized guinea pig antibody + Carcinoembryonic antigen (CEA) → Immobilized guinea pig antibody –CEA complex

The bead is washed free of serum and is treated with a solution of goat anti-CEA that has been

tagged with the enzyme horseradish peroxidase (HRP). During a second 2-hour incubation, the following reaction takes place:

Immobilized guinea pig antibody– CEA complex + Goat anti-CEA tagged with horseradish peroxidase → Immobilized guinea pig antibody CEA goat anti-CEA HRP– double complex

The bead with its immobilized double complex is then washed free of all unbound substances, and a solution containing hydrogen peroxide and o-phenylenediamine is added. During a 30-minute incubation, the horseradish peroxidase causes the hydrogen peroxide to react to form water and oxygen. The oxygen then reacts with phenylenediamine to produce a colored product, which is probably formed via the following reaction:

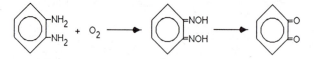

The colored product is measured at 492 nm. If suitable standards are analyzed, it is possible to develop a standard curve (Fig. A–10). The values for the sera of two patients, A and B, are included.

Fluorescent Polarization Immunoassay (FPIA)

In recent years methods for the analysis of antibodies have been developed around the use of fluorescent-tagged antigens. The procedures that have become common are competitive-binding methods that use polarized fluorescent measurements to measure the concentration of drugs. A good example of this approach is the analysis of phenytoin on the Abbott TD_X polarized fluorescent photometer.

Phenytoin is a drug that is used to control seizure disorders. Its concentration is generally measured in blood serum. In the TD_X method, the serum is treated with a surfactant in order to dissociate the phenytoin from serum proteins. The treated serum is then mixed with sheep antiphenytoin that is present in a fixed, limited amount. The mixture is then diluted with a solution of phenytoin that has the fluorescent tag fluorescein bonded to it. The following competitive reactions ensue:

Phenytoin + Sheep antibody directed against phenytoin → Sheep antibody– phenytoin complex

Phenytoin tagged with fluorescein + Sheep antibody directed against phenytoin → Sheep antibody tagged–phenytoin complex

The TD_X instrument then measures the concentration of sheep antibody fluorescein-tagged– phenytoin complex. The fluorescein-tagged com-

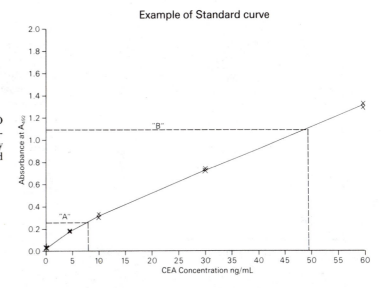

Example of Standard curve

Figure A–10. Example of standard curve—DO NOT USE IN PLACE OF STANDARD CURVE DETERMINED AT THE TIME OF ASSAY. (Courtesy of Abbott Laboratories, Abbott Park, Ill. Reprinted with permission.)

plex can be measured in the presence of the tagged phenytoin by the use of polarized excitation radiation on the fluorometer. The rapidly rotating fluorescein-tagged–phenytoin molecule loses the polarization of the excitation beam, and its fluorescent output can in effect be ignored. The large fluorescein-tagged complex rotates very slowly; thus the fluorescent polarization is not lost. The fluorescent output of the antigen-antibody complex passes through the polarization filter in the secondary filter system and is detected. (See Chapter 7 for a detailed discussion of the instrument.) As with all competitive-binding assays, the calibration curve for this method is nonlinear. The more phenytoin in the patient's serum, the less fluorescent intensity is seen.

SUBSTRATE-LABELED FLUORESCENT IMMUNOASSAY (SLFIA)

In direct contrast to the procedures that rely on an enzyme as a tag, the method discussed in this section relies on an enzyme substrate as a tag. These procedures are competitive-binding assays. Let's consider the determination of procainamide as our example.

Procainamide is an example of an antiarrhythmic drug. In the SLFIA analysis, procainamide is covalently bonded to the enzyme substrate umbelliferyl-β-D-galactoside. This tagged drug can be represented symbolically as follows:

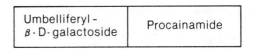

This is the tagged antigen in the method. When an antibody to procainamide is added, it binds the antigen (procainamide) in such a way that the substrate property of umbelliferyl-β-D-galactoside is lost. This is shown symbolically below.

When a patient's serum is mixed with tagged antigen, and the resulting solution is treated with

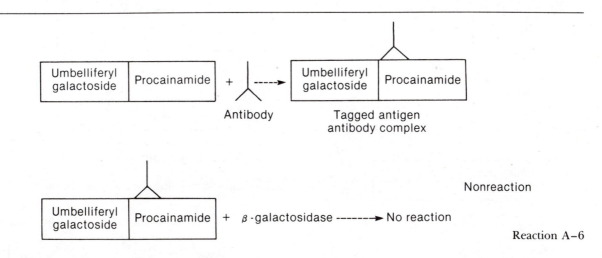

Reaction A–6

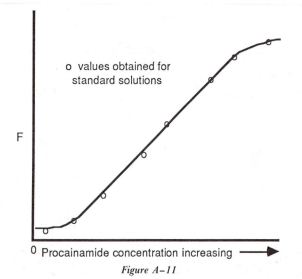

Figure A–11

Once the reactions have reached equilibrium, the enzyme β-D-galactosidase is added. This enzyme releases the tag from the uncomplexed, tagged antigen. Reaction A–9 results (see below.) The umbelliferone that is released fluoresces well when excited at 405 nm. The fluorescent emission is measured at 450 nm. Thus the status of Reaction A–8 can be established by the use of fluorescent spectroscopy. Because tagged antigen-antibody complex cannot react with the enzyme, the fluorescent intensity is directly proportional to the concentration of procainamide in the patient's serum.

As is the case with most competitive-binding methods, this one also has a nonlinear calibration curve (Fig. A–11). This curve would have to be established by running several standards.

a limited amount of antibody, the familiar competitive equilibrium is established (see Reactions A–7 and A–8 below). If the patient's serum contains a lot of procainamide, less of the tagged antigen is combined with antibody. If the patient's serum contains no procainamide, then most of the tagged antigen forms an antigen-antibody complex.

FLUORESCENT EXCITATION TRANSFER IMMUNOASSAY (FETIA)

In 1976, Ullman, Schwarzberg, and Rubenstein published the results of their research on fluorescent excitation transfer immunoassay (FETIA). The commercialization of their work as well as the work of Fischer, Choo Hsu, Daffern, and Cobb has led

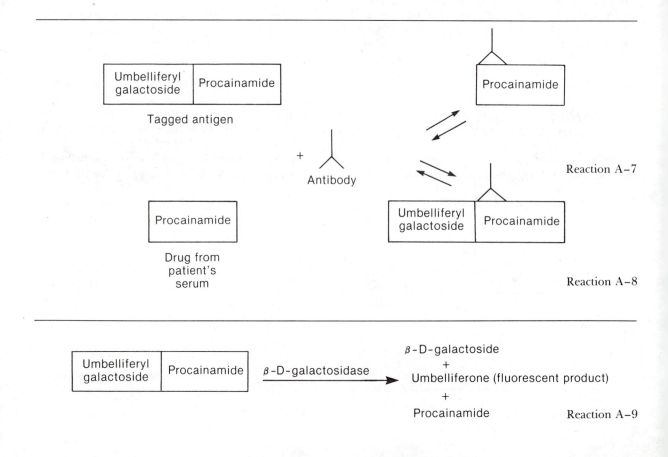

to the Syva Advanced System for the determination of immunoglobulins, thyroxine (T_4), triiodothyronine resin uptake (T_3U), and C-reactive protein assays. Let's consider the specifics of the IgA assay method.

The IgA determination is a good example of FETIA. The Syva FETIA methods are homogeneous competitive-binding assays. The patient's serum is mixed with a limited, fixed amount of human IgA that has been tagged with highly fluorescent fluorescein, which can be excited at about 470 nm and fluoresces at about 540 nm. The solution that results has the patient's IgA, the analyte, mixed with the fluorescein-tagged IgA. Next a solution of goat antibody that has been tagged with rhodamine is added. The rhodamine is called a quencher. The goat antibody has been raised against human immunoglobulin, IgA, and is present in a limited, fixed amount. Let's first consider the method as if it were an end-point procedure.

When the analyte IgA and the tagged IgA are placed in contact with goat antihuman IgA, a competitive reaction (bottom of page) ensues. The resulting solution is examined by fluorescent spectroscopy. The product does not fluoresce well. This

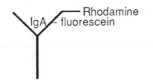

Quenched product

is true because the rhodamine absorbs the photon emitted by the fluorescein. Because of this absorption and the resulting loss of emission, the fluorescence of the solution is "quenched" to some degree. If there is no analyte IgA present, the fluorescence of the solution is greatly quenched. This is true because most of the IgA-fluorescein will form a complex with antibody and the rhodamine that is

bonded to the antibody will quench the fluorescence. On the other hand, if much analyte IgA is present, the solution fluoresces a great deal because the analyte IgA ties up a lot of the antibody. This leaves less antibody to react with the IgA-fluorescein in the mixture.

If the fluorescence of several standards is determined, and if these values are plotted against the IgA concentrations in the standards, a calibration curve will result. One would expect the curve to have the shape depicted in Figure A–12. This curve

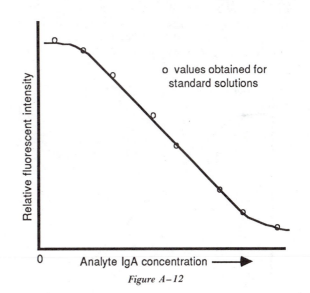

Figure A–12

could be used to find the concentration of IgA in the serum of a patient.

Now that we have discussed the method as if it were an end-point procedure, let's consider the actual method.

In the actual method, the reactions are not given sufficient time to come to equilibrium. Instead, the rate of the disappearance of fluorescent intensity is measured. The rate of disappearance of

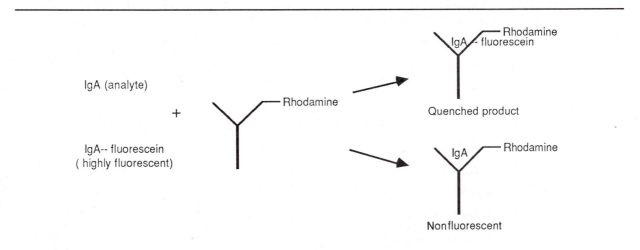

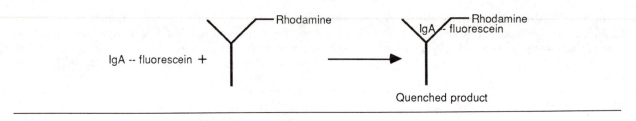

Quenched product

fluorescence is directly proportional to the rate of the reaction (see above). The rate of this reaction is directly proportional to the relative concentrations of IgA analyte and IgA-fluorescein. If no analyte IgA is present, the rate of the preceding reaction is at a maximum value. On the other hand, if a lot of analyte IgA is present, the rate of the preceding reaction is slower. This is due to the fact that the antibody is present in limited amount and that the two types of IgA must compete for antibody.

If the concentration of analyte IgA were plotted against the rate of disappearance of fluorescence for several standards, the calibration curve shown in Figure A–13 would result. The microprocessor on the Syva fluorometer uses this type of curve to transfer the rate of change in fluorescence data on patients to actual analyte concentrations.

The term fluorescent excitation transfer immunoassay is an unfortunate choice of words. This technique would be less ambiguously named fluorescent emission transfer immunoassay or, better yet, fluorescent quenching immunoassay. Hindsight is always better than foresight.

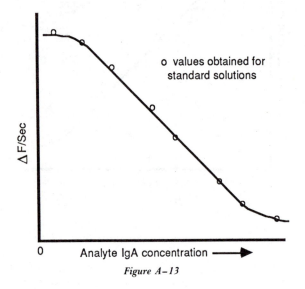

o values obtained for standard solutions

Figure A–13

REFERENCES

Lab Management, Oct; *21* (13), 1983.

Walter B., Greenquist A. C., Howard W. E. 3d. Solid-phase reagent strips for detection of therapeutic drugs in serum by substrate-labeled fluorescent immunoassay. Anal. Chem., May; *55* (6): 873–878, 1983.

Problems

1. Immunoglobulin (IgG) is determined by nephelometric immunoassay. A patient's serum, when reacted with an anti-IgG preparation, produces a relative scattering value of 37 units. Consider the data for seven standards and a blank.

IgG Concentration (g/dl)	Relative Scattering Intensity
Blank	0
0.4	4.0
0.8	14.3
1.2	31.0
1.6	49.0
2.0	67.0
2.4	82.0
2.8	90.0

What is the patient's IgG level?

2. Immunoglobulin (IgA) is determined by rate nephelometry. If a patient's serum produces a rate of 45.0 arbitrary rate units per second, what is his or her IgA concentration? Consider the data on the standards.

IgA Concentrate (g/100ml)	Reaction Rate in Arbitrary Units
0	0
0.1	4.7
0.2	21.0
0.3	38.5
0.4	57.0
0.5	74.0
0.6	90.0
0.7	98.0

3. Human chorionic gonadotropin (hCG) can be analyzed by radioimmunoassay. Using Figure A–7 as a standard curve, determine the hCG level in a patient's serum if the count rate is 33,500 cpm.

4. T_4 (l-thyroxine) can be analyzed by radioimmunoassay. If a competitive assay is done and the following data result, what will be the T_4 level in a patient whose serum produces a count rate of 37,000 cpm?

T_4 Concentration (mg/dl)	Count Rate (cpm)
0.1	48,000
0.5	44,750
1.0	41,700
2.0	36,500
5.0	32,500

5. Procainamide can be measured by EMIT methodology. The change in absorbance per 30 seconds is used in the methodology. The following standards are run.

Procainamide (µg/ml)	Concentration (mg/l)
0	Blank
1	Standard 1
2	Standard 2
4	Standard 3
8	Standard 4
16	Standard 5

The rate/30 sec for the blank is called ΔA_0. The rate/30 sec is determined for each of the five standards. The data are plotted in the next figure.

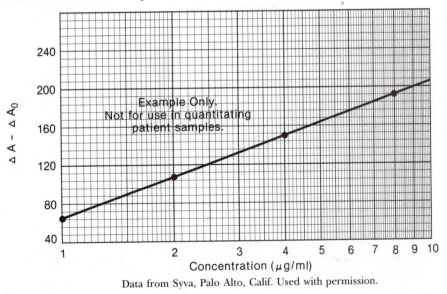

Data from Syva, Palo Alto, Calif. Used with permission.

If $\Delta A - \Delta A_0$ for a patient's serum is 130, what is the concentration of procainamide in the serum?

Appendix B

IONIC ACTIVITY, ACTIVITY COEFFICIENTS, AND IONIC STRENGTH—AN ABBREVIATED EXPLANATION

If a 0.1 molar hydrochloric acid solution is very carefully prepared, and if a pH electrode and a reference electrode are immersed in the solution, it can be shown experimentally that the pH of the solution is 1.081. This value is in direct contrast to the value 1 that is obtained by assuming that the following reaction is 100 per cent complete.

$$HCl + H_2O \xrightarrow{100\%} H_3O^+ + Cl^-$$

Equation B-1

Similarly, when a sodium ion–selective electrode is immersed in a 0.1 molar sodium chloride solution, a pNA value of 1.108 is obtained. This value is 0.108 higher than a novice might expect. A third example is the case of a calcium ion–selective electrode (ISE). A calcium ISE would yield a pCa of 1.4 for a 0.1 molar calcium nitrate solution. This is of course 0.4 higher than one might expect. Concern about the difference between the experimental numbers and the theoretical numbers prompted careful studies of ionic solutions and potentiometric electrodes. These studies produced a clarification of the problem. Let's consider the solutions first.

IONIC SOLUTIONS

When an ionic solute is placed in water, the solute dissolves and reacts with the solvent to yield several products. The relative proportion of these products is strongly concentration dependent. Let's use hydrogen chloride as an example; its behavior is very typical. When hydrogen chloride is placed in water, the principal reaction that it undergoes is

$$HCl_{(g)} + H_2O \rightarrow H_3O^+ + Cl^-$$ Equation B-2

This is the only reaction that HCl undergoes in solutions more dilute than 10^{-4} mol/liter. At higher concentrations a second reaction becomes increasingly important.

$$HCl_{(g)} + H_2O \rightarrow H_3O^+ \text{-} \text{-} \text{-} Cl^-$$ Equation B-3
$$\text{ion pair}$$

The product of this reaction is an ion pair. Ion pairs are not molecules in the ordinary sense of the word, but they are not ions either. They are aggregates of ions that are held together by electrostatic attraction. They exist in solution and are in dynamic equilibrium with the free ions from which they are made. This equilibrium is shown in Equation B–4.

$$H_3O^+ \text{-} \text{-} \text{-} Cl^- \rightleftarrows H_3O^+ + Cl^-$$ Equation B–4

Thus a hydrogen chloride solution consists of hydronium ions and chloride ions, and, depending on concentration, it may contain ion pairs.

ACTIVITY AND ACTIVITY COEFFICIENTS

Now that you have seen what happens to ions in solution, I will define the concept of activity. The activity of an ion is the molarity of the ion, excluding ion pairs. Using hydrochloric acid as our example, the activity of hydrogen ion (H_3O^+) is

$$(H_3O^+) = [H_3O^+]_A$$ Equation B–5

where $[H_3O^+]_A$ is the actual molarity of H_3O^+ ions in the solution,
and (H_3O^+) is the activity of hydrogen ion.

A second useful equation is an outgrowth of Equations B–2 and B–3.

$$[HCl]_I = (H_3O^+) + [H_3O^+ \text{-} \text{-} \text{-} Cl^-]$$ Equation B–6

where $[H_3O^+ \text{-} \text{-} \text{-} Cl^-]$ is the molarity of the ion pair,
and $[HCl]_I$ is the number of moles of HCl initially added divided by the solution volume in liters.

This equation simply states that the amount of solute added must equal the sum of the solute reaction products.

As analytical scientists, we are generally interested in knowing the molar concentration, $[HCl]_1$, of the solute. Fortunately, the relationship between activity and $[HCl]_1$ is known. Using HCl as our example, we obtain

$$(H_3O^+) = \gamma_{H_3O^+} [HCl]_1 \quad \text{Equation B-7}$$

where $\gamma_{H_3O^+}$ is the activity coefficient of hydrogen ion.

The activity coefficient of an ion is the fraction of the solute originally placed in solution that forms simple hydrated ions. This can be stated in an equation.

$$\gamma_{H_3O^+} = \frac{(H_3O^+)}{(H_3O^+) + [H_3O^+ - - - Cl^-]}$$

$$\text{Equation B-8}$$

Research chemists have been able to estimate activity coefficient values for many ions. I have plotted the activity coefficients for hydrogen ion and calcium ion in Figure B-1 (see Kielland in references).

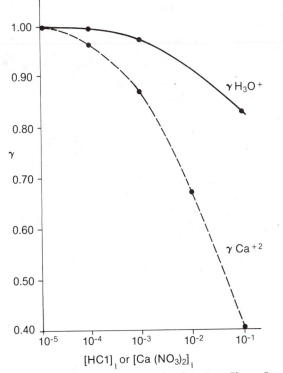

The values provided in the figure were estimated at 25°C. The values are different at different temperatures.

Notice that calcium ion is more prone to ion-pair formation than is hydrogen ion. This is the only explanation of the fact that γ Ca^{+2} values are lower than $\gamma_{H_3O^+}$ values at all but the lowest concentrations. All polyvalent ions are more prone to ion-pair formation than are univalent ions.

POTENTIOMETRIC ELECTRODES AND ACTIVITY

Researchers have found that potentiometric electrodes are poised to ion activities and not to ion pairs. Let's use a pH electrode and hydrochloric acid as an example. A pH electrode is poised to H_3O^+ ions. The electrode is not poised to $H_3O^+ - - - Cl^-$. Consider a solution with $[HCl]_1$ equal to 0.1 mol/l. Figure B-1 can be used to find the activity coefficient of 0.1 molar HCl; it is 0.83. Using Equation B-7, we can calculate the hydrogen ion activity.

$$(H_3O)^+) = \gamma_{H_3O^+} [HCl]_1$$
$$(H_3O^+) = (0.83)(0.1) = 0.083 \text{ molar}$$

If this value is used to find the pH, we find

$$pH = -\log (H_3O^+)$$
$$pH = -\log (0.083)$$
$$pH = 1.081$$

The value obtained is the same as the experimental value discussed at the beginning of this appendix. Thus we see that the discrepancy discussed previously is due to the formation of ion pairs and the fact that potentiometric electrodes do not detect them.

In much of clinical analysis, physicians want data on electrolytes in units of mg/dl or in milliequivalents per liter (mEq/l). These values are calculated by the instrument microprocessors from a knowledge of activity coefficients. For example, sodium ion–selective electrode provides a measurement of (Na^+), the sodium activity. If we want to express the results in milliequivalents of total sodium per liter we can start with Equation B-7 expressed in terms of sodium ion.

$$(Na^+) = \gamma_{Na^+} [Na^+]_{Total}$$

$$[Na^+]_{Total} = \frac{(Na^+)}{\gamma_{Na^+}}$$

$$Na^+ \text{ in mEq/l} = \frac{(Na^+) \text{ mol}}{\gamma_{Na^+} \text{ l}} (1000 \text{ mEq/mol})$$

The value of γ_{Na^+} is determined experimentally by determining $[Na^+]_{Total}$ on a serum sample by flame photometry. This serum standard is analyzed by the sodium ion–specific electrode, and the instrument is adjusted to make the $[Na^+]_{Total}$ value equal to the value found by flame photometry. This experimental approach makes it possible to avoid a theoretical calculation of γ_{Na^+}.

Here is a second example. Physicians prefer calcium results expressed in mg/dl. Calcium ion–selective electrodes provide calcium activities, (Ca^{+2}). Once again, the activity coefficient can be used.

$$(Ca^{+2}) = \gamma_{Ca^{+2}} [Ca^{+2}]_{Total}$$

$$[Ca^{+2}]_{Total} = \frac{(Ca^{+2})\ mol}{\gamma_{Ca^{+2}}\ l}$$

$$[Ca^{+2}]_{Total} = \frac{(Ca^{+2})\ mol}{\gamma_{Ca^{+2}}\ l}\ 40.1\ \frac{g}{mol}\ 10^3\ mg/g\ \frac{l}{10(dl)}$$

$$[Ca^{+2}]_T\ in\ mg/dl = (Ca^{+2}) \frac{4.01 \times 10^3}{\gamma_{Ca^{+2}}}\ mg/dl$$

In actual practice, the constant $4.01 \times 10^3/\gamma_{Ca^{+2}}$ is determined experimentally and fed into the microprocessor on the instrument. This is done by determining $[Ca^{+2}]_T$ on a serum standard by atomic absorption spectroscopy. The microprocessor on the potentiometric instrument is adjusted to make the instrument read out the known $[Ca^{+2}]_T$ value for the serum standard. Once standardized, the potentiometric instrument can be used to find total calcium in mg/dl.

CONCLUSION

This appendix is not a complete work on ionic activity. It has been prepared with an eye to the student's immediate need to understand activity as it affects potentiometry. Students interested in learning more about this topic should consult a good quantitative analysis book or a physical chemistry text. The quantitative analysis text entitled *Fundamentals of Analytical Chemistry*, 4th edition, by D. A. Skoog and D. M. West, is my favorite.

A second point may be of interest. The ionic equilibria functions taught in freshman courses are really equations that should be expressed in terms of activity. Thus the acid ionization constant (K_a) for acetic acid is

$$K_a = \frac{(H_3O^+)\ (CH_3COO^-)}{[CH_3COOH]}$$

The solubility product (K_{sp}) of AgCl is

$$K_{sp} = (Ag^+)\ (Cl^-)$$

REFERENCES

Kielland J.J. Am. Chem. Soc., *59*:1675, 1937.
Skoog D.A. and West D.M.: Fundamentals of Analytical Chemistry, 4th Ed. Philadelphia, Saunders College Publishing, 1982.

In Chapter 1 the concepts of potential, current, and resistance were developed. Ohm's law, Kirchhoff's voltage law, and series resistive circuits were also introduced. This appendix supplements Chapter 1 by providing additional background information concerning the electronics in instrumentation.

PARALLEL RESISTIVE CIRCUITS

An example of a parallel resistive circuit is illustrated in Figure C–1. The current flowing from

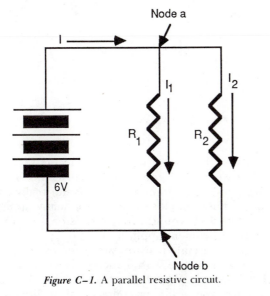

Figure C–1. A parallel resistive circuit.

the battery to node a must divide itself at the node. Part of the current travels through R_1 and part through R_2. The current divides itself between the two in a way that is in keeping with the ease of the paths. The least resistive path gets more of the current. With this qualitative explanation in mind, let's consider the quantitative approach.

The value for the currents I, I_1, and I_2 are best found by the application of Ohm's law. In Figure C–1 both resistors have 6 volts applied across them. From Ohm's law we can write

$$6 \text{ V} = I_1 R_1 = I_2 R_2$$

If we substitute the resistance values and solve for I_1, we find

$$I_1 = \frac{6 \text{ V}}{10^2 \ \Omega} = 6 \times 10^{-2} \text{ amp}$$

Solving for I_2, we find

$$I_2 = \frac{6 \text{ V}}{10 \ \Omega} = 6 \times 10^{-1} \text{ amp}$$

It is also clear that I is equal to the sum of $I_1 + I_2$. Thus I would be

$$I = I_1 + I_2 = 6.6 \times 10^{-1} \text{ amp}$$

The fact that the current going into a node is equal to the sum of the currents leaving the node is Kirchhoff's current law. This law will be helpful in later circuit analysis.

If the two resistors in Figure C–1 were replaced by a single resistor that would produce the same I value, the resistance of that resistor would be found by the following equation:

$$\frac{1}{R} = \frac{1}{R_1} + \frac{1}{R_2}$$

$$\frac{1}{R} = \frac{1}{10^2 \ \Omega} + \frac{1}{10 \ \Omega} = 10^{-2} \ \Omega^{-1} + 10^{-1} \ \Omega^{-1}$$

$$\frac{1}{R} = 1.1 \times 10^{-1} \ \Omega^{-1}$$

$$R = 9.1 \ \Omega$$

Thus the current, I, in Figures C–1 and C–2 would be equivalent. The resistance of any parallel network of resistors is equal to

$$\frac{1}{R_{network}} = \frac{1}{R_1} + \frac{1}{R_2} + \frac{1}{R_3} + \ldots$$

The circuit in Figure C–2 is said to be an equivalent circuit to the one in Figure C–1. Parallel resistive networks have a useful application in instruments; they are current dividers.

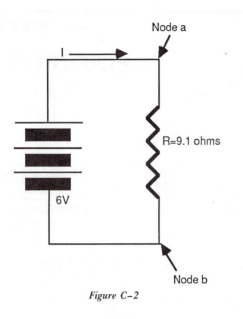

Figure C–2

VARIABLE RESISTORS

While we are discussing electrical resistors, it is also important to realize that variable resistors are manufactured and are widely used in instrumentation. The materials from which they are manufactured vary according to the desired range of resistance values, but all variable resistors function by sliding a contact down the length of the resistive material. The symbols used for such devices are shown in Figure C–3. Although the first symbol is

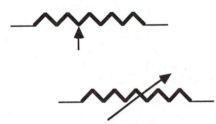

Figure C–3. Symbols used for variable resistors.

more descriptive, the second is more commonly used. Consider Figure C–4.

Figure C–4 depicts a variable resistor that has been prepared from 1000 cm of Nichrome wire of

sufficient diameter to provide 100 Ω of total resistance. The device also has a sliding contact, b. The device is a variable resistor because the length, L, from the equation

$$R = \rho \frac{L}{A}$$

can be varied by moving the sliding contact. In this way various resistance values between 0 and 100 Ω can be produced across terminals a and b.

The variable resistor in Figure C–4, although easy to understand, is impractical because of its size. Variable resistors of a more convenient size are easily produced. For example, if the Nichrome wire from the previous example were coiled and were supported in a circular configuration, the result would be a more compact variable resistor. Figure

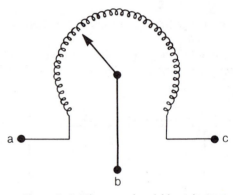

Figure C–5. Wire-wound variable resistor.

C–5 shows the resistor with the connections labeled as they were in Figure C–4. Because the sliding contact goes from loop to loop it is not really continuous, but it is close enough for many applications. The wire-wound variable resistor is common in some instrument applications.

Other variable resistors are made by replacing the Nichrome wire with a carbon layer on an insulated backing. This type of resistor can be built in an arc, and a sliding contact can be provided. Variable resistors such as these serve as the volume control on radios and are also used in noncritical instrument applications.

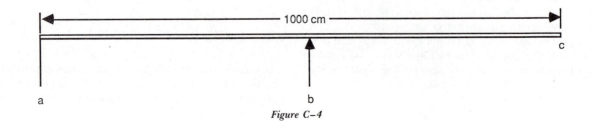

Figure C–4

VARIABLE VOLTAGE SOURCE

Using the variable resistor from Figure C–4 and a 10-volt battery, it is possible to see how a continuously variable voltage source can be made. Consider Figure C–6. The range of potentials that

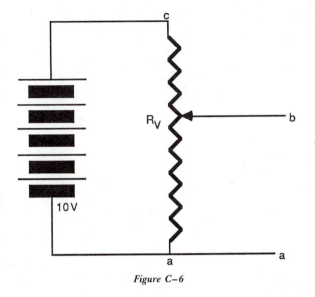

Figure C–6

the circuit could provide would cover 0 to 10 volts. As this type of circuit is applied a great deal in instrumentation, let us analyze it in some detail. First, because this is a series circuit, we know that the current through the circuit is the same at all points. The value of the current may be found from Ohm's law.

$$V = IR$$
$$I = V/R = 10 \text{ V}/100 \text{ }\Omega = 0.1 \text{ amp}$$

As the Nichrome wire is uniform in diameter, the equation

$$R = \rho \frac{L}{A}$$

from Chapter 1 becomes

$$R = kL \text{ where } k = \frac{\rho}{A}$$

If 1000 cm of the Nichrome wire produces 100 Ω, then k equals 0.1 Ω/cm. Thus, the resistance value for our variable resistor is

$$R = (0.1)L \text{ }\Omega$$

When the sliding contact is positioned at 250 cm from node a, the resistance subtended by ter-

minals a and b is 25 Ω. As the current through the variable resistor is 0.1 amp, the potential difference that is available between terminals a and b is 2.5 V.

$$V = IR$$
$$V = (0.1 \text{ amp}) 25 \text{ }\Omega = 2.5 \text{ V}$$

Thus by sliding the contact until it subtends an appropriate amount of resistance, any desired potential between 0 and 10 volts may be made available between terminals a and b.

It would be prudent at this time to do problems 1 through 4 at the end of this appendix (pp. 337–339). The topics ahead require mastery of the material presented up to this point, and the problems will help you develop the understanding that is required to continue.

WHEATSTONE BRIDGE

A resistive circuit that finds a great deal of use in instrumentation is the Wheatstone bridge circuit, which is used to measure unknown resistances. It finds application in potentiometric recorders, in some infrared radiation detection systems, and in some gas chromatography detection systems.

A Wheatstone bridge circuit (Fig. C–7) generally has two identical fixed resistors, R_1 and R_2, and a variable resistor, R_3. The variable resistor must be at least as high in resistance as the resistance of unknown magnitude, R_u, that is to be measured. The variable resistor must have a scale associated with its sliding contact so that the resistance value at any setting is known. A galvanometer of high sensitivity is used in the bridge as a current detector in order to sense the flow of current, I_5, and its direction. The galvanometer frequently has an internal resistance of about 10 Ω. In order to gain an appreciation of how the bridge measures resistance, let us consider the behavior of the currents in the circuit at three different settings of R_3.

Assume that in Figure C–8 the sliding contact of R_3 is subtending a resistance that is equal to R_u. When this condition exists, the resistance of R_1 + R_u is equal to $R_2 + R_3$. Under this condition, no path is less resistive than another. Thus I_1 must equal I_2; I_2 must equal I_3; and I_4 must equal I_1. If the currents are equal and if R_1 and R_2 are equal, there can be no potential difference between nodes a and b. If there is no potential difference between a and b, I_5 must equal zero. In this condition, the bridge is said to be balanced, and the galvanometer shows a zero response. When the galvanometer shows no current flow, R_3 and R_u must also be equal. The resistance that is read from the scale on R_3 is the resistance of R_u as well.

Let us assume that in Figure C–9 the sliding contact of R_3 is subtending a resistance that is

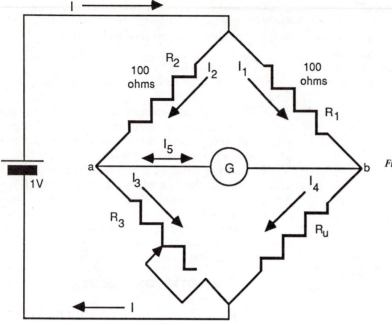

Figure C–7. A Wheatstone bridge circuit.

greater than R_u. When this condition exists, the current I_2, in finding the path of least resistance, splits at node a, and part of it flows through the galvanometer, causing a deflection on the meter that is proportional to I_5. This takes place because R_3 is more resistive than R_u. The greater the difference, $R_3 - R_u$, the greater the value of I_5. The name applied to I_5 is the unbalance current.

Next, consider the case in which R_3 is adjusted to a value lower than R_u (Fig. C–10). The current,

in finding the least resistive path, flows through the galvanometer, causing a deflection that is proportional to I_5, but the current travels in the opposite direction relative to the previous example. This can take place only when R_3 has a smaller resistance than R_u. The greater the difference, $R_u - R_3$, the greater the value of I_5.

In the actual use of a Wheatstone bridge, the technician adjusts the variable resistor R_3 while observing the galvanometer. If the current (I_5) flows

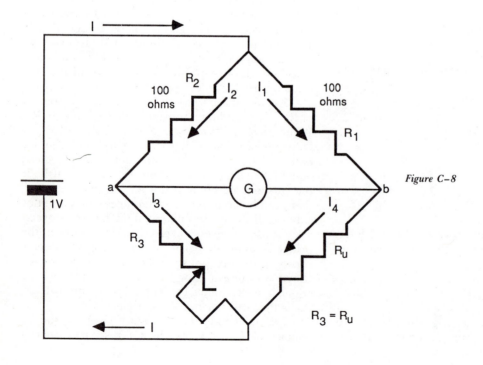

Figure C–8

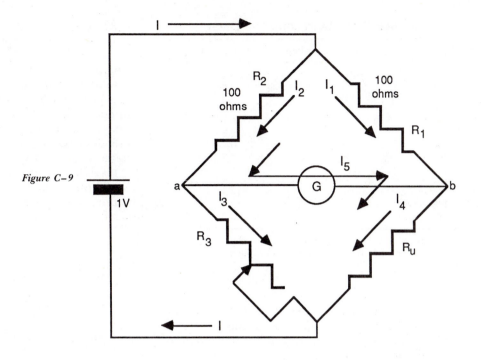

Figure C–9

from node a to node b, the variable resistor (R_3) is adjusted to a lower value. This continues until the galvanometer indicates 0 current flow. On the other hand, if I_5 flows from node b to node a, the variable resistor (R_3) is adjusted to a higher value. This is done until the galvanometer indicates 0 current flow. The value of R_u is then read from the scale on R_3.

One other significant feature of a Wheatstone bridge is that I_5 is directly proportional to the dif-ference, $R_u - R_3$, even when the bridge is out of balance. This feature accounts for the widespread use of the bridge in the detection circuits of instruments. Table C–1 shows the current values corresponding to the various degrees of unbalance for the circuit in Figure C–11.

The relationship between $R_u - R_3$ and I_5 is linear, as is the relationship between $R_u - R_3$ and the potential difference between a and b. Thus if R_u is proportional to some property that is being

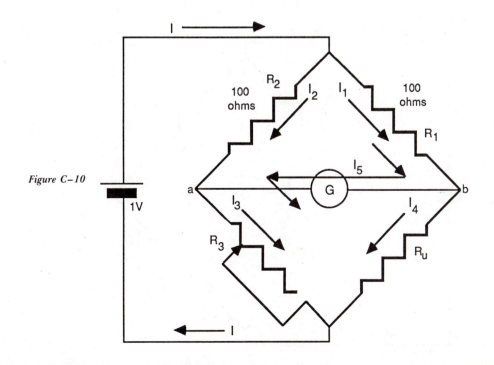

Figure C–10

TABLE C–1. Representative Data

$R_u(\Omega)$	$I_5(\mu A)$	Potential Difference ab (mV)
90	−245	− 26.3
92	−198	− 20.8
94	−143	− 15.5
96	−94	− 10.2
98	−46	− 5.1
99	−23	− 2.5
100	0	0
101	23	2.5
102	45	5.0
104	88	9.8
106	131	14.5
108	172	19.2
110	212	23.8

Data from: A. J. Diefenderfer, Principles of Electronic Instrumentation, 2nd Ed., Philadelphia, Saunders College Publishing, 1979.

measured, and the difference $R_u - R_3$ is proportional to the unbalance signal, then the measured property that affects R_u is directly proportional to the unbalance signal (Fig. C–12).

ALTERNATING VOLTAGE

In Chapter 1 and in the preceding pages of this appendix, circuits that are powered by batteries have been presented. Batteries are constant-voltage power sources; thus the focus of our discussion has been on circuits with constant steady-state currents only. They are called direct currents (DC), and batteries are, of course, DC power sources.

By contrast, there are some important applications involving alternating currents (AC) and voltages. Before discussing the circuits, let us consider the nature of alternating voltage.

In many countries of the world, electrical power that is produced for distribution is alternating in a continuous sine wave. In the United States the frequency of alternation is 60 cycles per second. Figure C–13 depicts one cycle of this power. The peak voltage (V_p) is 110 V, and the voltage peak-to-peak (V_{pp}) is 220 V. This cycle of voltage variation takes place 60 times per second. In many instrument applications, alternating voltages and currents of other frequencies are also important.

If an AC voltage source is connected to a resistor as in Figure C–14, the amount of power actually dissipated is not the same as the value obtained by the following equation:

$$Power = I_{peak}^2 R$$

The actual power dissipation is predicted by the equation

$$Power = I_{RMS}^2 R$$

where

$$I_{RMS} = \sqrt{\frac{I_{peak}^2}{2}}$$

The term I_{RMS} stands for root-mean-square current. The I_{RMS} value was invented to make the power

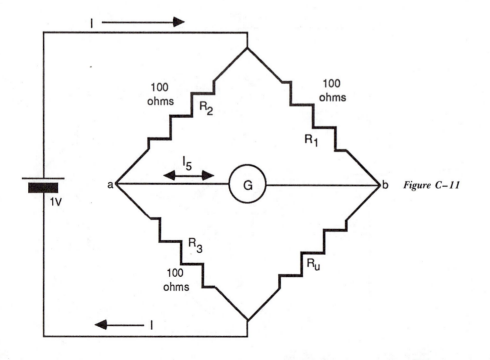

Figure C–11

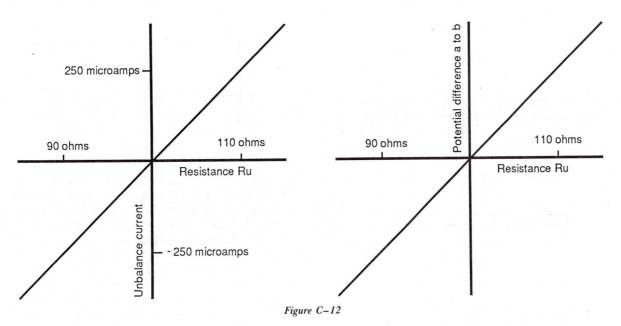

Figure C–12

equation work. It is RMS voltages that are measured with common AC voltmeters. The relationship between V_{RMS} and V_p is expressed by the following equation:

$$V_{RMS} = \sqrt{\frac{V_p^2}{2}}$$

Ohm's law for AC resistive circuits is $V_{RMS} = I_{RMS} R$. The alternation of the voltage and current is in phase in circuits that have only resistive devices. This means that peak positive voltage and peak positive current occur simultaneously. Otherwise, the information presented about resistive circuits applies to both AC and DC circuits. With these facts in mind, let us learn about some new electronic components. After completing just a few more sections on electronic components, you will be ready to consider a really important instrumentation circuit, the DC power supply.

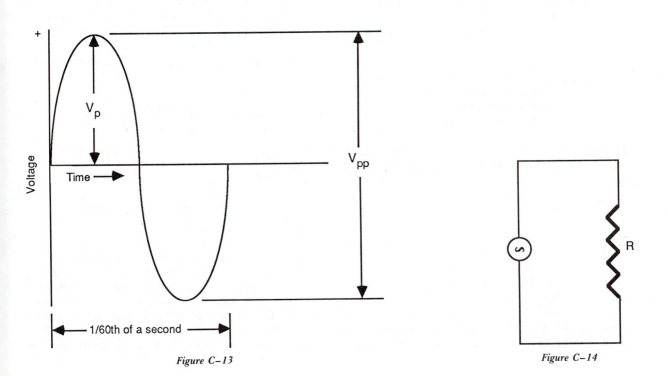

Figure C–13

Figure C–14

CAPACITANCE

A capacitor consists of two metal sheets that are separated by an insulator. Capacitors are manufactured from many materials; one example is the common oiled-paper type, in which two sheets of aluminum foil are separated by a sheet of oil-saturated paper. The two sheets of foil do not touch each other. Consider Figure C–15.

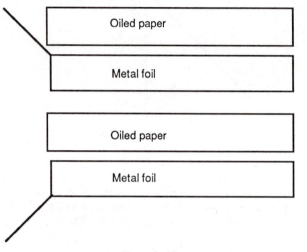

Figure C–15

If the sandwich depicted in the figure is laid flat on a table and is then rolled up into a tight cylinder, a capacitor that looks very much like a commercial capacitor would result. Its capacitance can be calculated by the following equation:

$$C = \frac{\epsilon A}{d}$$

C is the capacitance in farads (F)
A is the area of one of the aluminum sheets
d is the thickness of the oiled paper
ε is the dielectric constant of the oiled paper

The dielectric constant is a constant for a given material. It is a measure of the polarizability of the dielectric material. The symbol for a capacitor in an electric circuit is shown in Figure C–16.

Figure C–16. Symbol for a capacitor.

One of the most significant properties of the capacitor is its ability to sort alternating current from direct current. The reason that the capacitor

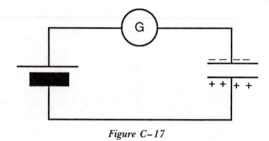

Figure C–17

can separate AC from DC can be seen by considering the circuit in Figure C–17. When the battery is connected, electrons from the battery flow down the wire through the galvanometer to the first metal foil of the capacitor. While the first foil is charging with electrons, electrons from the second metal foil will be driven back to the battery. This occurs because the electrons on the second foil feel the electrical field of the electrons on the first foil. The galvanometer will show a current for a few seconds or less while the capacitor is being charged but will then give a reading that the current equals 0. The charging process stops when the voltage across the capacitor is the same as the potential difference of the battery. If the charged capacitor is removed from the circuit, it will remain charged for many days unless the copper hook-up wires of the capacitor are "shorted" together. When the leads are shorted, the electrons from the negatively charged plate flow through the leads and neutralize the positive charge on the other plate. This is called discharging a capacitor.

Thus in DC circuits a capacitor is effectively a break in the circuit once steady-state conditions have been reached. In the steady-state condition, DC current cannot pass through a capacitor.

Consider the behavior of a capacitor in an AC circuit (Fig. C–18). Note that an AC ammeter must

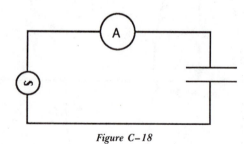

Figure C–18

be used to measure current, since a galvanometer is strictly a DC measuring device. When the battery is connected, the AC voltage source alternately charges one foil of the capacitor, then the other. The foil that is receiving the electrons at a given time induces a movement of electrons away from the second foil. This movement is due to the electrical field projected by the electrons. In this way the current is at its maximum value when the foil is just starting to charge with electrons. The instant the foil is fully charged with electrons from the AC power source, the current drops to 0. As the foil reaches peak charge, the current changes direction. This causes the charge on the first foil to decrease inductively. Thus the capacitor will allow an AC current to appear to "pass through" its dielectric center. The induced charge of opposite sign always appears on the other plate even though no electrons actually move across the dielectric center.

A capacitor does offer a "resistance" to the so-called flow of current through its dielectric core. The "resistance" to the "flow" of AC current through a capacitor is called the capacitive reactance (X_C) and is measured in ohms. The value of X_c is predicted by the following equation:

$$X_C = \frac{1}{2\pi fC}$$

where f is the frequency in Hertz (H_z)
and   C is the capacitance in farads (F)

For calculations of current and voltage in capacitive circuits, Ohm's law becomes $V = IX_C$. Except for the frequency dependence, X_C is like resistance.

The equivalent capacitance of several capacitors in parallel (Fig. C–19) is additive. Thus,

$$C_{equivalent} = C_1 + C_2 + C_3$$

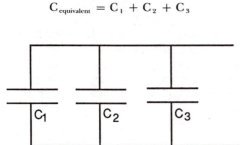

Figure C–19. Capacitors in parallel.

The equivalent capacitance of several capacitors in series (Fig. C–20) is given by the following equation:

$$\frac{1}{C_{equivalent}} = \frac{1}{C_1} + \frac{1}{C_2} + \frac{1}{C_3}$$

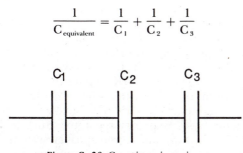

Figure C–20. Capacitors in series.

Air-core

Figure C–21. Symbol for an air-core inductor. (Reprinted from: A.J. Diefenderfer, Principles of Electronic Instrumentation, 2nd Ed., Philadelphia, Saunders College Publishing, 1979.)

INDUCTANCE

When a length of insulated wire of resistance, R, is tightly wound around a mandrel so that a compact spool of wire results, the electrical device that is produced is called an inductor. Inductors are also called chokes or coils. The name used depends only on the purpose the inductor is serving. The inductor described, in contrast to those wound on iron cores, is further classified as an air-core inductor. The symbol for an air-core inductor is depicted in Figure C–21, and the symbol for an iron-core inductor is shown in Figure C–22.

The useful property of an inductor has its origin in the interaction of a magnetic field with an electrical conductor. When a wire is placed in a changing magnetic field, an induced voltage is produced in the wire. The reason for this is beyond the scope of this text, but the relationship between

Iron-core

Figure C–22. Symbol for an iron-core inductor. (Reprinted from: A.J. Diefenderfer, Principles of Electronic Instrumentation, 2nd Ed., Philadelphia, Saunders College Publishing, 1979.)

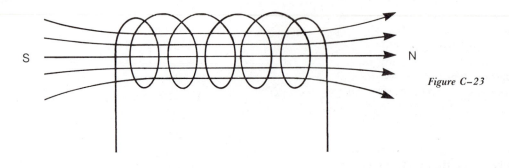

Figure C-23

changing magnetic fields and induced voltages in wires is a well-known experimental fact.

Let's consider the inductor and its magnetic field. When a constant DC passes through the inductor, a constant magnetic field is produced (Fig. C-23). The lines of force of such a field tend to concentrate in the center of the inductor. If a larger DC is passed through the device, a larger magnetic field is produced. In the steady state, the magnetic field is constant so there is no induced voltage.

Consider the behavior of the inductor when the current through the device is alternating (Fig. C-24). The connection of an AC voltage source across the inductor results in a flow of alternating

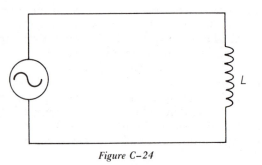

Figure C-24

current through the inductor. The alternating current causes the direction of the magnetic field as well as the field strength to vary with time. This varying magnetic field causes an induced potential difference to develop in the inductor. At any given instant, the induced potential difference is in direct opposition to the applied potential difference. For this reason inductors are called chokes. It is this difference between the two voltages that actually allows the flow of alternating current in the circuit. The choking action is not 100 per cent effective in destroying the applied voltage. The larger the number of turns of wire in the device, the more effective it is in choking the applied AC voltage.

Aside from the resistance of the wire from which the inductor is made, there is a second current-retarding quality that alternating currents en-

counter in their passage through inductors. The retarding effect that is felt specifically by alternating currents is called the inductive reactance (X_L) and is given by the following equation:

$$X_L = 2\pi fL$$

where   L is inductance measured in henrys
and   f is frequency measured in Hertz (H_z)

The value of L is controlled by the size, shape, and core material of the inductor. Note that X_L increases with frequency (f).

Inductors have resistance as well as reactance; the two current-resisting qualities may be combined to give what is called the impedance (Z). For an inductor,

$$Z = \sqrt{R^2 + X_L{}^2}$$

The impedance may be thought of as AC resistance, but it is frequency dependent, of course. In the series circuit in Figure C-24, we can calculate the current if the applied voltage is known and the impedance calculated. In such a circuit, Ohm's law becomes

$$V = IZ$$

A very significant fact that one may see by examining the impedance equation is that the higher the frequency of an alternating current, the more impedance the inductor exhibits. Thus an alternating current of a given value produces a much greater IZ drop than the corresponding IR drop that a direct current would produce in the inductor. Stated more simply, the inductor acts as a resistor of low resistance to a direct current but behaves like a resistor of high resistance when an alternating current passes through. In this way the inductor is a slightly imperfect AC-DC separator.

A point of interest can now be made. A capacitor and inductor properly wired make a good high frequency–low frequency separator. Such a circuit could be used in your stereo speaker system. Your speakers probably have a small speaker for high-frequency reproduction and a large speaker for low-

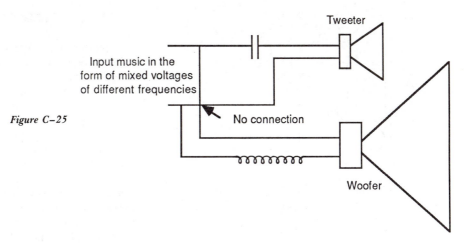

Figure C–25

Input music in the form of mixed voltages of different frequencies

Tweeter

No connection

Woofer

frequency reproduction. A frequency sorter would be helpful in routing the high-frequency current to the small speaker and the low-frequency current to the large speaker. Figure C–25 illustrates how it could be done. The capacitor with its high capacitive reactance for low frequency tends to keep low frequency away from the tweeter, while the inductor and its low resistance to low-frequency current allows the low-frequency currents to reach the woofer. The high impedance of the inductor for high frequencies tends to keep the high notes away from the woofer. The capacitor's low reactance for high frequencies allows the high notes to reach the tweeter. Now, back to instrumentation.

Filter Chokes. If an inductor is wound onto a soft iron core, the iron intensifies and concentrates the magnetic field. Such an inductor has a fairly high L value and hence a fairly high inductive reactance and impedance. For a 10-henry choke at 120 Hz, the inductive reactance is about 16,000 Ω. This is a high resistance to the flow of a 120-Hz current.

Also note that the frequency is equal to 0 for DC, so the choke has no inductive reactance for the passing of DC current. It does have a low resistance, however.

On the basis of its high resistance to AC and its low resistance to DC, the choke is used to help separate AC that is superimposed on DC. Filter chokes are widely used in DC power supplies that convert 110-V AC to DC potential differences.

Transformers. When two inductors are placed in close proximity, the magnetic field generated in one may induce a voltage in the other. This is particularly true in the case of iron-core inductors when both inductors are wound on the same iron core. This type of electrical component is called a transformer. Iron-core transformers are used widely in the conversion of inappropriate AC voltage values to values that are more useful. Transformers that increase voltages are called step-up transformers,

and those that decrease voltages are called step-down transformers. The symbol for a transformer is illustrated in Figure C–26.

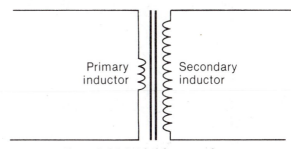

Primary inductor

Secondary inductor

Figure C–26. Symbol for a transformer.

The voltage that is to be transformed is applied to the primary inductor. The voltage that is induced in the secondary inductor is predicted by the following equation:

$$Vs = \frac{Ns}{Np} Vp$$

where Vs stands for AC voltage induced in the secondary inductor

Ns stands for the number of turns of wire in the secondary inductor

Np stands for the number of turns of wire in the primary inductor

Vp stands for the AC voltage applied to the primary inductor

The ratio Ns/Np is called the turns ratio and is larger for a step-up transformer and smaller for a step-down transformer.

Transformers are widely used in obtaining voltages used in instruments. This is an appropriate point for you to return to the problems at the end of the appendix. Do problems 5 through 19.

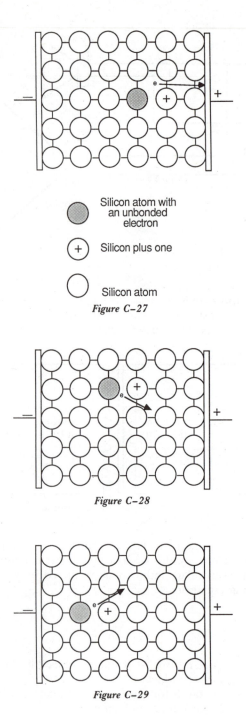

Silicon atom with an unbonded electron

(+) Silicon plus one

○ Silicon atom

Figure C–27

Figure C–28

Figure C–29

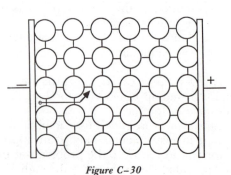

Figure C–30

SEMICONDUCTION

Metals are electrical conductors because the valence electrons of metals are so loosely bound that they can easily move under the influence of an applied potential difference. Insulators do not conduct current because the valence electrons of insulators are so tightly held in covalent bonds that they cannot conduct current even when rather large potential differences are applied. In contrast to metals and insulators is a group of elements and compounds called semiconductors. As the name implies, their conductivities are in between the two extremes of metals and insulators. The valence electrons of common semiconductors are thought to be involved in covalent bonds, but the bonds are relatively weak and, under favorable conditions, can be disrupted enough to allow a current to flow. Although there are many semiconductor materials in use, germanium and silicon are the most important.

Germanium and silicon are metalloids from the carbon family; their electronic configurations are Ar $3d^{10}4s^24p^2$ and Ne $3s^23p^2$ respectively. In their semiconductor forms, each germanium and silicon atom is bonded to four others by weak covalent bonds. Even at room temperature a few bonds break from thermal energy. In the absence of an electrical potential difference, these broken bonds soon reform. On the other hand, if the germanium or silicon crystal has electrical contacts applied to its ends, and if a potential difference is applied, the crystal conducts a small electrical current. The current is caused by the movement of electrons. (Fig. C–27). The electrons from the broken bonds are thought to be partly responsible for the conduction of current in an intrinsic semiconductor. The germanium or silicon atom that remains with the unbonded electron is called a hole and can be thought of as half of a covalent bond. An electron from an adjacent bond often fills the hole and, in the process, generates a new hole (Fig. C–28). In the next instant, the process is repeated (Fig. C–29). Finally an electron from the negative terminal enters the germanium or silicon and fills the hole (Fig. C–30).

In this way holes "carry" current in a direction opposite to the flow of the electron current. The holes appear to move toward the negative terminal. When a hole arrives at the negative terminal, an electron from the external circuit fills it. Although no positive-current carriers really exist, this "movement" of holes makes it appear as if a positive charge is moving toward the negative electrode. This type of conduction is called the intrinsic conduction.

The thermal generation of hole-electron pairs is a temperature-dependent process; the hotter the semiconductor, the greater the number of hole-electron pairs generated per unit time. The greater

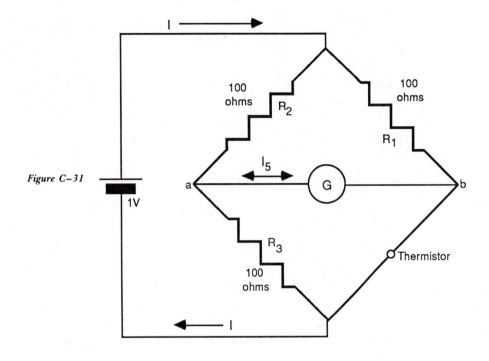

Figure C–31

the number of pairs, the greater the conductivity of the semiconductor. This increase in conductivity with temperature makes possible the measurement of temperature with small beads of semiconductor. The small beads are about 10^{-2} inches in diameter and are called thermistors. If a thermistor is made one arm of a Wheatstone bridge, the unbalance signal can be related directly to the temperature of the thermistor (Fig. C–31).

N-Type Semiconductors

For most applications, the conductivity of pure germanium or silicon is too low to be of value. In order to improve conductivity, about 10 ppb of an impurity can be added; the process of adding the impurity is called doping. If a small amount of an element such as arsenic is added to the pure silicon or germanium, the resulting crystal is called an N-type semiconductor. The arsenic atoms prefer to form five covalent bonds, but when they are placed in a silicon or germanium matrix, only four bonds can be formed. This leaves one electron per arsenic atom that is unbounded and is free for conduction. Thus in an N material, electrons are the main current carriers. Consider Figure C–32.

If an electron from an arsenic atom near the positive terminal is given up, that arsenic atom will take on a +1 charge. The +1 charge can attract an electron from a nearby arsenic atom, converting causing the second arsenic atom to have +1 charge, while returning the first arsenic atom to neutral charge. In this way electrons can move across an N-type semiconductor with relative ease. This process is illustrated in Figure C–33.

The intrinsic conduction of the semiconductor is also present but is small compared with the conduction caused by arsenic electrons. For most of our discussion of semiconductor devices, the intrinsic conduction can be overlooked.

P-Type Semiconductors

If a crystal of germanium or silicon is doped with an element such as indium, a P-type semiconductor is formed. Indium prefers to form only three covalent bonds. Thus when indium is forced into a germanium or silicon crystal lattice, a hole exists at the site of each indium atom (Fig. C–34).

If electrical contacts are applied to such a crystal, electrons from the negative terminal can come to an indium atom near the negative terminal and fill the hole. The indium atom is converted to an indium anion (Fig. C–35).

The excess electron on the In$^-$ can travel to the next hole. In the process, the next In is converted to In^{-1}, and the old In^{-1} is converted to In.

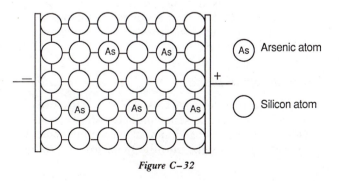

As — Arsenic atom

○ — Silicon atom

Figure C–32

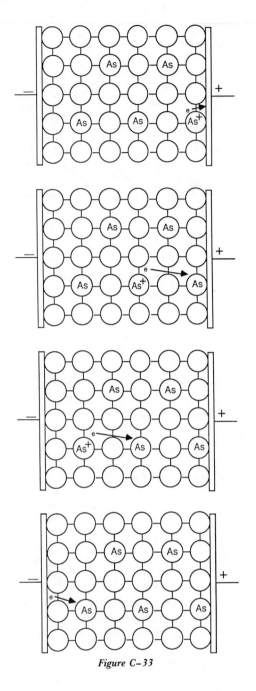

Figure C-33

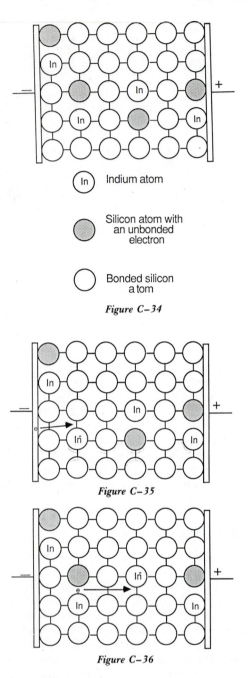

In   Indium atom

Silicon atom with
an unbonded
electron

Bonded silicon
a tom

Figure C-34

Figure C-35

Figure C-36

The electron has also moved closer to the positive contact (Fig. C-36). In this way electrons can move from hole to hole and cross the P-type semiconductor with relative ease.

The Junction Diode

If a heavily doped N crystal is joined to a heavily doped P crystal, a semiconductor diode results. Diodes are widely used and are very important. In order to gain insight into the behavior of a diode,

consider Figure C-37. None of the arsenic or indium atoms are ionized in the crystals before the crystals are joined. On the other hand, when the sections are joined to form a junction, the unbonded electrons form arsenic atoms on the P-N junction, cross over the junction, and fill holes in the indium atoms (Fig. C-38). Because of this fact, a charge separation develops across the junction. This separation of charge makes the diode a useful device because it allows current to pass through the diode in only one direction. For example, if the N

N   P

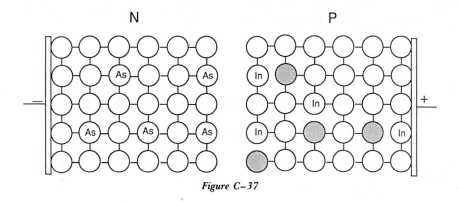

Figure C–37

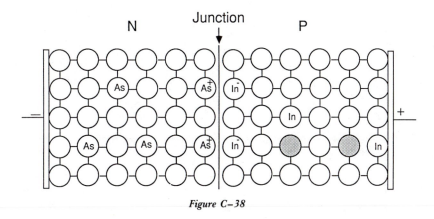

Figure C–38

crystal is made the negative terminal and the P crystal the positive terminal, electrons from the negative terminal flow through the N crystal to the junction, where they neutralize the charge on arsenic +1 ions. The electrons on the junction indium ions fill holes in the crystal nearer the positive electrode. Electrons then cross the junction and fill holes left behind. In this way the device becomes conductive.

When an applied potential tries to force electrons into the P crystal, the junction barrier potential is increased. The charge separation becomes greater, and no current can flow. Thus, the diode is conductive only when electrons enter the N crystal. If electrons try to enter the P crystal, the barrier potential stops the current flow. The symbol for a diode is shown in Figure C–39. The fact that the

Electrical current flow ——▶

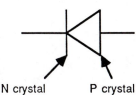

N crystal   P crystal
Figure C–39. Symbol for a diode.

diode passes current in only one direction makes it useful for the conversion of alternating voltage and current to pulsating direct current and voltage. Consider Figure C–40. The waveform from the AC

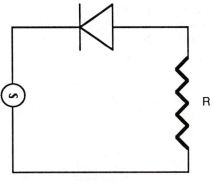

Figure C–40

source is a sine wave (Fig. C–41). The waveform across R is half of a sine wave (Fig. C–42). The diode conducts only during the half·cycle in which the electrons enter the N crystal. The output waveform in this case is called pulsating DC. Thus the diode can convert AC inputs to pulsating DC outputs.

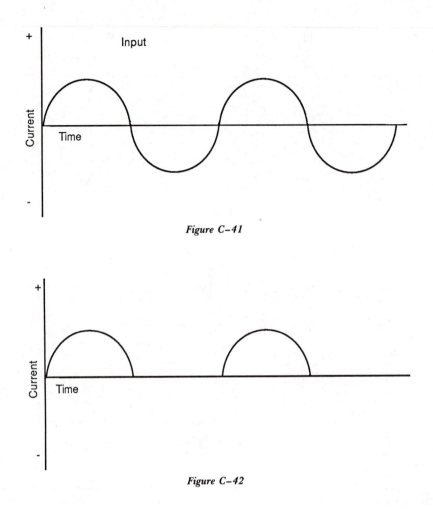

Figure C–41

Figure C–42

AC TO DC CONVERSION—RECTIFIER CIRCUITS

While many instruments use DC voltage in their operation, most do not use batteries as the source of this voltage. In most cases the AC voltage from a power company is changed to an appropriate AC voltage with a transformer and is then rectified to pulsating DC with the help of one or more diodes. There are several common ways of using diodes to do this; consider Figures C–43 to C–45.

Figure C–43 is a half-wave rectifier. Figures C–44 and C–45 are full-wave rectifiers. The bridge circuit in Figure C–45 is common because a less expensive transformer can be used. All three circuits employ diodes to convert AC to pulsating DC. Try to trace the current through the diodes as the AC input changes from positive to negative; remember that electrons are conducted only when they enter the N crystal. After you have tried to trace the current, read on.

Let's consider the current flow in the circuit

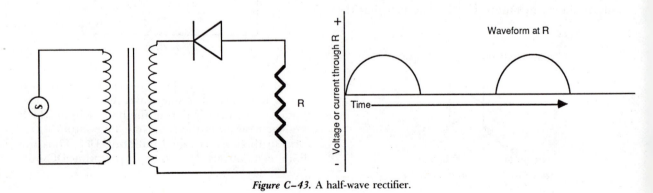

Figure C–43. A half-wave rectifier.

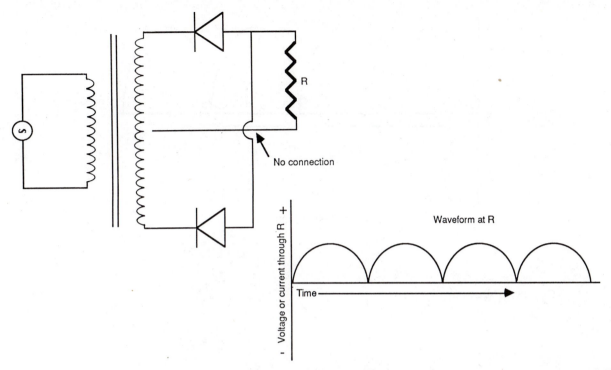

Figure C–44. A full-wave rectifier.

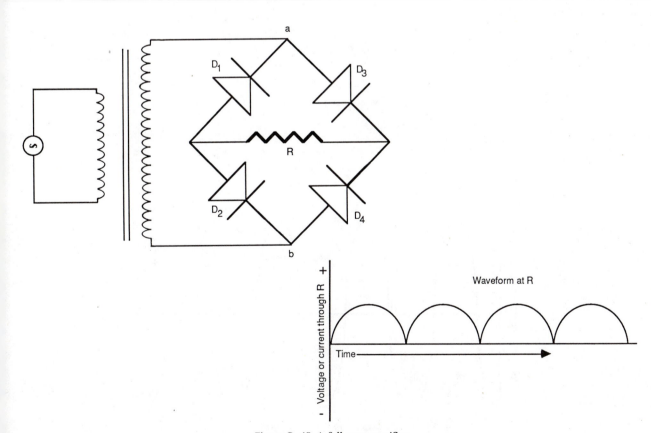

Figure C–45. A full-wave rectifier.

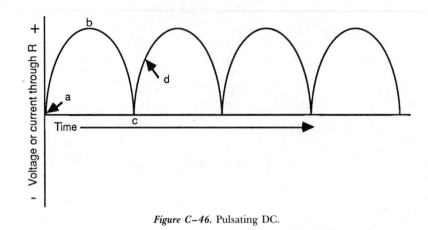

Figure C–46. Pulsating DC.

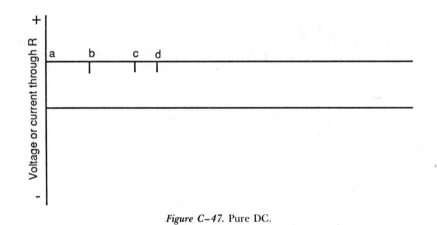

Figure C–47. Pure DC.

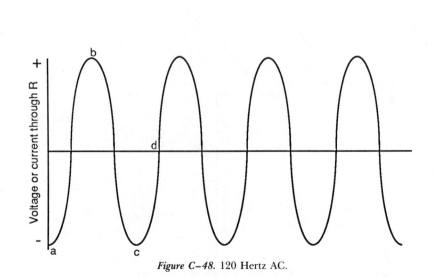

Figure C–48. 120 Hertz AC.

shown in Figure C–45. At a given instant, node a has an excess of electrons, while node b is electron deficient. This potential difference causes the electrons to move through the conductive path that is made up of diode 1, the resistor R, and diode 4. The other possible path that is made up of diode 3, the resistor R, and diode 2 is not conductive in this direction. In the next moment, node b has an excess of electrons, while node a is deficient in electrons. This potential difference causes the electrons to move through the conductive path that is made up of diode 2, the resistor R, and diode 3. The path that consists of diode 4, the resistor R, and diode 1 cannot conduct in this direction. Thus, the direction of current flow through R is the same at all times. The AC is converted to pulsating DC.

Power Supply Filter Networks

As is illustrated in Figures C–43 to C–45, the output from a rectifier circuit is pulsating DC and is a long way from the highly refined DC needed in many instrumental applications. The pulsating DC that is illustrated in Figure C–46 is really the sum of two less complicated waveforms (Figs. C–47 and C–48).

Figure C–47 shows the pure DC needed in many instruments, whereas the waveform depicted in Figure C–48 is a 120-Hz AC waveform that we must get rid of. The two waveforms, when combined, create the pulsating DC in Figure C–46. Carefully consider points a, b, c, d in the figures. The waveform in Figure C–47 is called the DC component, whereas the waveform in Figure C–48 is called the AC component of pulsating DC.

As has been discussed, capacitors and inductors have properties that allow them to separate AC components from DC components. There are several ways to do the separation. The instrument that needs DC power is represented in our diagrams by the resistor R_I. One way to separate the AC from the DC components is shown in Figure C–49. The capacitor in this circuit prevents the DC component from passing through but allows the 120-Hz AC component through with ease. Thus the desired DC component goes to our instrument, while the AC component reaches node a and must divide. The path through the instrument is resistive (2000 Ω), whereas the path through the capacitor is easy. In this way most of the AC component is routed away from the instrument.

If, for some reason, the AC component that does pass through R_I is too great, a second capacitor can be added (Fig. C–50). The second capacitor provides a second chance for the AC component to avoid passing through R_I. The AC component prefers going through a 100-ohm capacitor to going through a 1000-ohm instrument. The DC component cannot pass the capacitor at all so it must go to power the instrument.

Chokes are sometimes used in place of R(Fig. C–51). Because of its high impedance for the AC component and low resistance for the DC component, the choke is a very desirable component for this application.

A DC Power Supply

It is now appropriate to combine the components to make a DC power supply. Consider the circuit diagram in Figure C–52.

Study this circuit carefully. Most instruments have DC power supplies. Although there is some variation among instruments, they all involve a rectifier and some sort of filter network. The power supply of an instrument is a likely place for electrical failure.

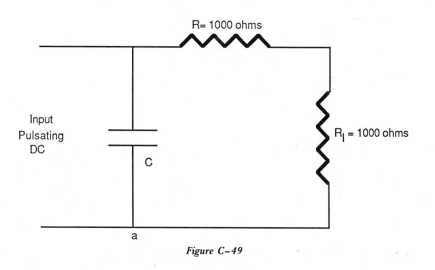

Figure C–49

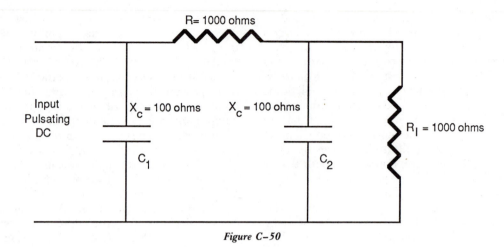

Figure C–50

FIELD EFFECT TRANSISTORS

When the voltage of a current-sensitive source must be amplified and measured, the best way to limit the current flow is to build an amplifier that has a very high internal resistance. Such an amplifier can be made with a device called a field effect transistor (FET). A field effect transistor, along with its electrical symbol, is depicted in Figure C–53.

The N-type semiconductor is connected to a battery or to a DC power supply that produces a flow of current from the source end of the crystal to and out the drain. This part of the circuit is illustrated in Figure C–54.

If a second battery, V_{GS}, is connected to both gates and the source, so that the gate-source junctions are reverse biased, the device can be used to measure the voltage, V_{GS} (Fig. C–55). Recall that when a P-N junction is formed, electrons from the N material cross the junction and form negative ions in the P material. This process leaves positive ions in the N material. This charge separation forms

during the manufacturing process and is called the barrier potential.

When the battery V_{GS} is connected with reverse bias, its electrons try to flow into the gates but cannot proceed because of the barrier potential. In fact, the higher the voltage of V_{GS}, the greater the barrier potential becomes. The higher the barrier potential, the more negative charges collect at the junction and the more positive ions form in the N material on the other side of the junction. In this way a band of positive ions forms in the region near the gate, and its thickness varies with the magnitude of V_{GS}. In the meantime, V_{DS} is driving electrons through the N material, but the conduction of electrons through the N material is restricted by the bands of positive charge near the gates. As the positively charged ion in an N-type semiconductor cannot help with current conduction, the current, I_D, is dependent on the magnitude of the voltage, V_{GS}. In fact, I_D can be adjusted to zero if V_{GS} is great enough. Consider Figure C–56. This figure illustrates the fact that the value of I_D is inversely related

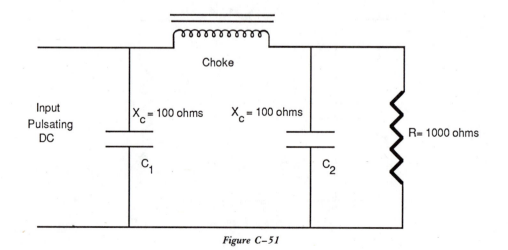

Figure C–51

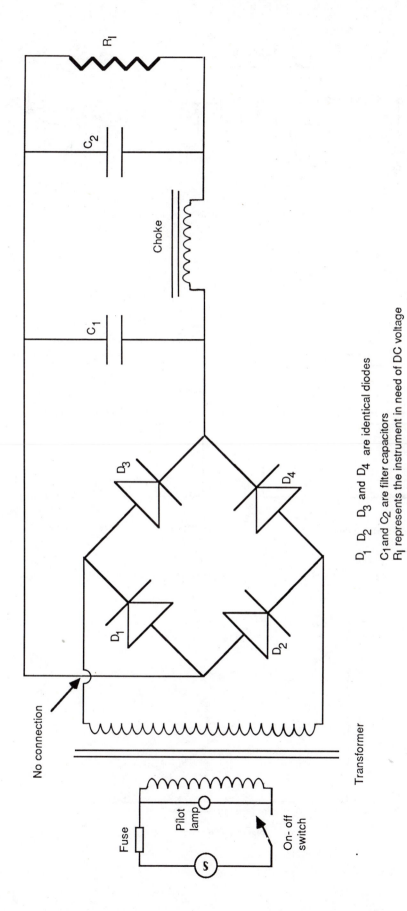

D_1 D_2 D_3 and D_4 are identical diodes

C_1 and C_2 are filter capacitors

R_l represents the instrument in need of DC voltage

Figure C-52

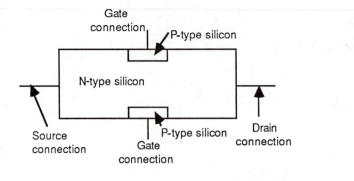
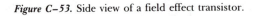
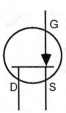

Electrical
symbol

Figure C–53. Side view of a field effect transistor.

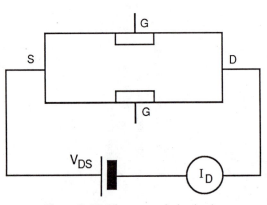
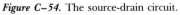

Figure C–54. The source-drain circuit.

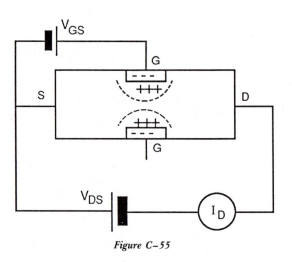

Figure C–55

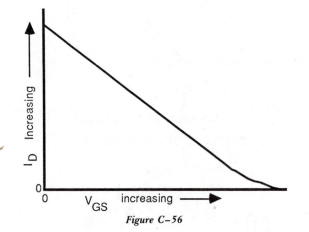

Figure C–56

to the value of V_{GS} and that the relationship is linear over quite a range. Another very significant fact is that there is no current flow in the source-gate circuit. Thus, the voltage of voltage sources that are perturbated easily by current flow can be measured using a field effect transistor. This application of the FET is the most important application of the device in chemical instrumentation.

Problems

1. Consider the following circuit:

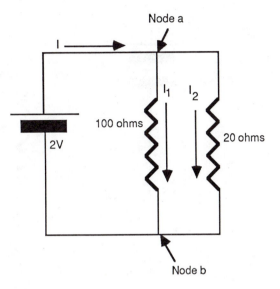

(a) What is the voltage a to b?
(b) What is I_1?
(c) What is I_2?
(d) What is I?

Note that the current divides in a way that is in keeping with the relative ease of the paths.

2. Consider the following circuit:

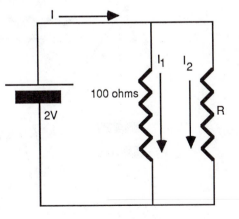

What must be the value of R so that $I_2 = 3 I_1$?

3. Consider the following circuit:

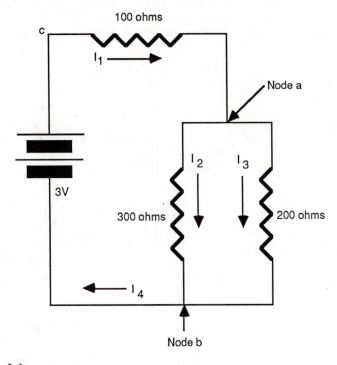

(a) What is I_1?
(b) What is I_2?
(c) What is I_3?
(d) What is I_4?
(e) What is the voltage across node a and node b?
(f) What is the voltage c to a?
(g) What is the voltage c to b?

4. Consider the following circuit:

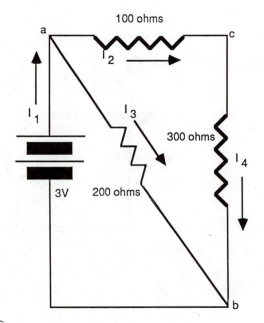

(a) What is I_1?
(b) What is I_2?
(c) What is I_3?
(d) What is I_4?

(e)   What is the voltage across node a and node b?
(f)   What is the voltage c to a?
(g)   What is the voltage c to b?

5.   What is the I_{RMS} current for a 3-amp peak current?

6.   What is the power dissipated in a 100-ohm resistor by a 3-amp peak current?

7.   What is the capacitive reactance of a 10-μF capacitor for DC current, that is, frequency equals 0?

8.   What is the capacitive reactance of a 10-μF capacitor for current with a frequency of 120 Hz?

9.   What is the capacitive reactance of a 10-μF capacitor for current with a frequency of 1000 Hz?

10. Consider the following circuit:

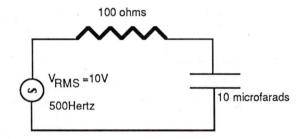

100 ohms

V_{RMS} =10V

500Hertz

10 microfarads

What will be the current$_{RMS}$?

11. Calculate the inductive reactance of a 2-henry inductor for current of 120 Hz.

12. Calculate the inductive reactance of a 2-henry inductor for current of 1000 Hz.

13. Calculate the impedance for a 2-henry inductor that has a DC resistance of 20 ohms. Assume the current is 120 Hz.

14. Calculate the current flow in the following circuit:

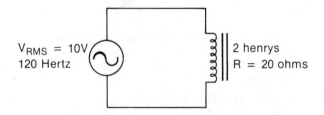

V_{RMS} = 10V
120 Hertz

2 henrys
R = 20 ohms

15. Referring to the figure in problem 14, if the AC power source were replaced with a 10-V DC power source, what would the value of the resulting current be?
Compare the results from problems 14 and 15. The values show that the inductor is much more resistive to AC current flow.

16. Consider the following circuit:

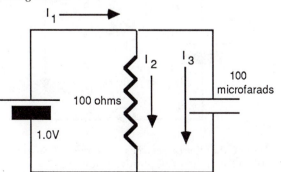

I_1

I_2 I_3

100 ohms

100 microfarads

1.0V

(a)   What is the peak positive value for I_1?
(b)   What is I_2?
(c)   What is the peak positive value for I_3?

17. Consider the following circuit:

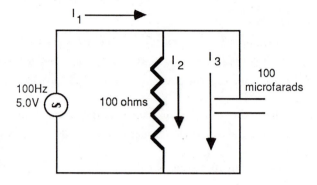

(a) What is I_1?
(b) What is I_2?
(c) What is I_3?

18. Consider the following circuit:

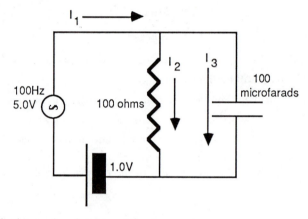

Assume that the internal resistance of the battery is zero.
(a) What is the peak positive value for I_1?
(b) What is the peak positive value for I_2?
(c) What is the peak positive value for I_3?

19. Consider the following circuit:

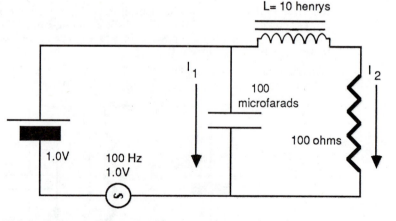

Assume that the battery has zero internal resistance.
(a) What is the peak positive value of I_1?
(b) What is the peak positive value for the AC voltage I_2?

Consider the following circuit when solving problems 20 through 24:

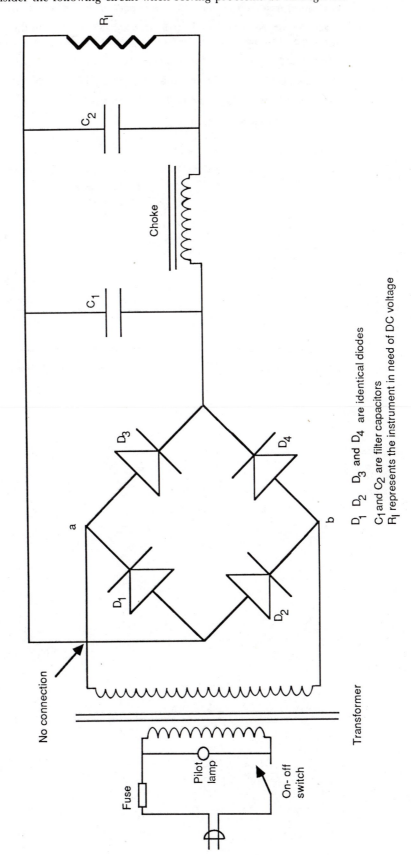

D_1 D_2 D_3 and D_4 are identical diodes

C_1 and C_2 are filter capacitors

R_l represents the instrument in need of DC voltage

20. Trace the movement of an electron arriving at node a from the transformer all the way back to node b. Assume that the electron is part of the DC component.

21. Trace the movement of an electron arriving at node b from the transformer all the way back to node a. Assume that the electron is part of the AC component.

22. What would happen if the fuse burned out?

23. What would happen if the choke burned a wire in two and developed infinite resistance?

24. What would happen if capacitor C_1 shorted out?

25. Consider the following circuit:

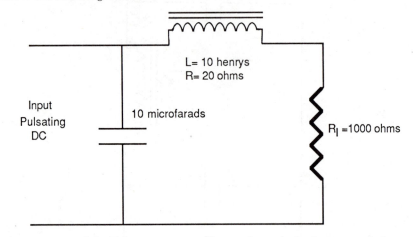

What percentage of the AC component will pass through the instrument? Assume that the AC component is 120 Hz.

Answers to Problems

Chapter 1

1. $12 \ \Omega$

2. $10^{11} \ \Omega$

3. 10^{-3} amps

4. (a) $3 \ \Omega$
   (b) 0.75 watts
   (c) A one-watt resistor should be used.

5. (a) 3.3×10^{-3} amps
   (b) 0.33 volts
   (c) 0.67 volts
   (d) 1.0 volts
   (e) 1.0 volts
   (f) 1.7 volts
   (g) 2.0 volts
   (h) $I_2 = I_1 = 3.3 \times 10^{-3}$ amps
   (i) 3.3×10^{-3} watts

6. (a) 1.0×10^{-2} amps
   (b) 2.0 volts
   (c) 1.0 volts
   (d) 3.0 volts
   (e) 3.0 volts
   (f) 4.0 volts
   (g) 6.0 volts
   (h) $I_2 = I_1 = 1.0 \times 10^{-2}$ amps
   (i) 3×10^{-2} watts

7. 10 volts

8. 10 volts

9. 2 volts

Chapter 3

1. (a) $A = 1.0$
   (b) $A = 0.629$
   (c) $A = 0.234$
   (d) $A = 2.0$
   (e) $A = 3$
   (f) $A = 0.0645$

2. (a) $A = 1.00$
   (b) $A = 0.629$
   (c) $A = 0.234$
   (d) $A = 2.00$

(e) $A = 3.0$
(f) $A = 1.06$

3. (a) $T = 79.4\%$
   (b) $T = 28\%$
   (c) $T = 10.0\%$
   (d) $T = 1.00\%$
   (e) $T = 6.31\%$
   (f) $T = 88.9\%$

4. 0.15

5. 0.51

6. (a) 1.15×10^{-5} mol/l
   (b) 1.49 ppm
   (c) 0.149 mg/dl

7. (a) 1.76×10^{-6} mol/l
   (b) 3.13×10^{-1} g/ml
   (c) 3.13×10^{-2} mg/dl

8. 0.187

9. 1.38×10^4 l/mol cm

10. 1.40×10^4 l/mol cm

11. 4.31×10^3 l/mol cm

12. (a) 8.91×10^{-6} mol/l
    (b) 0.141 mg/dl
    (c) 1.41×10^{-4} % by weight

13. 5.21 mg/dl

14. 2.01×10^{-5} mol/l

15. 2.5 mg/dl

16. $C_A = 3.43 \times 10^{-5}$ mol/l
    $C_B = 4.76 \times 10^{-5}$ mol/l

17. (a) λ_4 would have the greatest sensitivity.
    (b) λ_3 would be a good wavelength if the greatest sensitivity were not required.
    (c) λ_1

18. $C_R = 5.52$ ppm
    $C_Y = 6.26$ ppm

19. 5.13 mg/dl

20. 6.71 mg/dl

Chapter 4

1. 530 nm, 44%, 120nm

2. 725 nm, 25%, 90 nm

3. bluish green (see Table 4–2)

4. for M = 1, 1520 nm
   M = 2, 759 nm
   M = 3, 505 nm
   M = 4, 381 nm

5. glass C

6. glass B

7. for M = 1, 1000 nm
   M = 2, 500 nm
   M = 3, 333 nm

8. glass B

9. for M = 1, 13.8 degrees at 400 nm, 28.6 degrees at 800 nm
   for M = 2, 28.6 degrees at 400 nm, 73.4 degrees at 800 nm

10. 45.94 degrees

11. glass E

12. Everything in the M = 2 region from 45.94 degrees to 73.4 degrees is contaminated by diffraction from the M = 3 region.

13. glass F

Chapter 5

1. high result (falsely elevated result)

2. falsely elevated result

3.   low result

4.   very low result

5.   low result

6.   The correct result would be obtained within the limitations of random error.

7.   falsely elevated result

8.   falsely elevated result

9.   falsely elevated result

10.   falsely elevated result

11.   low result

12.   high result

13.   high result

14.   low result

15.   low result

16.   high result

17.   high result

18.   low

19.   A = 3

20.   slightly low; almost correct

Chapter 6

1.   224 mg/dl

2.   C = 111A

3.   335 units

4.   386 units

5.   amphetamine

6.   methaqualone

Chapter 7

1.   2.45 ppm

2.   pH has a profound effect on fluorescent output.

3.   Temperature has a profound effect on fluorescent output.

4.   7.8 mg/dl

5.   2.3 mmol/l

6. (b) 52.5 mg/dl

Chapter 8

1.   elevated

2.   elevated

3.   depressed

4.   yes (correct)

5.   correct

6.   correct

7.   correct

8.   correct

9.   3.43 ppm is the apparent concentration.

10.   Chemical interference. The new value, 5.3 ppm, should be more accurate.

11. (c) 18.7 ppm

Chapter 9

1.   0.94 mmol/l

2.   elevated

3.   depressed

4.   high

5.   48 mmol/l

6.   The resulting value would be low because of chemical interference.

7.   correct

8.   Add La^{+3} to the samples and standard. The La^{+3} would remove the chemical interference.

9.   Include 14 mEq/l NaCl in the standards. This would give the standards the same NaCl concentration as the samples.

10.   1.9 ppm

11.   The result would be high because of the ionization effect.

12.   2.96 ppm

13.   88 mmol/l

14.   1.0 ppm

15.   2.14 ppm

Chapter 12

1.   370 mg/dl, 92 mg/dl

2.   4.75%, 2.65%

3.   13 mg/dl, 37.4 mg/dl

Chapter 13

1. (a) 0.115 volts
   (b) $Cu^{+2} + 2Cl^- + 2Ag \rightarrow Cu + 2AgCl$
   (c) the silver–silver chloride electrode
   (d) the copper electrode

2. (a) 0.532 volts
   (b) $2Ag^+ + 2Cl^- + 2Hg \rightarrow 2Ag + Hg_2Cl_2$
   (c) the calomel electrode
   (d) the silver electrode

3. (a) 0.578 volts
   (b) $Ag^+ + Cl^- \rightarrow AgCl$
   (c) the silver–silver chloride electrode
   (d) the silver electrode

4. (a) 0.222 volts
   (b) $2AgCl + H_2 \rightarrow 2Ag + 2H^+ + 2Cl^-$
   (c) the hydrogen electrode
   (d) the silver–silver chloride electrode

5. (a) 0.268 volts
   (b) $Hg_2Cl_2 + H_2 \rightarrow 2Hg + 2H^+ + 2Cl^-$
   (c) the hydrogen electrode
   (d) the calomel electrode

6. (a) $pCu = \dfrac{2(0.115 - E)}{0.0591}$
   (b) the silver–silver chloride electrode
   (c) the copper electrode
   (d) the silver–silver chloride electrode

7. (a) $pAg = \dfrac{0.550 - E}{0.0591}$
   (b) calomel electrode
   (c) the silver electrode
   (d) the calomel electrode

8. (a) $pAg = \dfrac{0.519 - E}{0.0591}$
   (b) the silver–silver chloride electrode
   (c) the silver electrode
   (d) the silver–silver chloride electrode

9. (a) $pCl = \dfrac{0.529 - E}{0.0591}$
   (b) the silver electrode
   (c) the silver–silver chloride electrode
   (d) the silver–silver chloride electrode

10. (a) $pH = \dfrac{E - 0.250}{0.0591}$
    (b) the calomel electrode
    (c) the hydrogen electrode
    (d) the hydrogen electrode

11. (a) 0.177 volts
    (b) the electrode at the lower concentration

12. (a) $pH = \dfrac{E}{0.0591}$
    (b) the electrode in which $H^+ = 1.00$
    (c) the electrode in which $H^+ = ?$
    (d) the indicating electrode

13. (a) 0.714 volts
    (b) 0.313 volts

14. (a) −0.344 volts
    (b) 0.782 volts
    (c) 0.693 volts

15. E constant = 0.023 volts
    pH = 6.82

16. E constant = 0.0619 volts
    pNa = 1.88
    Na concentration is 1.32×10^{-2} molar.

17. pCa = 0.25 molar

18. Na concentration is 7.9 mmol/l.

Chapter 15

1. 64.7 mEq/l
2. 96.4 mEq/l
3. It would nearly double.
4. 1.63×10^{-2} Eq/l; if the acid is strong, pH = 1.79.

Chapter 17

1. primidone, phenobarbital
2. ethosuximide, primidone, carbamazepine
3. 3 mg/dl
4. 1.2 mg/dl
5. 1.4 mg/dl
6. 0.51
7. 9.3 ppm
8. 1.03
9. 145 ppm
10. 0.06 mm
11. 0.067
12. 0.14 mm
13. 0.31

Chapter 19

1. acetaminophen
2. propoxyphene

Chapter 21

1. 0.448 molal
2. 0.332 molal

Chapter 22

1. 1.42 mg
2. 2.26 mg

Appendix A

1. 1.35 g/dl
2. 0.34 g/dl
3. 3.9 mg/dl
4. 1.85 mg/dl
5. 2.9 mg/ml

Appendix C

1. (a) 2 V
   (b) 2×10^{-2} amp
   (c) 1×10^{-1} amp
   (d) 1.2×10^{-1} amp

2. 33.3 ohms

3. (a) 13.6 ma
   (b) 5.5 ma
   (c) 8.2 ma
   (d) 13.6 ma
   (e) 1.64 V
   (f) 1.36 V
   (g) 3.0 V

4. (a) 22.5 ma
   (b) 7.5 ma
   (c) 15 ma
   (d) 7.5 ma
   (e) 3.0 V
   (f) 0.75 V
   (g) 2.25 V

5. 2.1 amps RMS
6. 449 watts
7. infinity
8. 1.33×10^2 ohms
9. 15.9 ohms
10. 7.59×10^1 ma RMS
11. 1508 ohms
12. 1.26×10^4 ohms
13. 1508 ohms
14. 6.63 ma RMS
15. 0.5 amp

16. (a) 1×10^{-2} amp
    (b) 0
    (c) 1×10^{-2} amp

17. (a) 0.36 amp peak positive
    (b) 0.05 amp
    (c) 0.31 amp peak positive

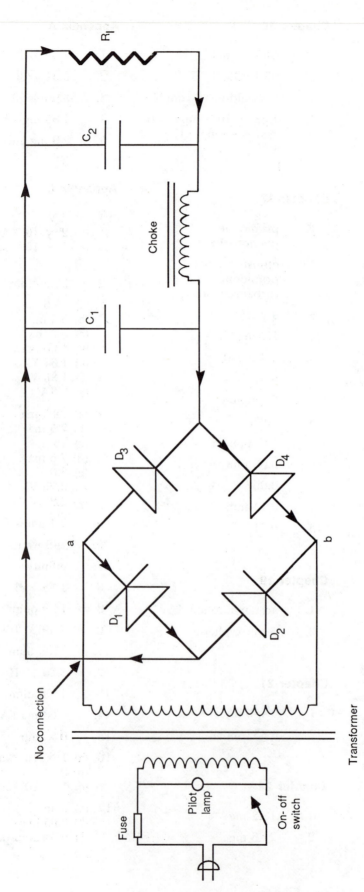

Answer to problem 20, Appendix C.

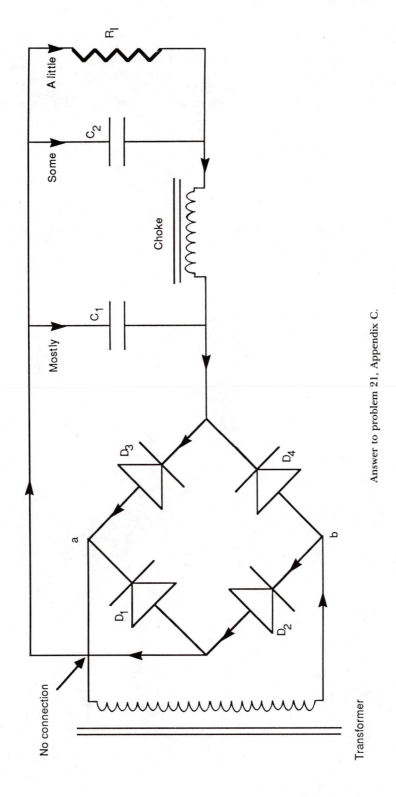

Answer to problem 21, Appendix C.

18. (a) 0.37 amp peak positive
    (b) 0.06 amp peak positive
    (c) 0.31 amp peak positive

19. (a) 0.063 amp peak positive
    (b) 8.6×10^{-3} amp peak positive

20. See figure on page 346.

21. See figure on page 347.

22. The pilot light would go out, and no voltage would exist across nodes a and b.

23. The pilot light would go out, and the voltage would still exist across nodes a and b. There would not be a voltage across R_1.

24. Excessive current would be allowed to flow through C_1. This would burn out some diodes or blow the fuse or both.

25. 1.53%

TRADE NAMES AND MANUFACTURERS

Note: The trade names of many of the products mentioned in this book are alphabetized in the following index. The manufacturer of the product is listed after the trade name.

Abbott, Abbott Laboratories

ACA V, E.I. du Pont de Nemours & Company, Incorporated

AG, Bio-Rad Laboratories

Amberlite, Rohm & Haas Company

Ames, Ames Company

Aminex, Bio-Rad Laboratories

ASTRA, Beckman Instruments, Incorporated

AutoAnalyzer, Technicon Instruments Corporation

AutoAnalyzer I, Technicon Instruments Corporation

AutoAnalyzer II, Technicon Instruments Corporation

BACTEC, Johnston Laboratories, Incorporated

Bakelite, Union Carbide Corporation

Beckman, Beckman Instruments, Incorporated

Carbopack, Supelco, Incorporated

CARBOWAX, Union Carbide Corporation

Celite, Johns-Manville Corporation

Chemstrip, Boehringer Mannheim

Chemwipes, Kimberly-Clark Corporation

Chromosorb, Johns-Manville Corporation

Chrono-log, Chrono-log Corporation

Clinical Assays, Travenol Laboratories, Incorporated

Coleman, Perkin-Elmer Corporation

Coleman Junior, Perkin-Elmer Corporation

Coulter Counter, Coulter Electronics, Incorporated

Cutie Pie, Warrington Laboratories, Incorporated

DADE, American Hospital Supply Corporation

DEMAND, Olympus Corporation of America

Diaion, Mitsubishi Chemical Company

Dowex, Dow Chemical Company

EKTACHEM, Eastman Kodak Company

ELT, Ortho Diagnostic Systems, Incorporated

EMIT, Syva Company

FETIA, Syva Company

Gamma Coat, Travenol Laboratories, Incorporated

Gas-Chrom, Alltech Associates, Incorporated

H6000, Technicon Instruments Corporation

Hematrak, SmithKline Beckman Corporation

Imac, Akzo Chemical Company

Kastel, Montedison Company

KDA, American Monitor Corporation

Lambda Array 3840, Perkin-Elmer Corporation

Lewatit, Mobay Chemical Company

Librium, Hoffmann-La Roche, Incorporated

Micro KDA, American Monitor Corporation

Multistat, Allied Instrumentation Laboratory

Multistix, Ames Company

N-Multistix, Ames Company

Oriel, Oriel Corporation

ORION, Orion Research Incorporated

ORTHO, Ortho Diagnostic Systems, Incorporated

OV-17, Ohio Valley Specialty Chemical Company

PARALLEL, American Monitor Corporation

Particle-enhanced turbidimetric immunoassay, E.I. du Pont de Nemours & Company, Incorporated

Permutit, Permutit Company

PETINIA, E.I. du Pont de Nemours & Company, Incorporated

Placidyl, Abbott Laboratories

Platelet Aggregation Profiler (PAP), Bio/Data Corporation

Pyrex, Corning Glass Works

Reichert, Reichert Scientific, Incorporated

Rexyn, Fisher Scientific Company

Seralyzer, Miles Laboratory

SMA, Technicon Instruments Corporation

SMAC, Technicon Instruments Corporation

SMAC II, Technicon Instruments Corporation

SP, Supelco, Incorporated

STRATUS, American Hospital Supply Corporation

Syva Advanced System, Syva Company

TD_x, Abbott Laboratories

Technicon, Technicon Instruments Corporation

Thorazine, Smith Kline &French Laboratories

Thrombokinetogram, Bio/Data Corporation

TKG, Bio/Data Corporation

Varian, Varian Techtron Pty. Ltd.

Wofatit, Wolfen Dye Factories

Zeocarb, Permutit Company

Zerolit, Permutit Company

INDEX

Note: Page numbers followed by (F) refer to illustrations; page numbers followed by (t) refer to tables.